S.S.R Kumar Challa (Ed.)
Chemistry of Nanomaterials

Also of interest

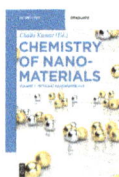

Chemistry of Nanomaterials
Volume 1: Metallic Nanomaterials (Part A)
Challa (Ed.), 2018
ISBN 978-3-11-034003-7, e-ISBN 978-3-11-034510-0

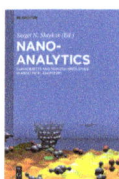

Nanoanalytics
Nanoobjects and Nanotechnologies in Analytical Chemistry
Shtykov (Ed.), 2018
ISBN 978-3-11-054006-2, e-ISBN 978-3-11-054024-6

Nanoscience and Nanotechnology
Advances and Developments in Nano-sized Materials
Van de Voorde (Ed.), 2018
ISBN 978-3-11-054720-7, e-ISBN 978-3-11-054722-1

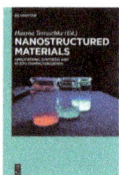

Nanostructured Materials
Applications, Synthesis and In-Situ Characterization.
Terraschke (Ed.), 2019
ISBN 978-3-11-045829-9, e-ISBN 978-3-11-045909-8

Physical Sciences Reviews.
e-ISSN 2365-659X

Chemistry of Nanomaterials

Volume 2: Metallic Nanomaterials (Part B)

Edited by
S.S.R Kumar Challa

DE GRUYTER

Editor
Dr. S.S.R Kumar Challa
Harvard University
Integrated Mesoscale Architectures for Sustainable Catalysis (IMASC)
Rowland Institute of Science
100 Edwin H. Land Blvd
Cambridge MA 02142
USA
challa@fas.harvard.edu
challakumar@gmail.com

ISBN 978-3-11-063660-4
e-ISBN (PDF) 978-3-11-063666-6
e-ISBN (EPUB) 978-3-11-063698-7

Library of Congress Control Number: 2018960714

Bibliographic information published by the Deutsche Nationalbibliothek
The Deutsche Nationalbibliothek lists this publication in the Deutsche Nationalbibliografie; detailed bibliographic data are available on the Internet at http://dnb.dnb.de.

© 2019 Walter de Gruyter GmbH, Berlin/Boston
Typesetting: Integra Software Services Pvt. Ltd.
Printing and binding: CPI books GmbH, Leck
Cover image: Science Photo Library / Ella Maru Studio

www.degruyter.com

Contents

Jiawei Zhang, Huiqi Li, Zhiyuan Jiang and Zhaoxiong Xie

Elena Piacenza, Alessandro Presentato, Emanuele Zonaro, Silvia Lampis,
Giovanni Vallini and Raymond J. Turner

List of contributing authors

Suparna Mukherji
Centre for Environmental Science and
Engineering (CESE)
IIT Bombay, Powai, Mumbai, India
mitras@iitb.ac.in
Chapter 1

Sharda Bharti
Centre for Environmental Science and
Engineering (CESE)
IIT Bombay, Powai, Mumbai, India
Chapter 1

Gauri Shukla
Department of Bioscience and Bioengineering
(BSBE)
IIT, Bombay, Powai, Mumbai, India
Chapter 1

Soumyo Mukherji
Department of Bioscience and Bioengineering
(BSBE)
IIT, Bombay, Powai, Mumbai, India
Chapter 1

Jiwen Hu
Guangzhou Institute of Chemistry
Chinese Academy of Sciences
Guangzhou, P. R. China, 510650,
hjw@gic.ac.cn
Chapter 2

Pei Zhang
Guangzhou Institute of Chemistry
Chinese Academy of Sciences
Guangzhou, P. R. China, 510650,
Chapter 2

Shudong Lin
Guangzhou Institute of Chemistry
Chinese Academy of Sciences
Guangzhou, P. R. China, 510650,
Chapter 2

Zhiyuan Jiang
State Key Laboratory of Physical Chemistry of
Solid Surfaces
Collaborative Innovation Center of Chemistry
for Energy Materials
Department of Chemistry,
College of Chemistry and Chemical
Engineering
Xiamen University
Xiamen 361005, China
zyjiang@xmu.edu.cn
Chapter 3

Jiawei Zhang
State Key Laboratory of Physical Chemistry of
Solid Surfaces
Collaborative Innovation Center of Chemistry
for Energy Materials
Department of Chemistry,
College of Chemistry and Chemical
Engineering
Xiamen University
Xiamen 361005, China
Chapter 3

Huiqi Li
State Key Laboratory of Physical Chemistry of
Solid Surfaces
Collaborative Innovation Center of Chemistry
for Energy Materials
Department of Chemistry,
College of Chemistry and Chemical
Engineering
Xiamen University
Xiamen 361005, China
Chapter 3

Zhaoxiong Xie
State Key Laboratory of Physical Chemistry of
Solid Surfaces
Collaborative Innovation Center of Chemistry
for Energy Materials
Department of Chemistry,

https://doi.org/10.1515/9783110636666-201

College of Chemistry and Chemical
Engineering
Xiamen University
Xiamen 361005, China
Chapter 3

Jun Yang
State Key Laboratory of Multiphase Complex
Systems,
Institute of Process Engineering, Chinese
Academy of Sciences
Beijing 100190, China
jyang@ipe.ac.cn
Chapter 4

Linlin Xu
State Key Laboratory of Multiphase Complex
Systems,
Institute of Process Engineering, Chinese
Academy of Sciences
Beijing 100190, China
Chapter 4

Saptarshi Mukherjee
Department of Chemistry
Indian Institute of Science Education and
Research Bhopal
Bhopal Bypass Road, Bhauri, Bhopal
462 066, Madhya Pradesh, India
saptarshi@iiserb.ac.in
Chapter 5

Nirmal Kumar Das
Department of Chemistry
Indian Institute of Science Education and
Research Bhopal
Bhopal Bypass Road, Bhauri, Bhopal
462 066, Madhya Pradesh, India
Chapter 5

Elena Piacenza
Microbial Biochemistry Laboratory,
Department of Biological Sciences
University of Calgary,
2500 University Dr. NW, Calgary

AB T2N 1N4, Canada
elena.piacenza@ucalgary.ca
Chapter 6

Raymond J. Turner
Microbial Biochemistry Laboratory,
Department of Biological Sciences
University of Calgary,
2500 University Dr. NW, Calgary
AB T2N 1N4, Canada
turnerr@ucalgary.ca
Chapter 6

Alessandro Presentato
Environmental Microbiology Laboratory
Department of Biotechnology
University of Verona
Strada Le Grazie 15, 37134
Verona, Italy
Chapter 6

Emanuele Zonaro
Environmental Microbiology Laboratory
Department of Biotechnology
University of Verona
Strada Le Grazie 15, 37134
Verona, Italy
Chapter 6

Silvia Lampis
Environmental Microbiology Laboratory
Department of Biotechnology
University of Verona
Strada Le Grazie 15, 37134
Verona, Italy
Chapter 6

Giovanni Vallini
Environmental Microbiology Laboratory
Department of Biotechnology
University of Verona
Strada Le Grazie 15, 37134
Verona, Italy
Chapter 6

Suparna Mukherji, Sharda Bharti, Gauri Shukla
and Soumyo Mukherji

1 Synthesis and characterization of size- and shape-controlled silver nanoparticles

Abstract: Silver nanoparticles (AgNPs) have application potential in diverse areas ranging from wound healing to catalysis and sensing. The possibility for optimizing the physical, chemical and optical properties for an application by tailoring the shape and size of silver nanoparticles has motived much research on methods for synthesis of size- and shape-controlled AgNPs. The shape and size of AgNPs are reported to vary depending on choice of the Ag precursor salt, reducing agent, stabilizing agent and on the synthesis technique used. This chapter provides a detailed review on various synthesis approaches that may be used for synthesis of AgNPs of desired size and shape. Silver nanoparticles may be synthesized using diverse routes, including, physical, chemical, photochemical, biological and microwave -based techniques. Synthesis of AgNPs of diverse shapes, such as, nanospheres, nanorods, nanobars, nanoprisms, decahedral nanoparticles and triangular bipyramids is also discussed for chemical-, photochemical- and microwave-based synthesis routes. The choice of chemicals used for reduction and stabilization of nanoparticles is found to influence their shape and size significantly. A discussion on the mechanism of synthesis of AgNPs through nucleation and growth processes is discussed for AgNPs of varying shape and sizes so as to provide an insight on the various synthesis routes. Techniques, such as, electron microscopy, spectroscopy, and crystallography that can be used for characterizing the AgNPs formed in terms of their shape, sizes, crystal structure and chemical composition are also discussed in this chapter.

This article has previously been published in the journal *Physical Sciences Reviews*. Please cite as:
Mukherji, S., Bartha, S, Mukherji, S. Synthesis and Characterization of Size and Shape-Controlled Silver Nanoparticles. *Physical Sciences Reviews* [Online] **2018**, 3. DOI: 10.1515/psr-2017-0082

https://doi.org/10.1515/9783110636666-001

Graphical Abstract:

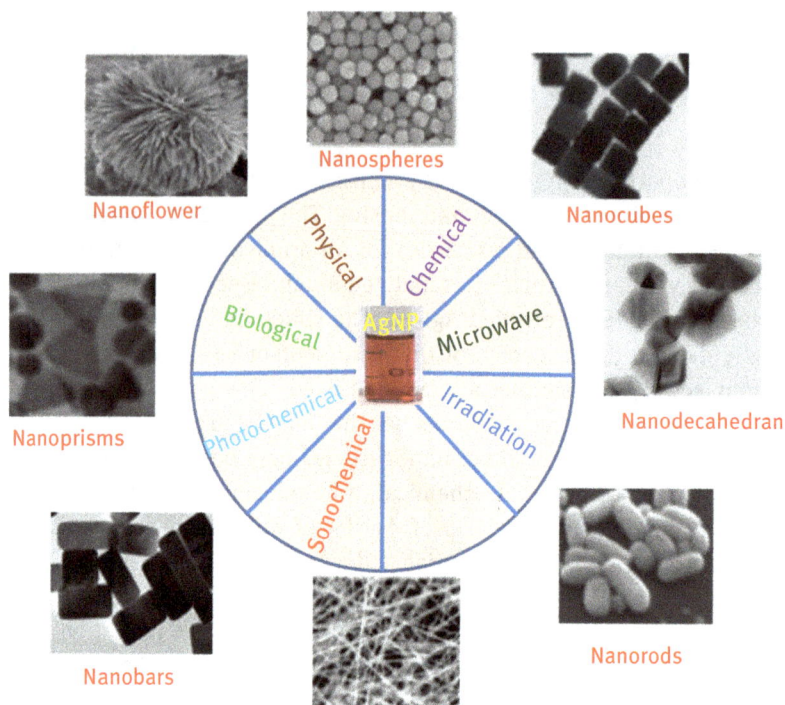

Nanospheres

Nanoflower

Nanocubes

Physical

Chemical

Biological

AgNP

Microwave

Nanoprisms

Photochemical

Irradiation

Sonochemical

Nanodecahedran

Nanobars

Nanorods

Keywords: silver nanoparticles, synthesis routes, nanospheres, nanoplates, nanoprisms, nanorods

1.1 Introduction

Nanoparticles (NPs) can be defined as small structures where one of the dimensions is of nanometre size range, i. e. 1–100 nm. NPs inherently possess a high surface area to volume ratio, which imparts higher reactivity to the NPs as compared to their bulk metallic form. The localized surface plasmon resonance (LSPR) frequency varies with the morphology of the NPs and the dielectric nature of the surrounding environment. The size and shape of NPs impart unique physical, electronic, catalytic and optical properties and determine their potential application [1–4]. Decrease in the metal core size changes the energy-level structure such that the continuous band of energy levels change to discrete energy levels, resulting in strong size-dependent physical and chemical properties. Among all metallic NPs studied, silver NPs (AgNPs) have been studied most widely. Aspects such as, synthesis routes, optical properties and catalytic properties and their use in various applications have been extensively explored. The physical, chemical, catalytic, optical, magnetic and electronic properties vary with the size and shape of the NPs, their interactions with stabilizers and the surrounding media and the synthesis method employed. The application areas for AgNPs include chemical sensing, biological sensing, pollutant removal through catalysis and antimicrobial applications [5–7]. Numerous researchers have focussed on size- and shape-controlled synthesis of AgNPs for various applications.

A number of excellent reviews are available on synthesis and application of AgNPs. They have focused on diverse synthesis routes [8–11], application in various domains [2, 12–14] and synthesis of AgNPs of diverse shape [15]. Most of the reviews have focused on a particular domain with respect to the synthesis of AgNPs, such as, shape-controlled synthesis of AgNPs [16, 17], physical synthesis approaches [18], microbial synthesis [9, 19–21], bacteria-mediated synthesis [10, 22], plant-mediated synthesis [23], green synthesis [8, 24] or microwave -based synthesis [25], while some others have selectively reviewed a few of the synthesis routes for production of shape-controlled AgNPs [3, 11, 13, 15, 26–28]. None of the reviews available on silver NP synthesis have comprehensively discussed all the synthesis approaches in depth. This chapter discusses the synthesis approaches starting from the synthesis of spherical AgNPs to synthesis of shape-controlled AgNPs and also discusses how size-controlled AgNPs can be synthesized. A preview of the characterization techniques presented before discussion on synthesis allows better comprehension of the characteristics of the AgNPs synthesized using each process. An attempt is made to provide insight on the mechanisms involved in the various synthesis processes. This allows a more scholarly discourse on the various synthesis routes and mechanisms, and elucidates how shape- and size-controlled synthesis of silver NPs can be achieved through appropriate choice of energy source, chemicals used as precursor, reducing agent and capping agent as well as the selection of concentration and molar ratio of the chemicals. Figures and tables are used to illustrate the diverse synthesis options available for synthesizing AgNPs of the desired size and shape. Finally, safety considerations are also discussed in this review since not only does synthesis involve numerous hazardous chemicals, adequate precaution is needed to prevent inadvertent release of the AgNPs thus synthesized. The overall breadth and detailed coverage will enable scientists, who are not completely familiar with the domain, to understand and appreciate the various techniques that have evolved over decades of international research.

1.2 Characterization of silver NPs

To establish the success of any NP synthesis protocol, it is important to characterize the NPs synthesized in terms of shape, particle size distribution, solubility, morphology, surface area, crystallinity, pore size, dispersivity in solution, and aggregation behaviour. The commonly used characterization techniques include UV-Vis spectroscopy, transmission electron microscopy (TEM), scanning electron microscopy (SEM), energy dispersive spectroscopy (EDS), X-ray diffraction (XRD), Fourier Transform Infra-Red (FTIR) spectroscopy, dynamic light scattering (DLS) and zetasizer. The shape and size distribution of AgNPs is typically measured and visualized using electron microscopy techniques, such as, TEM and SEM. The SPR band or absorption spectra can be obtained using UV-Visible spectroscopy, where the wavelength corresponding to peak absorbance (λ_{max}) reflects the size of the AgNPs. For AgNPs, the peak is usually in the range of 400–450 nm [29]. These and other techniques that can provide further insights on the silver NPs synthesized are discussed in the following section.

1.2.1 UV-Vis spectroscopy

The UV-visible optical absorption spectrum and the λ_{max} value of metallic NPs depend on the type of metal, on the size and shape of the NPs and on the dielectric media it is dispersed in. It is an outcome of the LSPR phenomena caused by collective oscillation of electrons in specific vibrational modes in the conduction band near the particle surface in response to light. The UV-Vis spectrum is also affected by the free electron density and inter-particle interactions. This same phenomenon is responsible for the intense colour that characterizes a dispersion of colloidal AgNPs due to scattering and absorption of visible light. The latter is affected by the energy of the photons which can induce conduction electrons to oscillate by absorbing incident electromagnetic radiation. Apart from characterization of the AgNPs, this technique is commonly used to study the stability/aggregation tendency of AgNPs over time [30]. In unstable colloidal dispersions, the plasmon resonance peak shifts to longer wavelengths and broadens as the diameter increases due to aggregation [29]. Thus, aggregated NPs would exhibit red-shifted LSPR peak with respect to the peak of well-dispersed freshly synthesized NPs. The aggregation state of the NPs may also be revealed through determination of effective size of the NPs in solution through alternative techniques, such as, DLS. Silver NPs exhibit unique extinction peak corresponding to their size. Agnihotri et al. [29] reported SPR peaks in UV-Vis spectra at wavelength 393, 394, 398, 401, 406, 411, 420, 429, 449, and 462 nm corresponding to AgNPs of average size 5, 7, 10, 15, 20, 30, 50, 63, 85, and 100 nm, respectively (Figure 1.1a). Moreover, distinct colour was observed with change in size of the AgNP colloidal suspensions, as shown in Figure 1.1b.

The absorbance spectra of AgNPs is also reported to reflect alteration in shape of AgNPs. Such changes in UV-Vis-NIR spectra were illustrated during photochemical synthesis of Ag nanoprisms grown by illuminating small silver NP seeds (λ_{max} of 397 nm) with low-intensity light-emitting diodes (LEDs) [31]. As the seeds were converted to nanoprisms, the peak absorbance at 397 nm decreased over time and new

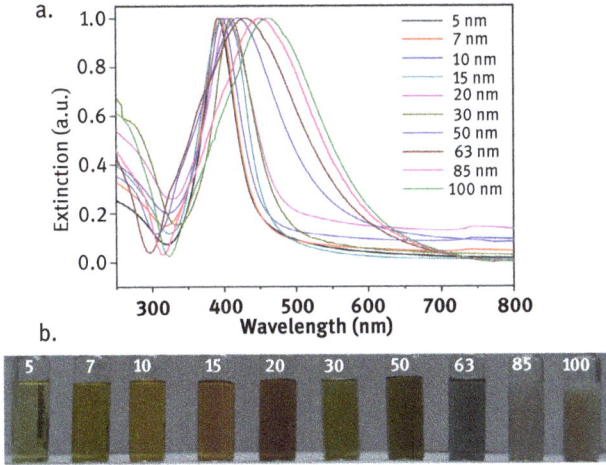

Figure 1.1: Variation in (a) UV-Vis absorption spectra and (b) distinctive colour of colloidal AgNP suspension of varying size (Reprinted from ref. [29], Published by The Royal Society of Chemistry).

peaks appeared at 1330 nm and 890 nm representing the LSPR of the nanoprisms formed. An intermediate stage of aggregated NPs was reflected by a red shift in the peak at 397 nm before the peaks due to nanoprisms appeared. Thus, UV-Vis-NIR spectra can reveal insights on the shape and size of NPs.

1.2.2 Transmission electron microscopy

High-resolution imaging of NPs below the wavelength of visible light is possible due to the fact that de Broglie wavelength of electrons is orders of magnitude lower compared to photons. Thus, if a beam of electrons is incident on NPs, fine details on shape, size and crystal structure of NPs is revealed in the transmission image that is formed. This technique is known as TEM. Further analysis of the image through image analysis software can provide information on uniformity in terms of shape and size, size distribution and aspect ratio thereby indicating success of a synthesis protocol in generating NPs of controlled size and shape. Imaging at various stages during the synthesis can illustrate the emergence of NPs of distinct shape as illustrated in Figure 1.2.

Figure 1.2a shows small quasi-spherical AgNP seeds and multiply twinned particles (MTPs) ~20 nm in size. Over 16 h, the seed and MTPs formed decahedral NPs with progressive increase in edge length. The edges were found to be irregularly etched (Figure 1.2f, Figure 1.2g) and were characterized by saw-toothed structures [32].

In addition to imaging, TEM can also be used for detection of diffraction ring pattern corresponding to a specific lattice plane/crystal structure. Selected area electron diffraction (SAED) pattern is generated due to scattering of electrons by

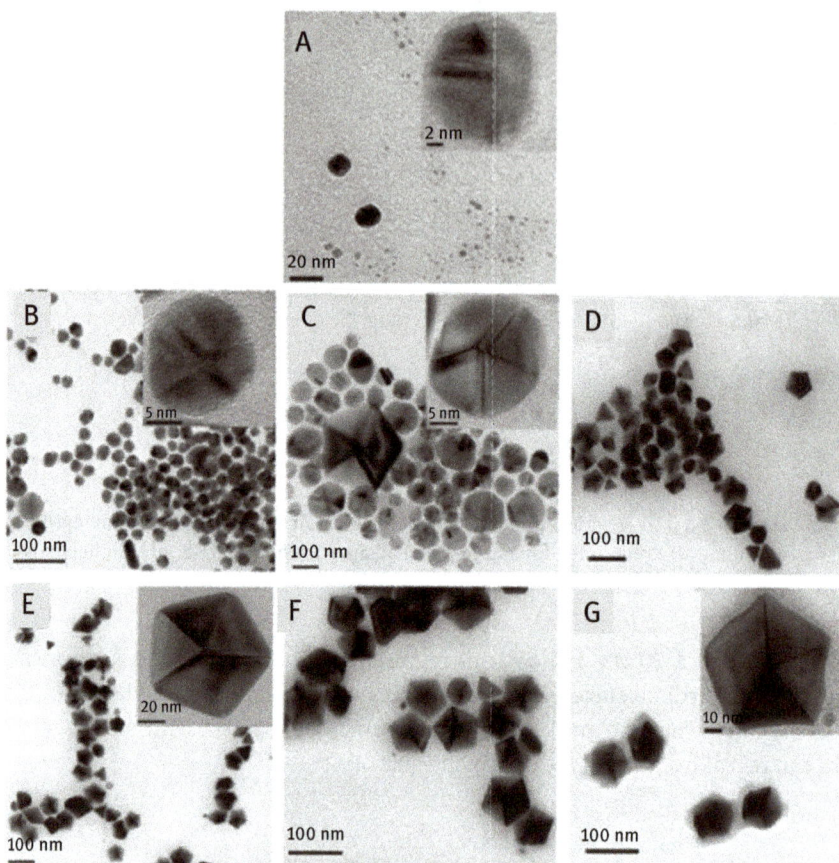

Figure 1.2: Time course of synthesis of decahedral AgNPs as revealed through TEM images at (a) 6 min, (b) 10 min, (c) 20 min, (d) 50 min, (e) 90 min, (f) 9 h and (g) 16 h (Reprinted with permission from ref. [32]. Copyright 2012, American Chemical Society).

the arrangement of atoms in a crystal. Thus, the SAED ring pattern over a selected area can be interpreted for determining the growth directions in a crystal and for phase identification. TEM image, SAED and dark-field images of icosahedral AgNPs synthesized via photo-assisted tartrate reduction method under UV (UV-B) irradiation over 48 h is shown in Figure 1.3 [33].

The brighter spots shown in Figure 1.3c can be grouped in concentric circles based on their distances from the centre. The radii of the circles were 2.67, 4.20, 5.79, 6.75, 7.39 and 8.38 nm^{-1} and the d-spacings determined from these radii were 0.374, 0.238, 0.172, 0.148, 0.135 and 0.119 nm, respectively. The d-spacing of 0.238, 0.148 and 0.119 nm corresponds to the {111}, {220} and {222} facets of AgNPs, respectively. Spot A was attributed to the {311} facets. These observations were also supported by the XRD spectrum [33].

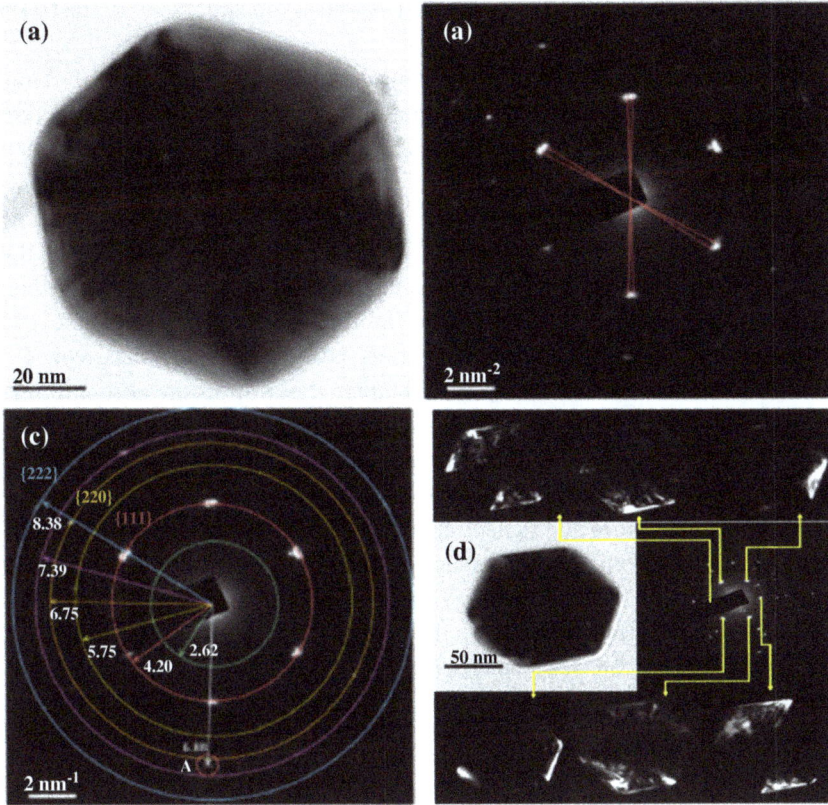

Figure 1.3: (a) TEM image (b) SAED pattern (c) assignments based on SAED (d) dark-field image and SAED pattern of an icosahedral AgNP (Reprinted with permission from ref. [33]. Copyright 2016, John Wiley and Sons).

1.2.3 Scanning electron microscopy and energy dispersive spectroscopy

SEM reveals the external morphology and texture and also provides insight on chemical composition of the surface and orientation of nanomaterial on the surface. Spatial variation in properties is illustrated through a two-dimensional image generated by scanning the specimen surface over a selected area using the signals from both secondary electrons (SE) and backscattered electrons (BSE). These electrons are generated from the surface of the specimen upon exposure to the electron beam irradiation. While inelastic collision and consequent scattering of the incident beam generates SEs, BSEs are produced by elastic collision followed by scattering. SEs, generally characterized by energy less than 50 eV, are effective in revealing the surface topography and morphology while BSEs are useful for demonstrating the contrasts that aids the visualization of samples containing multiple phases. BSEs can also be used to resolve atomic number contrast as well as topographical contrast with a resolution of more than 1 μm.

High-energy electron beam irradiation of a sample also generates an X-ray signal that can be employed to gain information regarding chemical composition of the specimen. The X-ray signal originates from below the surface of the specimen and provides elemental and chemical composition based on the characteristic X-ray signals and the characteristic energy peak. The technique is known as energy dispersive X-ray analysis (EDX or EDAX). The basic principle of EDX is that the detector detects the characteristic X-rays emitted from a specimen surface when bombarded with a focused electron beam. The intensity of characteristic X-rays emitted due to each element present on a surface is a function of its mass concentration or atomic fraction. EDX analysis is comprised of a spectrum with multiple peaks, where the peak position (eV) reveals the elements present in a sample. SEM images and EDS spectra of AgNPs immobilized on a surface is illustrated in Figure 1.4a–d. [34] While the SEM micrographs (Figure 1.4a–c) illustrate the size, shape, morphology, texture and surface coverage of AgNPs on the surface, the EDX spectra (Figure 1.4d) confirms the presence of silver in the specimen by exhibiting the characteristic peak of silver at 3 keV. The adjacent image reveals elemental mapping of silver and other elements on the surface. Apart from silver, presence of carbon, oxygen and silicon was also observed.

1.2.4 X-ray crystallography

X-ray crystallography/diffractometry (XRD) is a method used for determining crystal structure of NPs. Crystalline NPs cause a beam of X-rays to diffract into specific directions. Based on the angle and intensity of the diffracted beam, a three-dimensional electron density map is produced for a crystal. The XRD spectra generated is a manifestation of the electron density map, such that, insights can be obtained regarding the position of atoms in a crystal. Moreover, XRD can yield information on chemical bonds, their disorders and various other useful information. The diffraction peak position and pattern helps in identification of the crystalline phases present and the mean crystallite size of NPs may be determined using Scherrer's formula: $D = k\lambda/(\beta \cos\theta)$ where, D = size of the NP, k = 0.89 (shape factor used for spherical NPs), λ = 1.54056 nm (X-ray wavelength for Cuk_α), β is the line broadening, i. e., full width at half the maximum intensity (FWHM) expressed in radians, and θ is the angle obtained from 2θ value corresponding to an XRD peak. Peak broadening is inversely proportional to the size of the NPs; hence, significant peak broadening and correspondingly lower intensity of signals are found with smaller-sized NPs [35]. For AgNPs, a high-intensity diffraction peak at a 2θ value of 38.18º is commonly observed and it represents the (111) lattice plane of face-centred cubic (fcc) silver [36].

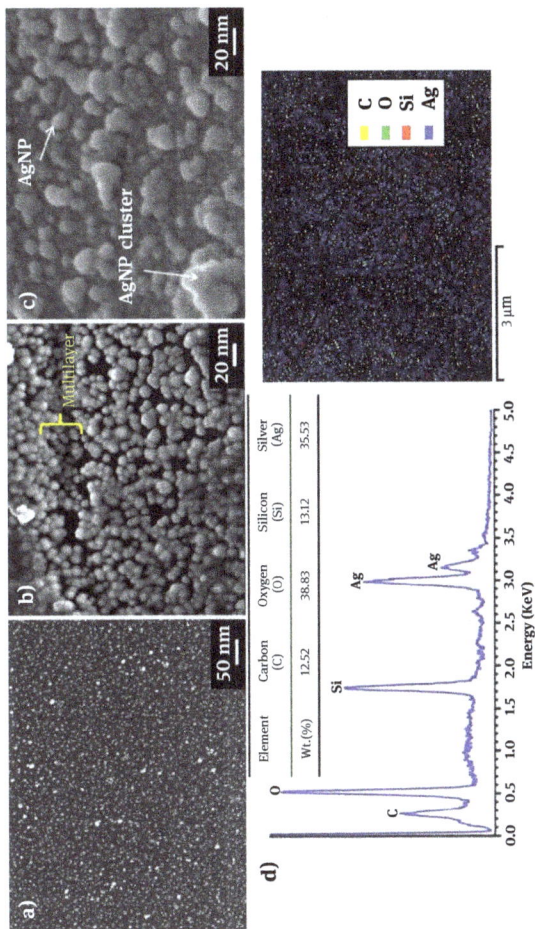

Figure 1.4: AgNPs immobilized on a glass substrate (a–c) FEG-SEM images, (d) EDX spectra (left image) and elemental mapping (right image) over an area (Reprinted from ref. [34]. Published by The Royal Society of Chemistry).

1.2.5 Fourier transform infra-red spectroscopy

In FTIR spectroscopy, as a sample is irradiated with infrared radiation, a part gets absorbed while the remaining is transmitted. The spectra generated based on molecular absorption and transmission represents a molecular fingerprint of the sample. It reveals unique molecular structures based on the functional groups present. This technique can be used to ascertain various unknown materials based on characteristic peak corresponding to a chemical species. It can also be used for determining the amount of components in a mixture depending on the vibrational spectra of the surface functional groups present in the sample. Its utility is illustrated through Figure 1.5 which depicts the presence of diverse functional groups in a leaf extract of *Skimmia lauriola* (A) and significant reduction in the abundance of various functional groups in the residual extract (B) after it was utilized for synthesis of silver NPs [37].

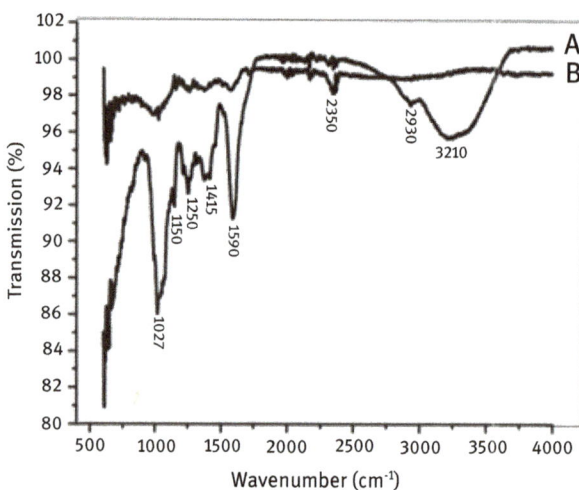

Figure 1.5: FTIR spectra of leaf aqueous extract of *S. laureola* before (A) and after (B) its use in AgNP synthesis. (Reprinted with permission from ref. [37]. Copyright 2015, Elsevier).

The various absorption peaks at wavenumbers, 1027, 1150, 1250, 1415, 1590, 2930 and 3210 cm^{-1} in the leaf extract corresponds to aliphatic amine (C–N stretch), alkyl halide (–CH X), aromatic amine (C–N stretch, C–C stretch), amine (N–H bend), alkane (C–H stretch) and carboxylic acid (O–H stretch), respectively. Reduction in these functional groups in the residual extract indicates the role of these biomolecules in AgNP synthesis.

1.2.6 X-ray photoelectron spectroscopy

X-ray photoelectron spectroscopy (XPS), also referred as electron spectroscopy for chemical analysis (ESCA) is useful for surface characterization in terms of its chemical state. XPS analysis is performed over average depth of approximately 5 nm. It is useful for determining the presence and chemical state of various elements associated with AgNPs synthesized as illustrated in Figure 1.6.

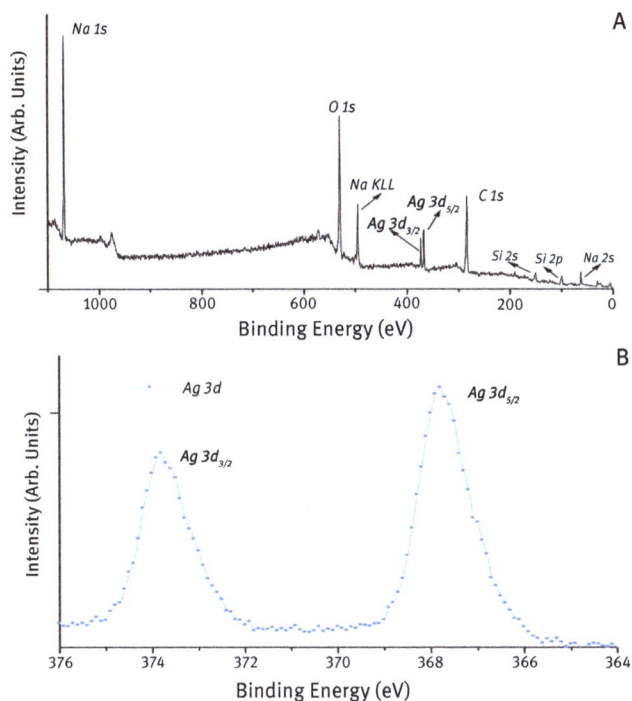

Figure 1.6: XPS spectra of AgNPs (A) Overall spectra and (B) Spectra due to Ag only (Reprinted with permission from ref. [38]. Copyright 2009, American Chemical Society).

The XPS spectra for AgNPs shown in Figure 1.6A exhibit characteristic peaks for Ag. The Ag (3d) region is characterized by a doublet with two peaks at binding energy of 373.8 and 367.9 eV (Figure 1.6B). These peaks may be attributed to spin-orbit coupling of Ag $3d_{3/2}$ and Ag $3d_{5/2}$. The XPS spectra also reveals characteristic peaks due to presence of various other elements in the AgNPs, such as, Na (1s) at 1070 eV, O (1s) at 530 eV, Na (Auger $KL_{23}L_{23}$) at 495 eV and C (1s) at 284.5 eV that may be attributed to the presence of residual reducing and stabilizing agent associated with the AgNPs. No Ag_2O peaks are observed for the XPS spectra shown in Figure 1.6 [38]. In another study, the XPS spectra of AgNPs demonstrated peaks at 374.5 eV and 368.4 eV corresponding to Ag $3d_{3/2}$ and Ag $3d_{5/2}$, respectively [34]. Deconvolution of the XPS spectra for Ag was found to reveal three additional peaks corresponding to binding energy of 367.7, 368.8 and 374.2 eV, confirming the presence of a small fraction of

AgO and Ag$_2$O along with metallic silver [34]. Thus, XPS analysis may reveal valuable information regarding the chemical nature of the surface.

1.2.7 Dynamic light scattering and zeta potential

Hydrodynamic size distribution and stability of NPs are important characteristics that should be determined after NPs are synthesized. DLS, also known as photon correlation spectroscopy (PCS), is widely used for determining the size distribution of NPs in a colloidal suspension. DLS works on the fact that NPs in colloidal dispersion move randomly due to Brownian motion and hence cause scattering of the incident light. The diffusive transport of NPs through a liquid of low Reynold's number can be explained by Stokes Einstein equation ($D = k_B T/6\pi\eta R$, where, D denotes the translational diffusion coefficient, k_B is the Boltzmann's constant, R is the hydrodynamic radius of the particle and η is the dynamic viscosity of the solvent). The scattering of incident light is proportional to the 6th power of the radius of the NPs. When the size of NPs is less than 1/10th of the wavelength of incident light, elastic scattering referred as Rayleigh scattering occurs. In contrast, when the size of the NPs exceeds this threshold, inelastic and anisotropic scattering, referred as Mie scattering occurs. The scattering of light is dependent on various factors such as concentration, shape and size of the particles as well as the refractive index of the solvent. The scattering of light from individual NPs may get affected by other particles due to random Brownian motion of all the NPs in solution. At higher concentration, the inter-particle interactions also become significant as the number of collisions increase; however, the average path length travelled by the particles between successive collisions decrease [39].

Scattering of incident laser light is detected in DLS. The intensity of the scattered light show fluctuations due to the random motion of the NPs. Statistical analysis of such intensity variations can yield the diffusion coefficient, which is then used for estimating the hydrodynamic radius R as per Stokes Einstein equation. Hydrodynamic radius is typically much larger than the radius of the NPs determined through TEM. DLS measures size distribution of NPs is a colloidal dispersion based on intensity of scattered light whereas in TEM a direct measure of particle size is obtained from the image of the NP and size distribution may also be obtained through image analysis of a sufficiently large number of NPs. In addition, while DLS measures the hydrodynamic radius, TEM captures images of NPs based on the projected surface area by the incident electrons transmitted through the sample in dry condition, under high vacuum. The advantage of determining size distribution using DLS is that it provides the size distribution based on measurement of millions of NPs. In contrast, TEM gives size distribution based on only a few hundred NPs. Hence, DLS can be a preferred choice when determination of size distribution and polydispersity is a primary objective [39]. For non-spherical NPs, the hydrodynamic radius obtained is a hypothetical value, and in some cases, e. g., cylindrical NPs, the length, diameter and aspect ratio can also be derived from the hypothetical radius.

Presence of charge (+/−) on the surface of NPs impart stability to the NPs and prevents aggregation over time due to electrostatic repulsion between particles carrying like charges. This is achieved using appropriate stabilizing agents during the synthesis of NPs as discussed in later sections. The charge on the surface of NPs is indicated by the zeta potential, since the exact potential on the particle surface, i. e., Nernst potential cannot be directly measured. Zeta potential of NPs is the potential at the plane of shear/slipping plane. An electric double layer (EDL, comprising of the Stern layer and Gouy Chapman layer) enriched in oppositely charged ions forms around charged particles in solution. When a charged particle moves in an electric field, a part of the EDL up to the slipping plane moves with it. The potential developed on the slipping plane, i. e., zeta potential, can be determined based on electrophoretic mobility. Often DLS instruments have provision for determination of zeta potential also. During electrophoresis, NPs scatter the incident laser beam. The scattered beam from mobile NPs have different frequency than the incident laser light and the frequency shift is proportional to the velocity of the NPs. Zeta potential can be determined based on this velocity. It may be emphasized that although zeta potential reflects the particle surface charge, it is also affected by water chemistry, i. e., pH and ionic strength of the solution (I). Depending on the ionic strength, the value of zeta potential may change, i. e., the zeta potential becomes less negative for negatively charged particles as "I" increases, even when the particle surface charge remains unaltered [40]. NPs with magnitude of zeta potential exceeding 25 mV is reported to demonstrate a high degree of stability [41]. When AgNPs synthesized are characterized by low zeta potential values, they are found to aggregate spontaneously due to Van der Waal interactions. Thus, determination of zeta potential gives insight on stability of the NPs formed.

1.3 Synthesis of silver NPs

Primarily two approaches are used for fabrication of metal NPs, i. e., top-down approach and bottom-up approach (Figure 1.7). The former deals with the synthesis of NPs using a bulk material, in which NPs are directly generated from the bulk material using techniques such as milling, pyrolysis and photolithography [42]. The size of NPs produced using these approaches are between 10 and 100 nm. The surface structures commonly show imperfections that adversely affect the physical and chemical properties of the NPs. Thus, top-down approach cannot be gainfully applied in most applications. Bottom-up approach involves the construction of complex clusters and NPs starting from individual atoms, molecules or clusters by employing processes, such as, chemical reactions coupled with nucleation and growth processes [42].

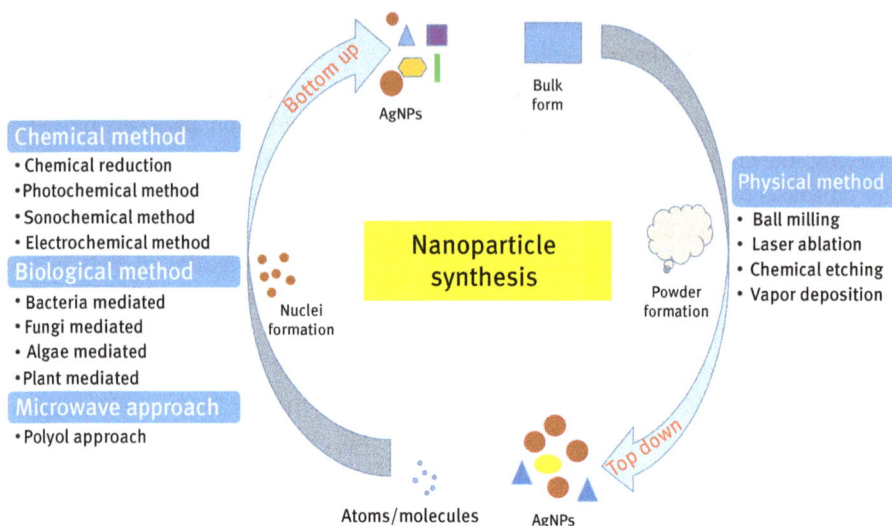

Figure 1.7: Nanoparticle synthesis: top-down approach versus bottom-up approach.

1.3.1 Top-down approaches

The top-down synthesis route involves synthesis of nanomaterials by breaking down the bulk material using various physical forces. The physical energy for size reduction may be applied as: mechanical energy that is utilized in processes, such as, ball milling, crushing or grinding; electrical energy that is utilized in processes, such as, electrical arc discharge method, high-energy laser ablation method, or as photon beams; and thermal energy, such as in pyrolysis. A detailed discussion is provided in the subsequent sections.

1.3.1.1 Physical method
The predominant physical processes are classified as mechanical processes and vapour-based processes. Mechanical methods involve high-energy ball milling techniques, which can be achieved in mills such as planetary, vibratory, rod or tumbler-based mills. Vapour-based techniques include pyrolysis, physical vapour deposition, electric arc discharge and laser ablation. In pyrolysis, an organic precursor (liquid/gas) is forced through a vent at elevated pressure and burnt. Oxidized NPs may be recovered from the ash generated. In the evaporation/condensation method, metal NPs are usually synthesized in a tube furnace maintained at atmospheric pressure. The material of interest is vaporized into the carrier gas from the precursor placed in a boat held within a furnace. Diverse types of NPs including AgNPs have been successfully synthesized using the evaporation/condensation

technique. The advantage of physical methods includes uniform size distribution of NPs and high purity of the NPs formed [43].

1.3.1.1.1 Ball milling/attrition

The attrition/ball milling technique consists of a container filled with hardened steel or tungsten carbide balls and air or inert gas, where the material of interest is fed as flakes maintaining a specific mass ratio of balls to the desired material. The container is rotated at a high speed around a central axis, where the desired material is forced to move towards the wall and pressed against it. The factors that can be controlled to produce NPs with desired characteristics include varying the speed of rotation, duration of milling to grind the material and the atmospheric medium, which give rise to NPs of size range of a few nm to few tens of nm. In case of melt mixing method, the NPs formed are arrested in glass, an amorphous solid lacking symmetric arrangement of atoms/molecules. The basic principle is that when metals are cooled at a very high cooling rate (10^5–10^6 K/s), the metals can form amorphous solid metallic glasses. The mixing of molten streams of metals at very high velocity with turbulence leads to the formation of NPs.

Khayati and co-workers [44] synthesized silver nanopowder of average size 28 nm in a high-energy planetary ball mill by mechanochemical reduction of Ag_2O (5–40 µm) using 40 mol% excess graphite (10–50 µm) as reducing agent. Excess graphite acted as a diluent and facilitated maximum interaction between the precursors during synthesis. The parameters affecting synthesis using ball mill machine were: rotation speed of the disc of diameter 350 mm (250 rpm), rotation speed of hardened chromium steel vials of 90 mm diameter having 150 ml capacity (450 rpm), ball (20 mm diameter, made up of hardened carbon steel) to powder weight ratio (20:1) and time of milling (0–22 h) in dry milling condition with total powder mass of 9.75 g. With increase in milling time, the Ag_2O peak in XRD spectra gradually broadened and its height decreased and Ag peaks in the XRD spectra appeared between 3 and 6 h of milling [44]. Thus, the reduction reaction was initiated after a period of 3 h. Moreover, increase in intensity of the Ag peaks was obtained on further milling and a single phase of Ag was obtained only after milling for 22 h. The rate of transformation decreased at the end of milling process, i. e., beyond 22 h. Thus, the reduction process started with a nucleation phase and was followed by a growth phase. The activation energy for nucleation depends on the degree of deformation. In this case, the energy barrier got reduced due to the large number of imperfections generated by mechanical milling. Larger amount of nuclei was generated after 6 h of milling and hence synthesis of AgNPs was enhanced as large amount of reactants were available for reduction. Thus, the AgNP synthesis rate improved, such that 72% reduction of Ag_2O occurred during 6–16 h of milling.

Cryomilling is a variation of the ball milling technique, where ball milling is performed at a very low temperature (i. e., −160 ± 10°C) under an inert argon atmosphere. Conventionally, it can be achieved by mixing the silver powder with liquid nitrogen (LN) during ball milling to achieve a lower temperature. However, the mixing of LN with silver powder during ball milling process poses a challenge. Kumar et al. [45] synthesized highly pure spherical AgNPs (size: 4–8 nm, with impurities <35 ppb) using an advanced version of cryomill containing a single tungsten carbide (WC) ball over 7 h. The ball to silver powder (40 μm) ratio was maintained at 80:1. The surfactant-free AgNPs synthesized were found to be thermally stable and could withstand a temperature up to 350°C, whereas AgNPs synthesized through traditional ball milling technique are reported to coalesce at 200°C and form micron-sized silver particles within just 30 min of annealing. In contrast to conventional cryomilling, this custom-built cryomill did not require direct mixing of LN with the silver powder during ball milling. The custom-built cryomill had a special reservoir surrounding the milling chamber which was maintained at –160°C during the milling process. This approach could generate large quantity of AgNPs in a single ball mill i. e., ~60–70 g/day. The extremely low temperature used in cryomilling provided additional advantages. Cooling the silver powder and milling ball facilitated the formation of finer grain structure, accelerated the fracturing process and suppressed the recovery and recrystallization of particles such that size-controlled AgNPs (6 ± 2 nm) could be formed.

1.3.1.1.2 Physical vapour condensation

For NP synthesis, the vaporization method has been widely used. In this method, the target materials are evaporated by a heat source followed by rapid condensation. Based on the reactions, the vaporization process can be divided into physical and chemical methods. Physical vapour condensation (PVC) methods are used when the resultant NPs have the same composition as the target material. However, when NPs have to be of different composition than the target, chemical vapour condensation (CVC) method is used. In CVC, the vapour and other system components react with each other during the vaporization and condensation processes [46].

1.3.1.1.3 Electrical arc-discharge method

The electrical arc discharge method can produce high-purity NPs and is also characterized by high yield. It is a cost effective and environment-friendly method that utilizes a DC power supply and an open vessel containing either ultra-pure/deionized water, or alcohol [47, 48]. Tien et al. [47] synthesized AgNP suspension in water using DC arc-discharge between two silver electrodes immersed in pure water [49]. During arc-discharge, the Ag electrodes used were etched in the aqueous medium and the surface layer was vaporized due to the very high temperature (~1000°C) in the vicinity of the electrodes. Subsequently, the metal vaporized was condensed as AgNPs in the dielectric liquid. To effectively ionize the aqueous medium between the electrodes, a pulse voltage of 70–100 V was applied for 2–3 ms using the DC arc-discharge system and the voltage was later lowered to 20–40 V for around 10 ms.

AgNP formation in this system was facilitated by four effects, i. e., the strong electric field, a high temperature plasma, release of active silver atoms from the silver electrodes and release of hydrogen and oxygen atoms from water molecules. The AgNPs initially formed upon condensation were positively charged such that atomic oxygen was attracted to it and water molecules also formed hydrogen bonds with the oxygen on the surface (shown as dashed line in Figure 1.8). This resulted in a thermodynamically stable colloidal suspension of AgNPs, of average size 20–30 nm [49].

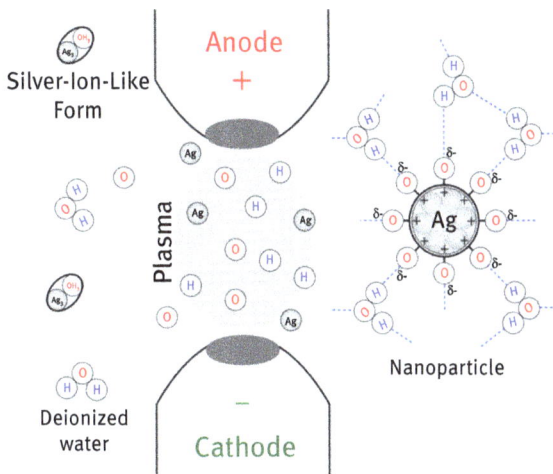

Figure 1.8: Plasma and electrical discharge contributing to the formation of a stable suspension of AgNPs (Reprinted with permission from ref. [49]. Copyright 2008, Elsevier).

Rashed [48] also synthesized AgNPs using this method by applying a stable voltage of 130 V between two silver electrodes immersed in deionized water maintaining a distance of ~1 mm between the electrodes. The short distance between the electrodes caused generation of electric spark which caused passage of ~35 mA electric current through the water. The silver atoms vaporised from the anode surface later condensed in water to form AgNPs of size ~8.4 nm through nucleation mechanism.

1.3.1.1.4 Laser ablation method

The laser ablation technique illustrated in Figure 1.9 involves generation of non-equilibrium vapour/plasma at the surface of a material using a strong laser pulse. It differs from the laser evaporation process where heating and evaporation of material occurs under equilibrium conditions. A high-power pulsed laser together with an ablation chamber [50] is used in the laser ablation device. The temperature of the target is elevated due to absorption of the high-power laser beam which causes vaporization of atoms from the surface of the target into the laser plume.

Ti:Sapphire or OPO laser

focusing lens

holder

target

distilled
water

quartz cell

magnetic
stirrer tip

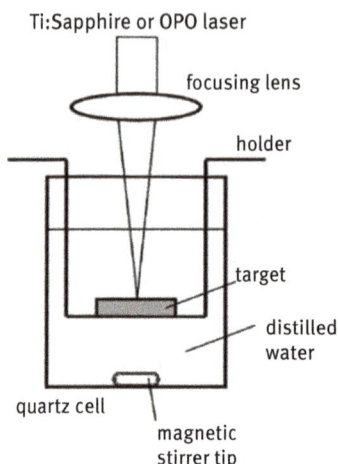

Figure 1.9: Schematic illustrating laser ablation in solution (Reprinted with permission from ref. [50]. Copyright 2002, Elsevier).

The vaporized atoms may either condense as clusters and particles without any chemical reaction or may react in the vaporized state to yield new materials. The condensed particles may be collected through deposition on a substrate or by passage through a glass fibre mesh-type filter.

Synthesis of spherical AgNPs have been illustrated using both nanosecond and femtosecond laser pulses using apparatus as illustrated in Figure 1.9. However, although femtosecond laser pulses yielded AgNPs with narrower size distribution (20–50 nm), its efficiency was found to be lower [50]. The ablation efficiency in air and water medium has also been compared for these two types of laser. While nanosecond laser ablation resulted in comparable efficiency irrespective of the medium, femtosecond laser yielded higher efficiency in air compared to water [51]. Apart from duration of laser pulse and use of air/water media, other factors that influence the ablation efficiency and characteristics of the AgNPs formed are wavelength and fluence of the laser, duration of the ablation process and the presence of surfactants in liquid media [43]. Only a sufficiently high laser fluence (power) can eject metal particles from the target. However, as fluence increases, the size of NPs formed also increases. Thus, the minimum threshold fluence is commonly used for generating small NPs. The wavelength of the laser and time duration of irradiation also controls the concentration and morphology of the NPs formed. NP concentration increases with increase in irradiation time until a saturation value is reached. In a highly concentrated colloidal suspension of NPs, light absorption is adversely affected, leading to saturation in concentration of the NPs.

The effect of laser wavelength on the size of AgNPs formed is illustrated in Figure 1.10 based on a study conducted by Tsuji et al. [50]. They found that AgNPs prepared using lower wavelength laser yielded AgNPs with narrower size distribution. The mean size of AgNPs formed was significantly different at

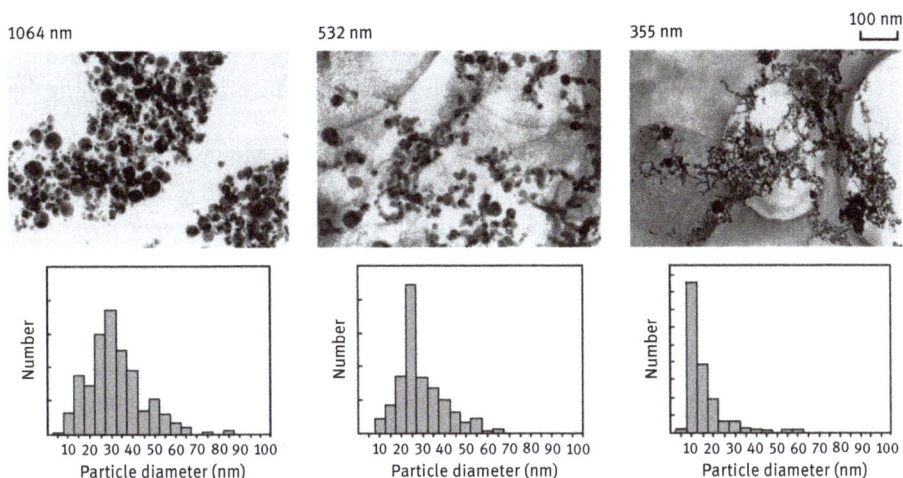

Figure 1.10: TEM images of silver particles in colloidal solution prepared by laser of varying wavelength (100, 1000X). (Reprinted with permission from ref. [50]. Copyright 2002, Elsevier).

532 nm (26 ± 11 nm) and 355 nm (12 ± 8 nm) (Figure 1.10). Alteration in laser wavelength affects the ablation efficiency, self-absorption and penetration depth of the laser beam such that the mean size and size distribution of NPs formed is altered. The characteristics of NPs formed by laser ablation in an aqueous medium is also affected by the type and concentration of surfactants added to the medium. Smaller NPs can be formed by increasing the surfactant concentration [43].

In a study by Mafune et al. [52], AgNPs were synthesized using laser ablation on a silver plate immersed in an aqueous solution containing a surfactant, e. g., sodium dodecyl sulphate (SDS). The surfactant was used to prevent aggregation of the NPs. It surrounded each NP thereby preventing them from coming in direct contact with each other. Synthesis of AgNPs was accompanied by irradiating the silver plate placed in 10 mL aqueous solution of SDS in a glass beaker. Irradiation was done using 532 nm Nd:YAG laser (operated at 10 Hz, using a lens with focal length 250 mm) and varying output power of 40, 55, 70 and 90 mJ/pulse. For output power of 90 mJ/pulse, the diameters of AgNPs synthesized were 16.2 ± 4.0, 14.9 ± 8.4 and 11.7 ± 5.3 nm for SDS concentration of 3, 10 and 50 mM, respectively. Thus, a gradual decrease in size of AgNPs was observed with increase in surfactant concentration. For AgNPs synthesized at constant SDS concentration of 10 mM with variable output power, the diameters of AgNPs formed were 7.9 ± 3.3, 10.7 ± 5.8 and 12.8 ± 4.1 nm for power output of 40, 55 and 70 mJ/pulse, respectively. AgNP synthesis using laser ablation was also attempted using dodecanethiol solution in heptane. Synthesis of AgNPs was revealed through colour change of the solution to light brown observed after 10 h and TEM micrographs revealed synthesis of AgNPs with size range 5–30 nm.

Dodecanethiol solution in heptane generated significantly lower yield of AgNPs compared to aqueous solution of SDS. Thus, the concentration and type of surfactant, solvent used and power output of the laser plays an important role in determining the yield and the characteristics of the AgNPs synthesized.

Bae et al. [53] fabricated AgNPs using pulsed laser ablation of a target silver plate placed in the bottom of a glass vessel containing NaCl solution. AgNPs were synthesized by laser irradiation of a silver plate with 1 mm spot size using Nd: YAG laser beam (wavelength: 1064 nm, 5 ns pulse). Effect of NaCl concentration (5–20 mM) on AgNP synthesis was also tested. Maximum yield of AgNPs was observed for 5 mM NaCl concentration and the AgNPs synthesized had average size of 26 nm (range: 5–50 nm). The absorbance due to LSPR (at λ_{max} = 400 nm) increased from 0.86 to 1.45 with increase in NaCl concentration from 0 to 5 mM; however, the FWHM decreased over this range. The increase in yield, smaller size of the AgNPs formed and greater aggregation tendency observed were attributed to the presence of chloride ions. Increase in absorbance at λ_{max} was not observed when NaCl was added after AgNPs were synthesized in pure water. However, AgNPs prepared in pure water exhibited better stability as compared to those synthesized in NaCl solution. Increase in NaCl concentration beyond 5 mM caused further decrease in stability of the AgNPs synthesized and promoted aggregation.

Silver NP synthesis through laser ablation in organic solvent was reported by Amendola et al. [54]. Synthesis of AgNPs was performed using a silver plate within a cell containing the solvent. Laser ablation was performed with 1064 nm Nd:YAG laser pulses (9 ns pulses for 10 min) with output power of about 10 J/cm². Various organic solvents such as, acetonitrile, N,N-dimethylformamide, tetrahydrofuran and dimethyl sulfoxide were used for AgNP synthesis. Spherical and crystalline AgNPs with average size of 1.9 ± 1.5 nm and 2.2 ± 2.5 nm were synthesized in acetonitrile and dimethylformamide, respectively. However, a few nm thick amorphous shell was observed surrounding the AgNPs synthesized in tetrahydrofuran. Average size of the core AgNPs was found to be 2.4 ± 1.1 nm, whereas the thickness of the amorphous shell varied from sub-nanometre to several nanometres. AgNP synthesis could also be achieved in dimethyl sulfoxide and the average size of AgNPs formed was determined as 3.9 ± 1.9 nm.

In a recent study by Sebastian et al. [55], AgNPs were synthesized by laser ablation using a silver plate placed within ethanol containing PVP as stabilizing agent. They also explored laser-induced transformation of spherical AgNPs. Initially spherical AgNPs were synthesized using 1064 nm Nd-YAG laser (9 ns pulses) as the source and laser ablation was conducted for 20 min in an ethanolic solution of PVP (6 M, MW 10,000). Stable spherical AgNPs (average size: 45 nm) having fcc crystal structure were obtained. The spherical AgNPs synthesized were subjected to structural transformation by irradiating the solution with 532 nm laser. After 45 min of irradiation with 532 nm laser, a colour change from light to dark yellow was observed in the colloidal AgNP solution. TEM micrographs revealed fragmentation of almost all larger AgNPs (size: 45 nm) into smaller fragments of size up to 4 nm. Upon irradiation using 532 nm laser for up to 9 h, the fragmented NPs (of size range 1–4 nm) assembled to produce stable Ag nanorods (average width 160 nm, length 8.7 µm and average aspect ratio: 54). However, since PVP restricted growth along specific crystallographic facets, it also facilitated transformation of the nanorods to nanowires. Hence, a mix of nanorods and nanowires was obtained upon ripening the PVP solution.

In general, top-down approaches for AgNP synthesis has various limitations, such as, consumption of large amount of energy that is required for raising the temperature around the source material. Correspondingly large time duration is required for achieving thermal stability. Moreover, most of the synthesis methods discussed in the preceding section are less amenable for synthesis of size- and shape-controlled AgNPs. Hence, bottom-up AgNP synthesis approaches are often preferred. These approaches are discussed in the following section.

1.3.2 Bottom-up approaches

Crystal growth is a bottom-up approach, where atoms, ions or molecules are assembled into the desired crystal structure on a growth surface. The size of the NPs synthesized using the bottom-up approach covers the full nano scale. Bottom-up approaches for NP synthesis improves the probability of achieving specific shape and size, reduces defects in the crystal structure and can provide greater homogeneity in chemical composition. Since NPs are synthesized by reducing the Gibbs free energy, they remain in a state that is closer to thermodynamic equilibrium [56]. Therefore, bottom-up approaches are found more useful and are widely used for synthesis of AgNPs.

As illustrated in Figure 1.11, two main reaction phases completes the formation of AgNPs in the bottom-up approach: nucleation, i. e., the formation of nuclei; and growth of NPs through coagulation and coalescence processes. During synthesis of NPs, aggregation (coarsening) through either Ostwald ripening or agglomeration over time is a spontaneous phenomenon since it reduces the surface to volume ratio. To synthesize NPs of controlled size, a key challenge is to overcome the large surface energy of the NPs and resist aggregation [57, 58]. The nature of the formulation and various process parameters, such as, temperature, reaction time, pressure and degree of mixing can be optimized to control the growth of NPs. The formulation typically contains diverse chemicals serving as precursor salt, reducing agent and stabilizing agent/capping agent. The composition of the formulation and concentration of various chemicals along with the order of addition may affect the process as discussed in later sections. The commonly used bottom-up methods include chemical reduction and biological reduction of a precursor salt. Biological

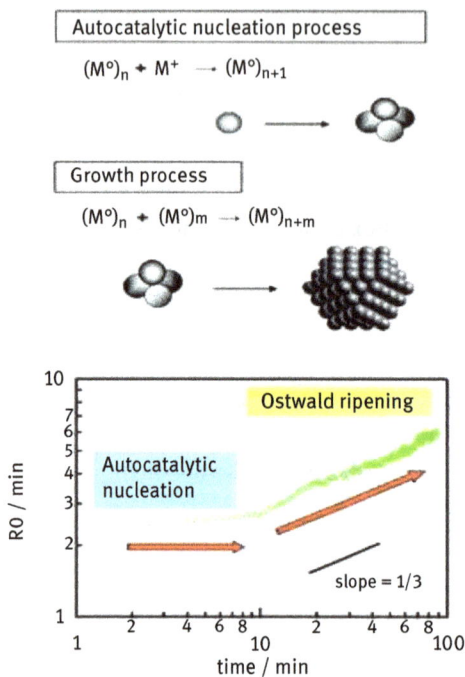

Figure 1.11: Change in particle radius and proposed mechanism for formation of silver nanoparticles (Reprinted with permission from ref. [57]. Copyright 2010, American Chemical Society).

reduction commonly relies on microbial enzymes, extracellular polymeric substances and other components that are released into the culture broth or involve the use of plant extracts [59, 60]. Chemical reduction [29, 61, 62] is commonly facilitated by heating while biological reduction may proceed under ambient conditions due to presence of enzymes that serve as biocatalysts. Although commonly coupled with conventional heating, chemical reduction is also sometimes coupled with alternative energy sources, and the processes are subsequently designated based on the energy source used. These include electrochemical methods [63], photochemical methods [64], sonochemical methods, Y-radiation-based methods [65] and microwave-based methods. A detailed description of various bottom-up synthesis approaches is provided in subsequent sections with specific reference to size- and shape-controlled synthesis of AgNPs.

1.3.2.1 Mechanism of bottom-up synthesis of shape- and size-controlled silver NPs

Most of the bottom-up techniques result in synthesis of spherical NPs. However, recently there is much interest in shape-controlled synthesis and researchers have successfully demonstrated synthesis of NPs with specific shape using chemical and photochemical method. Synthesis of AgNPs via bottom-up approach requires a metal precursor salt, a reducing agent and a stabilizing/capping agent. The most commonly used Ag precursor salt is silver nitrate ($AgNO_3$) [66] whereas, sodium borohydride ($NaBH_4$) [67], polyol (ethylene glycol, EG) [68] or sodium citrate (TSC) [69] is frequently used as reducing agent. Use of various stabilizers/capping agents, such as, TSC [70], cetyltrimethylammonium bromide (CTAB) [67], bis(p-sulfonatophenyl)phenyl phosphine dipotassium dehydrate (BSPP) [70], polyvinylpyrrolidone (PVP) [68], polyacrylamide (PAM) [71], poly((sodium styrenesulfonate) (PSSS) [72], chitosan [73] and SDS [74] have been reported. The nucleation and growth of NPs is governed by various reaction parameters, such as, reaction temperature, pH and type of precursor, reducing agent and stabilizing agent, precursor concentration, molar ratio of surfactant/stabilizer and precursor [75]. In general, nuclei formation is achieved by reduction of the metal precursor salt by the reducing agent. This may either be achieved directly by the reducing agent or the process may be assisted by providing energy in various forms including conventional heating, microwave heating and light energy. The nuclei generated undergoes growth process through Ostwald ripening followed by coalescence leading to the formation of NPs. However, as already highlighted, the NPs may aggregate over time. Stabilizing/capping agents are required to prevent aggregation of the NPs. Stabilization can be achieved either through electrostatic or steric repulsion. Electrostatic stabilization is usually achieved through anionic species, such as, citrate, halides, carboxylates or polyoxoanions that adsorb or interact with the AgNPs to impart a negative charge on the surface. The surface charge on the AgNPs produces an EDL resulting in coulombic repulsion between the NPs. Steric stabilization can be achieved by interaction of NPs with bulky groups, such as, organic polymers and alkylammonium cation that prevent aggregation through steric repulsion [43]. Addition of stabilizers or surfactants may also facilitate directed growth of NPs thereby generating NPs of specific shape. Selective adsorption of the surfactant/stabilizer to specific crystal facets constraints the growth along specific crystal axes resulting in the synthesis of NPs characterized by distinct shape [76]. The various approaches towards control of size and shape are discussed in the following sections.

1.3.2.1.1 Chemical reduction

Three major group of methods are reported for shape- and size- controlled synthesis of AgNPs based on the reducing agent employed, i. e., polyol -based synthesis, citrate-based synthesis and borohydride-based synthesis. The specific shapes that have been synthesized through these three broad group of methods are illustrated in Figure 1.12, Figure 1.13 and Figure 1.14.

Polyol-based synthesis has been extensively studied for achieving various shapes and sizes of AgNPs. Spherical AgNPs of average size 10–30 nm can be obtained by proper choice of PVP to $AgNO_3$ mass ratio (such as, 2:1) [77, 78] and is sometimes facilitated by using ammonia in the formulation. Wiley et al. [79] explored variations of the polyol method for synthesizing AgNPs of diverse shapes, such as, pentagonal

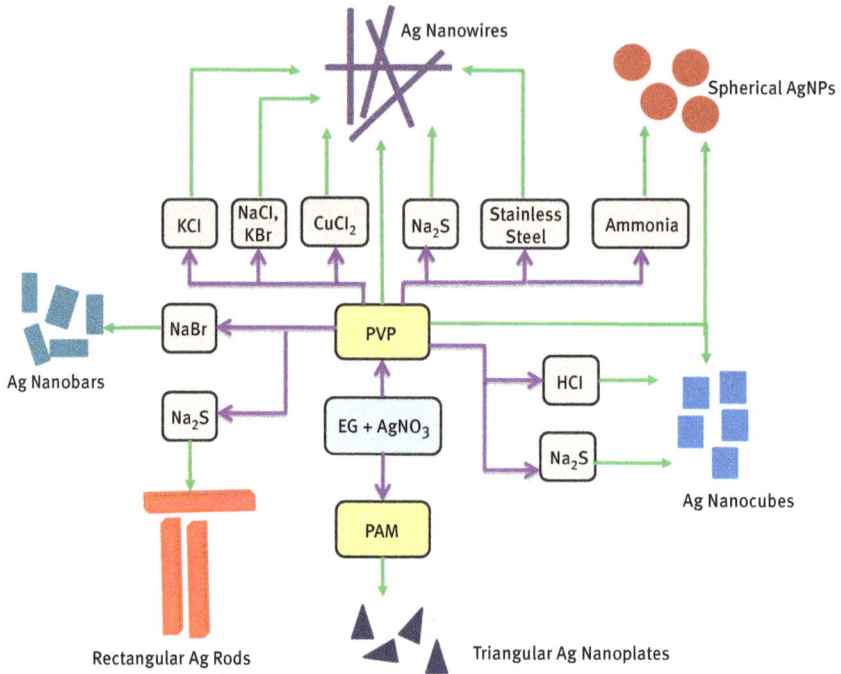

Figure 1.12: Schematic demonstrating AgNP synthesis via chemical route using polyol -based methods highlighting the various shapes that can be achieved through appropriate choice of reagents.

AgNPs, nanowires, cuboctahedra, nanocubes, nanobars, bi-pyramids and nanobeams. In a typical polyol method, EG, $AgNO_3$ and PVP are used as the polyol, salt precursor and polymeric capping agent, respectively. At elevated temperature, silver ions are reduced by EG to silver atoms and subsequently nuclei are generated. These nuclei grow to form multiply twinned, singly twinned or single-crystal seed as shown in Figure 1.15. These seeds grow to yield nanostructures of diverse shapes.

Crystallinity of the seeds plays a critical role in controlling the shape of AgNPs and Ag nanostructures [79]. In the presence of controlled amount of chloride and iron impurities, multiply twinned decahedral seeds grow to form nanowires whereas single crystal seeds grow to form silver nanocubes. When NaCl is replaced with NaBr and a specific NaBr concentration is employed, single crystal seeds grow into nanobars and seeds with a single twin plane grow to produce right bipyramids. In this approach, PVP facilitates synthesis of shape-controlled AgNPs by binding strongly with {100} crystal facets of the seed. This allows directed growth of AgNPs and yield AgNPs with specific shapes.

Presence of Na_2S in polyol-based synthesis can generate nanowires, nanocubes and rectangular nanorods. On addition of Na_2S, Ag_2S produced in the early stage helps to reduce the concentration of Ag^+ ions in the initial stage and thus controls the

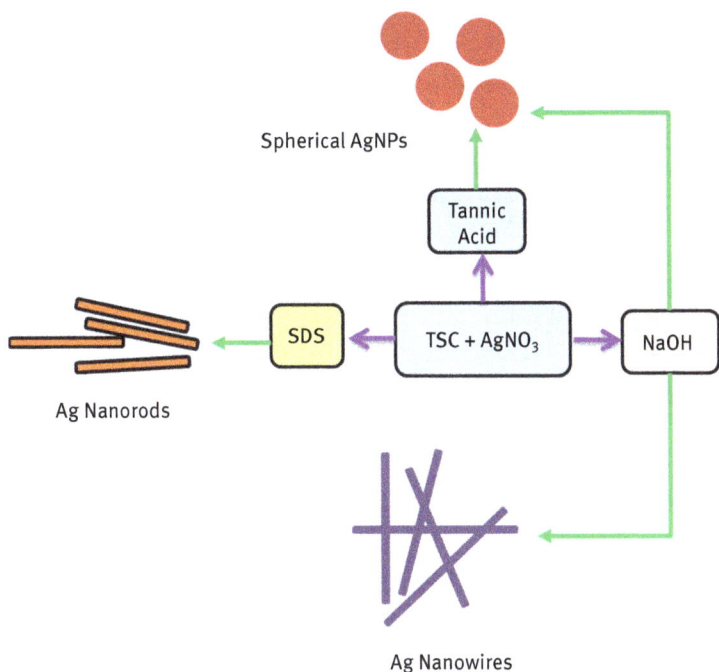

Figure 1.13: Schematic demonstrating AgNP synthesis via chemical route using citrate-based methods highlighting the various shapes that can be achieved through appropriate choice of reagents.

formation of silver seeds. In a successive reaction, Ag^+ ions from Ag_2S colloids are released into the solution [80]. This facilitates nanowire formation. The length and diameter of the nanowires can be adjusted by controlling the Na_2S concentration [80]. By lowering the surface energy of {100} facet of the seeds by controlling the PVP concentration and by controlling the concentration of sulfide ions (28–30 μM) at a temperature of 150–155°C, nanocubes can be obtained in less than 10 min [81]. Rectangular nanorods can be synthesized by maintaining molar ratio of PVP to $AgNO_3$ at 0.93 and by increasing the reaction time sufficiently[82]. Nanowires can be synthesized upon addition of $CuCl_2$ and KCl in polyol-based synthesis. This is facilitated by formation of AgCl nanocrystallites that reduces free Ag^+ ions during the initial stages of Ag seed formation. The subsequent slow release of Ag^+ ions helps in ensuring high yield of multiply twinned Ag seeds necessary for nanowire synthesis [83, 84]. However, the nanowires formed have narrower diameter when grown in the presence of bromide ions [85].

Using TSC as reducing agent (Figure 1.13), spherical AgNPs and Ag nanowire can be produced by controlling the pH of the reaction mixture with NaOH [69, 86]. Nanowires are formed at pH 6.5. Formation of spherical AgNPs is facilitated by rapid nucleation at solution pH slightly above neutral (7.7) followed by lowering the pH to 6.1 during the growth stage. SDS plays an important role in nanorod synthesis. SDS

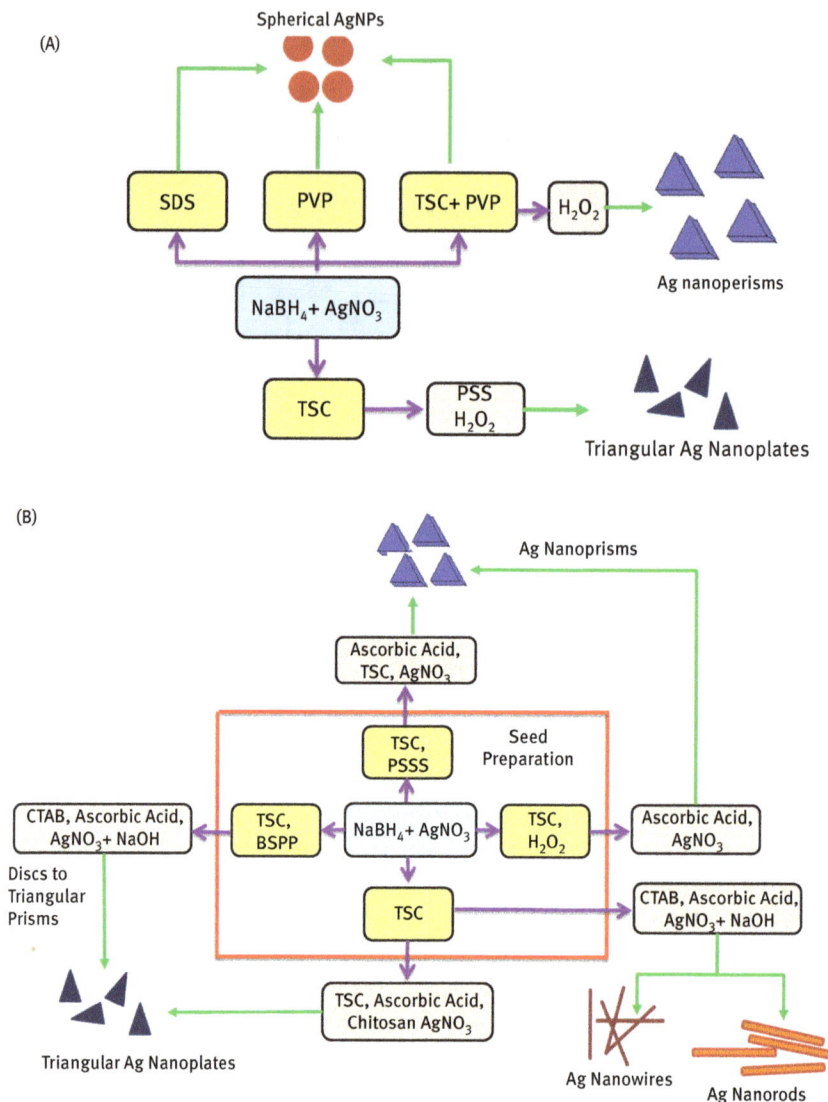

Figure 1.14: (A, B): Schematic demonstrating AgNP synthesis via chemical route using borohydride-based method highlighting the various shapes that can be achieved through appropriate choice of reagents.

micelles are produced in aqueous solution when its concentration exceeds the critical micelle concentration (CMC). The shape of the micelle depends on SDS concentration. As SDS concentration increases, the shape of the micelles changes from spherical to cylindrical. The AgNPs growing inside SDS micelles also takes the shape of the micelles, such that nanorods are formed at higher SDS concentration [74].

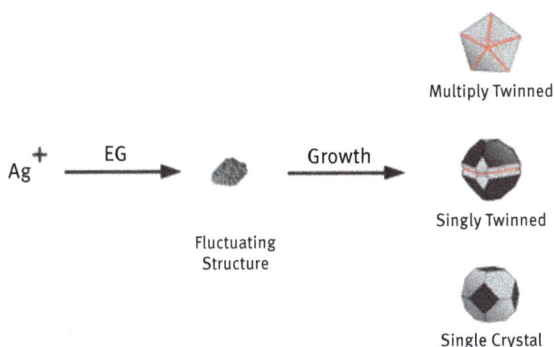

Figure 1.15: Formation of various types of seeds in polyol synthesis. (Reprinted with permission from ref. [79]. Copyright 2007, American Chemical Society).

Borohydride-based synthesis approach is also reported to produce a variety of shape and size (Figure 1.14). Spherical NPs less than 10 nm were obtained using PVP as stabilizer. However, addition of TSC along with PVP resulted in spherical NPs greater than 20 nm in size. Presence of SDS alone resulted in NPs of average size range 30–40 nm. Preferential binding of citrate on the (111) crystal facets of AgNP seeds can direct the formation of nanoprisms [87]. Binding of poly(styrene-4-sulfonate) (PSS) to (111) surface of AgNP seeds can lead to the formation of triangular nanoplates [88]. Chitosan also showed the same nature of binding [73].

1.3.2.1.2 Photochemical approach

In addition to chemical route of synthesis, photochemical methods have also been reported to yield anisotropic AgNPs such as, nanoprism, bipyramids, nanodisc, nanorods and nanodecahedron (Figure 1.16). To obtain size- and shape-controlled AgNPs via photochemical method, irradiation wavelength plays a vital role, in addition to all the reaction parameters discussed in the preceding subsection on the chemical reduction process. Synthesis of spherical AgNPs via photochemical route is usually a one-step process. However, to obtain anisotropic NPs, multiple synthesis steps are required. For nanoprism synthesis, Ag seed solution is prepared first and subsequently the seed solution is irradiated with light of appropriate wavelength. Commonly, bipyramids, nanodisc, nanorods and nanodecahedron synthesis process also involve two or more steps. Ag seeds prepared in the first step are grown in the second step using an appropriate growth solution (Figure 1.16). Sometimes multiple growth steps are carried out to produce size- and shape-controlled AgNPs.

Conventional photochemical reactions involve direct photoreduction of a metal precursor or reduction using photosensitizers as intermediate species [89]. TSC, sodium tartrate and PVP have been widely used as photoreduction reagents in the literature. Another class of light-driven, solution-based synthesis is plasmon-mediated reactions. In such reactions, incident light produces LSPR in metal NPs.

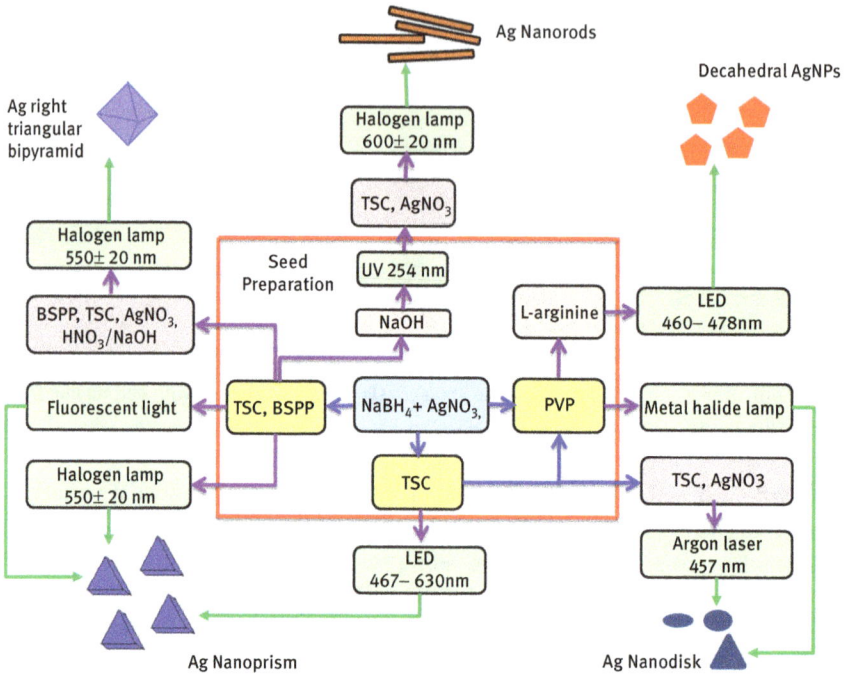

Figure 1.16: Schematic depicting two-step synthesis of AgNPs via photochemical method highlighting the various shapes that can be achieved through appropriate choice of reagents.

Langille et al. [90] studied photophysical effects of LSPR excitation responsible for growth of AgNPs in the plasmon-mediated synthesis approach. The essential components of this synthesis process are the small quasi-spherical AgNP seeds that may be obtained by chemical reduction using $NaBH_4$. These seeds, referred as plasmonic seeds, reduce the activation energy of shape transformation under visible light irradiation [33]. Visible light induces the reduction of Ag^+ to Ag^0 in presence of TSC and plasmonic seeds. Of the three carboxylic groups in citrate, one is free and contributes towards stabilization through electrostatic repulsion. The other two carboxylic groups bind to the surface of AgNP and cause fractional electron transfer due to their nucleophilic nature [91]. Thus, AgNPs can behave as electron storage and transfer medium, such that reduction of silver ions may occur on the surface of the NP.

Mirkin and co-workers [92] have studied photo-induced conversion of spherical AgNPs to triangular prisms. Spontaneous oxidative dissolution of small Ag particles can produce Ag^+ that may subsequently get reduced on the surface of Ag particles due to reduction by citrate under visible light irradiation [92]. Oxygen is essential for dissolving small AgNPs to produce Ag^+ and this step is critical for the formation of nanoprisms. Bis (p-sulfonatophenyl)phenylphosphine (BSPP) can form coordination complexes with Ag^+, such that the total concentration of Ag^+ in solution (including dissolved and complexed forms) is enhanced. These redox reactions facilitate the growth of triangular prisms. Later as the LSPR get red-shifted and increase beyond the maximum excitation wavelength (λ_{ex}) permissible, further growth is hindered. It has been shown that wavelength of incident light can control the edge length of nanoprisms [93]. Maillard et al. demonstrated that irradiation wavelength can also govern the shape of NPs [91]. For colloidal seed containing diverse shape and size, particles with the largest plasmon absorption cross-section at the laser wavelength used were found to grow at the highest rate. As particles grow, the reaction rate increases for shapes where the plasmons are in resonance with the irradiation wavelength used for synthesis [91].

Decahedral AgNPs can also be obtained by photochemical method. Simultaneous growth of platelets and decahedra is reported upon exposure to light. However, on prolonged irradiation, the platelets dissolve as they are less stable compared to decahedra and the released silver is integrated into the growing decahedral seeds. While the size of NPs is dependent on the irradiation wavelength and intensity, the shape is determined by various morphologies present in the seed and relative stability of these morphologies [94].

1.3.2.1.3 Microwave-assisted approach

Microwaves are a portion of the electromagnetic spectrum with frequency range of 300 MHz–300 GHz. The most commonly used frequency is 2.45 GHz. Although microwave heating is widely employed in various domains, its application in synthesis of NPs has not yet been explored extensively. The dielectric constant and dielectric loss of a medium affects its interaction with microwaves. When a dielectric medium is exposed to microwave irradiation, an internal electric field is generated. This field induces translational motion of electrons or ions and rotational motion of dipoles. The resistance to these induced motion attenuates the electric field and causes heating of the medium [95, 96]. Dipoles trying to follow the field causes heating that can drive reactions, and reactions may also occur due to electrical conduction causing polarization. For a good conductor, complete polarization may be achieved within a few seconds under the influence of 2.45 GHz microwave [25].

Microwave-based synthesis of AgNPs involve the reduction of silver ions using variable frequency microwave radiation instead of conventional heating methods. Synthesis of AgNPs using conventional heating methods, i. e., oil heating at high temperature (260°C) is time consuming (~24 h). In contrast, microwave-assisted synthesis provides rapid and uniform heating and thus greatly reduces the time of synthesis [97]. The main advantages of MW-based synthesis are: (i) rapid reaction kinetics; (ii) rapid temperature increase in the initial phase; (iii) localized high temperature at the reaction site and (iv) potential for formation of novel phases due to selective formation of one phase over another. For the same temperature and exposure time, MW-based synthesis yields a faster reaction and increases the yield of AgNPs [56, 95, 98]. When this technique is combined with other reduction techniques such as polyol method for NP synthesis, there is significant enhancement in kinetics of crystallization and this may lead to energy savings of up to 90%. Thus, microwave heating-based polyol process is ideal for the synthesis of AgNPs in environmentally benign closed systems [95]. The faster reaction kinetics reduces agglomeration in MW-assisted synthesis methods resulting in NPs with narrow-size distribution.

1.3.2.1.4 Biological synthesis of AgNPs

In extracellular biosynthesis by microorganisms, the spent media containing extracellular substances secreted by microbes facilitate synthesis of NPs. These extracellular biomolecules may include enzymes, proteins and amino acids, polysaccharides, vitamins and extracellular polymeric substances (EPS). These components together may serve as both reducing agent and stabilizing agent for synthesis of AgNPs from Ag precursor salt. The nitrate reductase enzyme has been demonstrated to play a key role in the reduction of Ag^+ ions [99]. Intracellular biosynthesis of AgNPs by the activity of this enzyme is illustrated in Figure 1.17. Nitrate is converted to nitrite by the nitrate reductase enzyme in the presence of reducing power provided by the reduced form of nicotinamide adenine dinucleotide (NADH). In *Bacillus licheniformis*, it has been demonstrated that the electrons released from NADH can drive the reduction of Ag^+ to Ag^0 and lead to the formation of AgNPs [99, 100]. Li et al. [101] also demonstrated synthesis of AgNPs by reductase enzymes secreted by *Aspergillus terreus*, using a similar NADH mediated mechanism as shown in Figure 1.18.

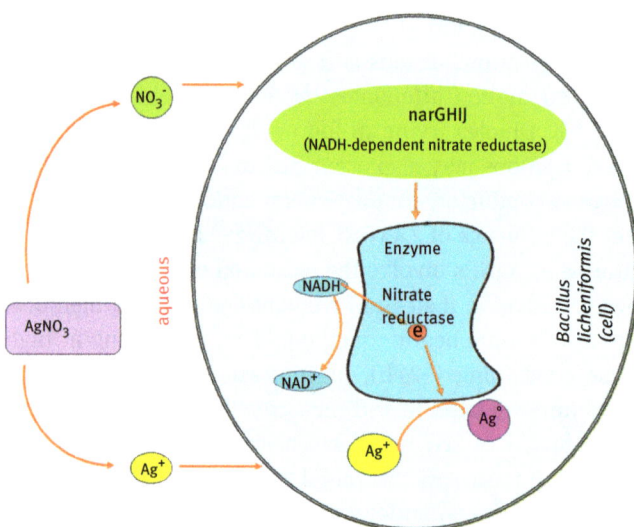

Figure 1.17: Possible mechanism for intracellular AgNP synthesis in *Bacillus licheniformis* mediated by NADH-dependent nitrate reductase enzyme (Reprinted with permission from ref. [99]. Copyright 2008, Elsevier).

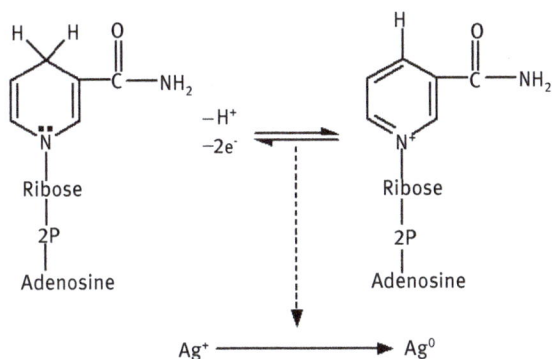

Figure 1.18: Schematic representation of biosynthesis of AgNPs by NADH-dependent reductase enzymes (Reprinted with permission from ref. [101]. Copyright 2012. Molecular Diversity Preservation International).

A direct evidence of this phenomenon was demonstrated by Kumar et al. [102] where nitrate reductase purified from the organism *Fusarium oxysporum* was used for AgNP synthesis in a test tube containing AgNO₃ and NADPH. Appearance of brown colour was observed in the reaction mixture thereby confirming the involvement of nitrate reductase towards biosynthesis of AgNPs in the size range 10–25 nm. Thus, these findings provide opportunities for large-scale biological synthesis of NPs through NADH or NADPH-coupled reactions [101, 102].

The detailed methodology for synthesis of size- and shape-controlled AgNPs using bottom-up approaches are discussed in the following section.

1.3.2.2 Chemical method

The most common method for synthesis of AgNPs is chemical reduction of silver salts by organic and inorganic reducing agents in the presence of a stabilizer. In addition to AgNO₃, silver sulfate (Ag₂SO₄) [103] and silver perchlorate [104] are also used as the Ag precursor salts. Use of reducing agents, N,N-dimethylformamide (DMF) [105], ascorbic acid [72], hydrazine [106], poly(EG)-block copolymers [107] and ammonium formate [108] have also been reported apart from TSC, NaBH₄ and polyol. CTAB, PVP and SDS are commonly used as stabilizers.

Three broad groups of methods for AgNP synthesis, based on reducing agents such as, polyol, citrate and borohydride have already been discussed under Section 1.3.2.1.1. Polyol-based synthesis involves heating a polyol with Ag salt precursor to generate metal colloids. Occasionally, polymeric capping agents are also introduced into the reaction mixture. In borohydride-based synthesis approach, various stabilizers such as citrate and PVP have been used for preventing the aggregation of AgNPs. Citrate can also behave as a reducing agent at a temperature of 90°C and above [29]. The weaker reductant citrate, can produce larger particles; however, controlling the size and shape with citrate alone is difficult [43, 109, 110]. Sometimes, PVP is also employed as reducing agent. Two-step synthesis process is carried out to obtain various shapes of AgNPs such as nanoprisms, nanoplates, nanorods and nanowires. In the first step, seeds are prepared using NaBH$_4$ as reducing agent. In the second step, these seeds are further grown into various morphologies of AgNPs. The synthesis protocols for obtaining AgNPs of desired size and morphology is discussed in the following sections.

1.3.2.2.1 Synthesis of silver nanospheres

A wide range of solvents and stabilizers are reported for the synthesis of silver nanospheres. The activity of reducing agent and nature of stabilizers determine the size of nanospheres.

In citrate reduction-based AgNP synthesis, reduction activity of citrate is reported to depend on the pH of the reaction mixture [86]. At high pH, spherical and rod-like silver NPs are observed due to fast reduction rate of the precursor. In contrast, at low pH, triangle or polygon-shaped silver NPs are observed due to slow reduction rate of the precursor. Hence, for shape-controlled synthesis of spherical silver NPs, a balance between the nucleation and growth processes needs to be achieved [86].

The Turkevich method, a typical synthesis approach, was used by Dong et al. [86] for synthesis of spherical AgNPs. In this method, 100 mL aqueous TSC (7.0 mM) solution was prepared using controlled amounts of citric acid and sodium hydroxide (NaOH) at room temperature. After adjusting the pH to the desired value using nitric acid or NaOH, the solution was continuously stirred and brought to a boil. Subsequently, 1.0 mL of 0.1 M aqueous AgNO$_3$ was added to this solution. The fast nucleation stage was carried out at a pH value slightly above neutral pH (7.7) and the growth stage was carried out by reducing the pH to 6.1. Using this method, the authors reported average size of spherical NPs to be 58 nm with ±11% standard deviation and presence of 12% non-spherical NPs [86].

Quasi spherical silver NPs were synthesized by Qin et al. [111] using ascorbic acid and citrate as the reductant and stabilizer, respectively. The pH of an aqueous solution (8 mL) containing ascorbic acid and TSC (at 6.0×10^{-4} M and 3.0×10^{-3} M, respectively) was adjusted over the range 6.0–10.5 using either citric acid (0.2 M) or NaOH (0.1 M). Subsequently, AgNO$_3$ (0.08 mL, 0.1 M) was introduced into the above solution maintained in a 30°C water bath while stirring continuously. The reaction was completed within 15 min. A decrease in average size of AgNPs from 73 to 31 nm was observed with rise in pH from 6.0 to 10.5. After aging for 2 h at 100°C, the AgNPs became more spherical due to intraparticle ripening. No significant change in average diameter of AgNPs was observed over time indicating good stability of the NPs [111].

Recently, Ranoszek-Soliwoda et al. [110] synthesized spherical AgNPs using TSC and tannic acid. A combination of these reagents yielded monodispersed

AgNPs with narrower size distribution compared to those synthesized using either TSC or tannic acid alone. It was observed that, a citric acid – TA complex was first formed and reduction of silver ions by this complex was responsible for AgNP synthesis. The reduction of AgNO$_3$ using a mixture of TSC and tannic acid at 100°C produced spherical AgNPs of average diameter 30 nm with a narrow size distribution [110].

Synthesis of spherical AgNPs with diameter ~54 nm ± 10 nm and narrow size distribution can be achieved using the polyol method (Figure 1.19). In a typical procedure described by Liang et al. [107], PVP (444 mg, MW 40,000) was added to 40 mL of PEG-600 with stirring at 80°C. PVP prevented agglomeration of the AgNPs and PEG was used as solvent. After the solution appeared transparent, AgNO$_3$ aqueous solution (1 mL, 0.5 M) was added as the Ag precursor, such that the PVP/AgNO$_3$ molar ratio was 8. After adding the precursor solution, the solution was heated at 260°C for 24 h.

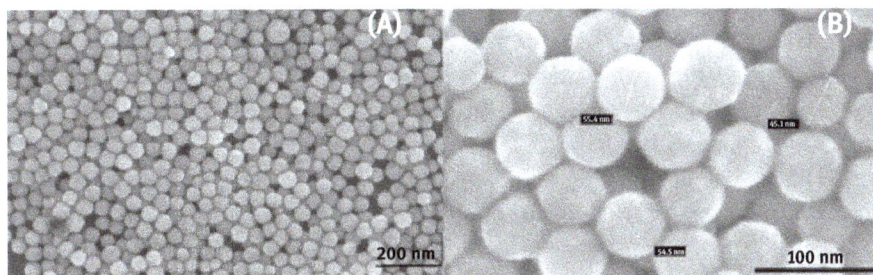

Figure 1.19: (A) Low- and (B) high-magnification SEM images of spherical AgNPs synthesized using polyol method with heating at 260°C and PVP/AgNO$_3$ ratio of 8. (Reprinted with permission from ref. [107]. Copyright 2010, American Chemical Society).

High-resolution TEM (HRTEM) images (Figure 1.20) of silver nanospheres formed using this method were found to be crystalline with their face indexed to {111} planes [107]. The four peaks observed in XRD corresponded to diffraction from (111), (200), (220) and (311) planes of fcc crystal phase (Figure 1.20C). At temperature lower than 260°C, the AgNPs were of irregular shape and demonstrated twinned structure. When PVP/AgNO$_3$ molar ratio was increased from 8 to 16, a wider size distribution of nanospheres ranging from 21 to 74 nm with average diameter 38 nm resulted. For PVP/AgNO$_3$ molar ratio of 4, nanospheres and some irregular-shaped NPs having diameter ~61 nm with a wide size distribution of 30–95 nm was observed [107].

Synthesis of size-controlled spherical AgNPs has also been reported using PVP and EG through a modified polyol method [77]. PVP (0.17–1.7 g, Mol. wt. = 40,000) was dissolved in 10 ml EG and heated to 160°C in an oil bath. Once a light-yellow colour appeared, AgNO$_3$ (0.17 g) dissolved in 10 ml EG was slowly added to the PVP-EG solution. The reaction was continued for 4 h. NPs with mean sizes 80 nm and 50 nm

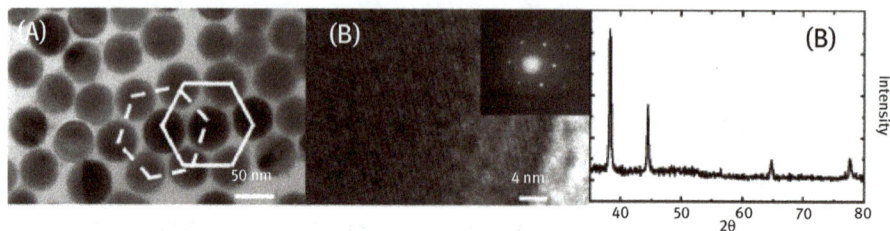

Figure 1.20: (A) TEM image and (B) HRTEM image and (C) XRD pattern of silver nanospheres synthesized by polyol method with heating at 260°C and PVP/AgNO$_3$ ratio of 8. The inset shows the SAED pattern of an individual AgNP (Reprinted with permission from ref. [107]. Copyright 2010, American Chemical Society).

were obtained with PVP/AgNO$_3$ mass ratio of 8:1 and 10:1, respectively. However, with further increase in PVP/AgNO$_3$ ratio to 15:1 and 20:1, the size of AgNPs did not change considerably. At higher PVP/AgNO$_3$ ratios, viscosity of the reaction mixture increased drastically. Hence, diffusion of Ag atoms in the reaction mixture was hindered, such that, the particle size remained almost unaltered. NPs of smaller mean size viz. 30 nm and 10 nm could be achieved by adjusting the pH of the AgNO$_3$ solution to 10 using ammonia solution and maintaining PVP/AgNO$_3$ mass ratio of 8:1 and 10:1, respectively. After adding ammonia, Ag(NH$_3$)$_2$$^+$ was formed by reaction between NH$_3$ and Ag$^+$. This complex is inherently more difficult to reduce compared to Ag$^+$, and thus produces AgNPs of smaller size. In a typical reduction reaction, (as shown in eq. (1.1)–(1.3)), the EG, after losing water, is transformed into acetaldehyde. Ag$^+$ or [Ag(NH$_3$)$_2$]$^+$ is subsequently reduced by acetaldehyde to Ag^0.

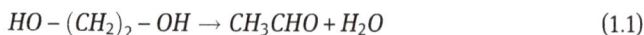

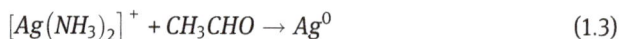

$$HO - (CH_2)_2 - OH \rightarrow CH_3CHO + H_2O \tag{1.1}$$

$$Ag^+ + CH_3CHO \rightarrow Ag^0 \tag{1.2}$$

$$\left[Ag(NH_3)_2\right]^+ + CH_3CHO \rightarrow Ag^0 \tag{1.3}$$

XRD and FTIR analysis of the synthesized AgNPs was performed to interpret the crystal structure of AgNPs and to confirm PVP coating on these NPs. The XRD spectrum of the PVP-capped AgNPs (typical size 30 nm) showed five characteristic diffraction peaks for silver at $2\theta = 38.2^0, 44.5^0, 64.7^0, 77.5^0$ and 81.8^0 corresponding to the {111}, {200}, {220}, {311} and {222} crystallographic planes of fcc Ag crystals, respectively (Figure 1.21). The results indicated that the synthesized NPs were comprised of metallic silver with fcc crystal structure [77].

To confirm the PVP coating on synthesized AgNPs, FTIR spectra of PVP- and PVP-coated AgNPs (Figure 1.22) having average diameter of 30 nm were compared by Zhao

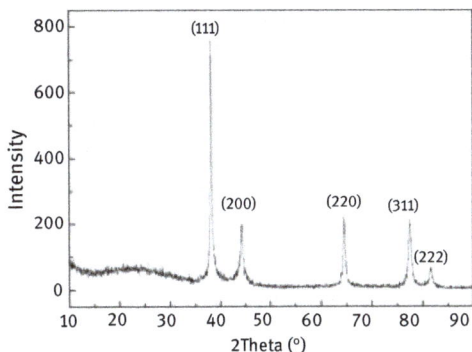

Figure 1.21: XRD spectrum of PVP-capped AgNPs synthesized using PVP and EG with PVP/AgNO$_3$ mass ratio of 8:1 at pH 10 (average size 30 nm) (Reprinted with permission from ref. [77]. Copyright 2010, Elsevier).

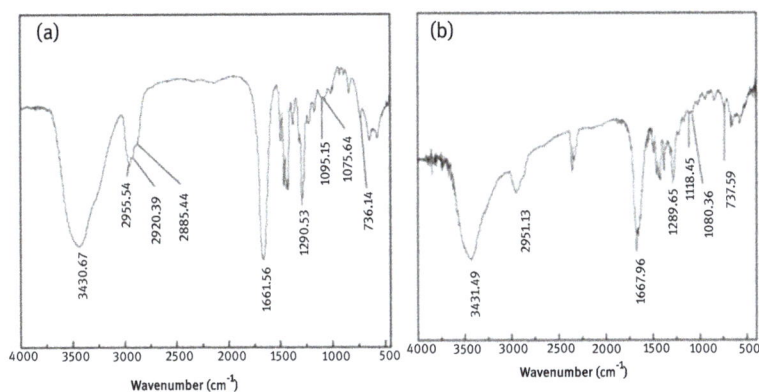

Figure 1.22: FTIR spectra of (a) PVP alone and (b) PVP-capped AgNPs synthesized using PVP and EG with PVP/AgNO$_3$ mass ratio of 8:1 at pH 10 (Reprinted with permission from ref. [77]. Copyright 2010, Elsevier).

et al. [77]. The asymmetric and symmetric stretching vibration peaks of C–N were shifted from 1095 and 1075 cm^{-1} in PVP to 1118 and 1080 cm^{-1}, respectively, in the PVP-capped AgNPs. The stretching vibration peak of –C = O was also shifted from 1661 to 1668 cm^{-1}. These peak shifts indicated that PVP may have adsorbed on to the surface of AgNPs through the N or O atoms [77].

Recently, Leng et al. [78] produced 20–30 nm spherical AgNPs by further modifying the polyol process in the presence of PVP. During synthesis, 35 ml of EG was heated at 160°C for 1 h under mechanical stirring at 800 rpm. EG solution (15 ml) containing 0.1 M AgNO$_3$ and 1 g PVP (Mol. wt. ~30,000) was introduced into this solution at the rate of 1 ml/min. After complete reduction, the solution was cooled to room temperature. Three successive centrifugation steps at 8000 rpm with anhydrous ethanol were carried to out to separate the AgNPs. Smaller-size AgNPs were produced at higher AgNO$_3$ concentration (0.25 M), with higher AgNO$_3$ to PVP mass ratio (i. e., 2:1, 1 g PVP). With increase in reaction temperature (100–180°C), the size range of the as-prepared AgNPs increased. The silver NPs synthesized using 2 g AgNO$_3$ and 1 g PVP at 160°C were comprised of fcc single crystals and the size ranged from 20 to 30 nm as illustrated through TEM and HRTEM. The d-spacing was 0.223 nm and XRD peaks corresponded to diffraction from (111), (200), (220), (311) and (222) planes of Ag crystals.

Change in reductant and stabilizers has an impact on size and monodispersity of AgNPs. In a spectrum of studies, PVP [112], PVP and TSC [62] and SDS [61] were used as stabilizers along with NaBH$_4$ as reducing agent. Spherical AgNPs of about 9 nm were synthesized by adding 10 ml of 2.0×10^{-2} M AgNO$_3$ dropwise into a solution comprised of PVP (10 ml of 0.375 M) and NaBH$_4$ (10 ml of 10^{-2} M) accompanied with vigorous stirring at room temperature. The solution was continuously stirred for more than 10 min after complete addition of AgNO$_3$ to obtain spherical AgNPs with mean size 9.0 ± 2.1 nm [112]. Dong et al. [62] produced spherical AgNPs of average size 21 nm using NaBH$_4$ as reducing agent and PVP and TSC as stabilizers. Freshly prepared NaBH$_4$ (0.5 ml, 50 mM) was added into an aqueous solution containing TSC (0.5 ml, 30 mM) and AgNO$_3$ (1 ml, 5 mM). After 30 sec, an aqueous solution of PVP (0.5 ml, 5 mg/ml, Mol. wt. = 40,000) was introduced into the mixture to obtain spherical AgNPs with average size of 21 nm [62]. Silver NPs (30–40 nm) were produced using SDS as a stabilizer by slowly adding AgNO$_3$ solution into NaBH$_4$ solution containing SDS [61]. After all the precursor solution was added, the contents were stirred for 1 h. In the presence of SDS, AgNO$_3$ was reduced by NaBH$_4$. The reaction is as follows:

$$Ag^+ + BH_4^- + 3H_2O \rightarrow Ag^0 + B(OH)_3 + 3.5H_2 \qquad (1.4)$$

For higher SDS/AgNO$_3$ weight ratios (5, 20 and molar ratio of NaBH$_4$/AgNO$_3$ = 4), absorption of SDS on the surface of AgNPs protected them from aggregation through steric effect. When the SDS/AgNO$_3$ weight ratios were low (0.5, 2 and molar ratio of NaBH$_4$/AgNO$_3$ = 4), the silver NPs were aggregated as indicated by broadening of the LSPR peak. Particle aggregation was also influenced by the NaBH$_4$/AgNO$_3$ molar ratio. Aggregation was observed with lesser amount of NaBH$_4$, i. e., for molar ratio of NaBH$_4$/AgNO$_3$ of 2 and SDS/AgNO$_3$ weight ratio of 2. By increasing the NaBH$_4$ concentration, aggregation could be minimized [61].

Synthesis of size-controlled spherical AgNPs (average size 5, 7, 10, 15, 20, 30, 50, 63, 85 and 100 nm) using NaBH$_4$ and TSC as primary and secondary reductant, respectively, and TCS as stabilizing agent was demonstrated by Agnihotri et al.[29]. A schematic of the method used is demonstrated in Figure 1.23. A mixture of NaBH$_4$ and TSC was heated to 60°C for 0.5 h under dark condition with stirring. After 0.5 h, AgNO$_3$ solution was introduced dropwise into the mixture and the temperature was increased to 90°C. Subsequently, 0.1 M NaOH was added to adjust pH to 10.5. The mixture was heated for another 20 min. In stage I, at 60°C reduction was

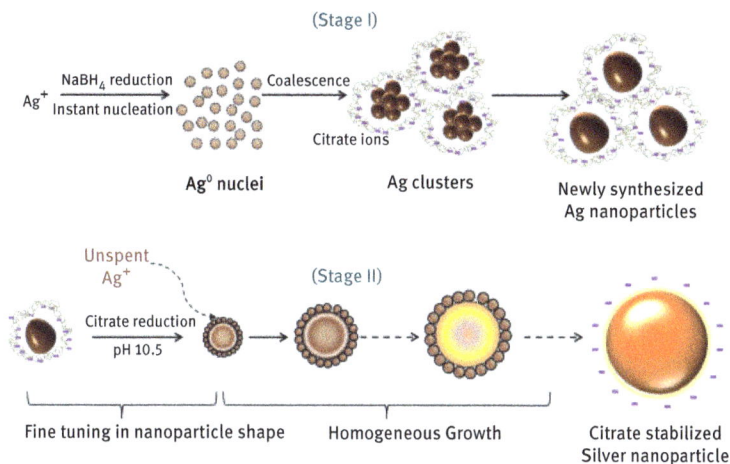

Figure 1.23: Schematic representation of size-controlled silver nanoparticles synthesized by the co-reduction method. (Reprinted from ref. [29]. Published by The Royal Society of Chemistry).

predominantly mediated by NaBH$_4$ and this resulted in production of an abundance of silver nuclei through reduction of Ag$^+$ and Ostwald ripening of the silver nuclei produced AgNPs through a secondary process. In stage II, AgNPs produced in the previous stage participated in growth process. At 90°C, the remaining Ag$^+$ ions were reduced by TSC. In stage I, TCS prevented the agglomeration of AgNPs. The NaBH$_4$ to TSC ratio was critical for controlling the nucleation and growth processes during the synthesis of AgNPs. It was concluded that for the various AgNP size ranges, 5–20 nm, 25–60 nm and 60–100 nm, reduction was caused primarily by NaBH$_4$, combined action of NaBH$_4$ and TSC and primarily by TSC, respectively. Reduction rate during the 2nd stage was controlled using an optimal pH of 10.5.

Apart from citrate and NaBH$_4$, other reducing agents and stabilizing agents have also been explored for synthesizing size-controlled spherical AgNPs. Spherical AgNPs of average diameter 15 and 43 nm were prepared by reducing a starch solution (1% w/v) of AgNO$_3$ (10^{-4} M) precursor salt by D(+)-glucose and NaOH, respectively [113]. The reaction was conducted by heating at 70 °C for 30 min. Another study employed the triblock copolymer pluronic P123 (poly(ethylene oxide)-poly(propylene oxide)–poly (ethylene oxide) PEO$_{20}$ PPO$_{70}$ PEO$_{20}$) and L64 (PEO$_{13}$ PPO$_{30}$ PEO$_{13}$) worked both as reducing as well as stabilizing agent and produced spherical AgNPs of 8 and 24 nm size, respectively, at ambient temperature [113].

Rivero et al. [114] synthesized AgNPs via chemical reduction route at room temperature using AgNO$_3$ precursor salt, dimethylaminoborane (DMAB) as reducing agent and poly(acrylic acid, sodium salt) (PAA) as protective agent. A wide range of colour, size and shape was produced by changing the concentration of either the protecting agent, PAA or the reducing agent, DMAB so as to alter their molar ratio at a fixed AgNO$_3$ concentration (3.33 mM) [114]. Increasing the PAA concentration with DMAB concentration held constant (0.33 mM) resulted in violet, blue or green solution predominantly comprised of hexagonal-, triangular- and rod-shaped AgNPs. Similar observation was made when the DMAB concentration was reduced (0.033 mM) with PAA held constant (10, 25 mM). However, lower PAA or elevated DMAB (6.66 mM) concentrations produced orange red-coloured solution of spherical AgNPs.

In another study, Zhang et al. synthesized AgNPs using ultrasonically treated aqueous solution of sodium dodecyl benzene sulphonate (SDBS) as both reducing and stabilizing agent. Either, AgNO$_3$ or silver acetate (2 mL, 0.05 M) was mixed with 8 ml ultrasonically treated SDBS aqueous solution (10 mL, 0.05 M) and the mixture was refluxed under magnetic stirring for 8 h at 100 °C in an oil bath. Using silver acetate and AgNO$_3$, the size ranges of AgNPs produced were 3–9 nm and 2–13 nm, respectively [115].

He et al. studied the effect of precursor concentration (AgNO$_3$), stabilizer concentration (PVP, Mol. wt. = 40,000 and β-cyclodextrin), NaOH, HCl, reducing agents (DMF, ethanol, 2-propanol and methanol) and presence of water on the size of AgNPs formed. A reduction in AgNP size was observed with increase in stabilizer to AgNO$_3$ molar ratio and with increasing amount of methanolic NaOH until an optimum level (7.767 × 10^{-4} M). A similar trend was also observed with methanolic HCl (7.767 × 10^{-7} M) using microwave irradiation method. The reducing agent had a significant influence on particle size. Also with increase in concentration of water (2–50%) in the reducing agent, the particle size increased linearly due to decrease in effectiveness of the reducing agent. Reduction in activity was highest for the most effective reducing agents, DMF [105]. When molar ratio of the stabilizer: AgNO$_3$ was increased from 1 to 150, with PVP as the stabilizer, the mean diameter of the particles decreased from 11.0 to 3.2 nm for conventional heating whereas with β-cyclodextrin as the stabilizer, the mean diameter of the particles decreased from 6.9 to 2.3 nm. Table 1.1 summarizes methods for synthesis of size-controlled spherical AgNPs using diverse synthesis protocols and reagents.

1.3.2.2.2 Synthesis of silver nanoprisms

Silver nanoprisms have been either synthesized directly using various reducing agents, such as, NaBH$_4$ and TSC or through the seed-mediated approach. Sometimes hydrogen peroxide is added to favour anisotropic growth required for nanoprism synthesis. The chemical synthesis of nanoprisms, generally involve transformation of colloidal silver NPs into nanoprisms. Therefore, the control over nanoprism thickness and size distribution by chemical route is a challenge.

Mirkin and co-workers [87] demonstrated a rapid route for silver nanoprism synthesis with control over nanoprism thickness using TSC, NaBH$_4$, PVP and hydrogen peroxide at room temperature. For initiating the synthesis, a mixture of aqueous solution of AgNO$_3$ (25 mL, 0.1 mM), TSC (1.5 mL, 30 mM), PVP (1.5 mL, 0.7 mM, Mol. wt. ~ 29,000) and hydrogen peroxide (60 μL, 30 wt%) was mixed vigorously in

Table 1.1: Synthesis of spherical silver nanoparticles using chemical method.

Reducing agent	Stabilizing agent/Additives	AgNO₃	Reaction conditions	Size (nm)	Reference
TSC: 100 mL, 7.0 mM		1 ml, 0.1 M	20 min, pH 7.7,6.1	58 ± 11	Dong et al. [86]
Ascorbic acid 0.6 mM	TSC 3.0 mM	0.08 mL, 0.1 M	15 min, pH 6.0–10.5	73–31[a]	Qin et al. [111]
TSC 4%; 4.2 g, TA 5%; 0.63 g	—	0.0165%, 95.15 g	100°C, 15 min	30	Ranoszek-Soliwoda et al. [110]
PEG 40 mL	PVP 444 mg [PVP/AgNO₃] 8	1 mL, 0.5 M	260°C, 24 h	54± 10 nm	Liang et al. [107]
EG	PVP 10 mL, 0.17–1.7 g [PVP/AgNO₃] 8,10	10 mL, 0.17 g	160°C, 4 h pH 10	30, 10	Zhao et al. [77]
EG 35 mL	PVP 15 mL,1g [PVP/AgNO₃] 0.5	0.1 M	160°C, >1 h	20–30	Leng et al. [78]
NaBH₄ 10 mL,10 mM	PVP 10 mL, 0.375 M	10 mL, 20 mM	RT, >10 min	9.0 ± 2.1	Li et al. [112]
NaBH₄ 0.5 mL, 50 mM	TSC 0.5 mL, 30 mM; PVP 0.5 mL, 5 mg/ml	1 ml, 5 mM	>30 min	21	Dong et al. [62]
NaBH₄, TSC	TSC		60°C, 30 min 90°C, 20 min	5–100[a]	Agnihotri et al. [29]
D(+)-glucose	Starch 1%	0.1 mM	70°C, 30 min	15	Shervani et al. [113]
NaOH	Starch 1%	0.1 mM	70°C, 30 min	43	Shervani et al. [113]
Pluronic L64	Pluronic L64	0.1 mM	RT	24	Shervani et al. [113]
P123	P123	0.1 mM	RT	8	Shervani et al. [113]
SDBS 8 ml, 0.05 M	SDBS	2 mL, 0.05 M	100 °C, 8 h	2–13	Zhang et al. [115]
DMF 100mL	PVP;β-cyclodextrin [Stabilizer:AgNO₃] 1–150	0.572 mM	CH, 60 min	11–3.2[a] 6.9–2.3[a]	He et al. [105]

[a]AgNPs were synthesized over this size range by varying the reaction conditions; RT implies synthesis at room temperature; CH implies conventional heating.

presence of air at room temperature. Later, $NaBH_4$ (100–250 µL, 100 mM) was injected into the mixture. By varying $NaBH_4$ concentration (0.3–0.8 mM), nanoprisms of varying edge length and thickness could be obtained, i. e., 31 ± 7 and 7 ± 1.5 nm for 0.3 mM $NaBH_4$; 32 ± 8 and 6.9 ± 1.2 nm for 0.5 mM $NaBH_4$; 35 ± 8 and 5.5 ± 0.6 nm for 0.67 mM $NaBH_4$; and 39 ± 8 and 4.3 ± 1.1 nm for 0.8 mM $NaBH_4$. With increase in $NaBH_4$, the tips of the nanoprisms became sharper with reduction in thickness and increase in the edge length.

Strong and weak reducing action of $NaBH_4$ and TSC, respectively, was exploited for triangular silver nanoprism synthesis by Dong et al. [116]. In a stepwise reduction method, at low temperature, the strong reductant, $NaBH_4$ was used to reduce $AgNO_3$ and form small spherical silver NPs. At high temperature, the weak reductant TSC was used for further reduction of the precursor. During synthesis, $AgNO_3$ (0.1 mM) and TSC (3.5 mM) were mixed to get 100 mL aqueous solution and the mixture was kept in an ice bath. $NaBH_4$ (5 µM) was added dropwise into this solution maintained at 4°C while stirring vigorously. After 30 min, the solution was heated to 70°C for ~48 h. Figure 1.24 shows TEM images demonstrating evolution of AgNPs leading to the formation of nanoprisms. At low temperature, within 1 min small spherical AgNPs having average size 4.7 nm (±29%) were seen. AgNPs having average size 3.1 nm (± 35%) were observed after 2 h of heating at 70°C. After 4 h, anisotropic larger NPs were seen. Triangular nanoprisms were observed at 6 h when the abundance of spherical AgNPs decreased. Transformation of spherical AgNPs into nanoprism within 2–20 h was attributed to consumption of the remaining precursor and dissolution of the small AgNPs (Ostwald ripening). From 20 to 48 h, uniform nanoprisms were observed due to Ostwald ripening. The nanoprisms synthesized were characterized by average thickness of 6.2 ± 0.7 nm and around 60% showed average edge length of 100 ± 20 nm. The ratio of two reductants ($NaBH_4$ and TSC) was critical for producing small spherical NPs and its subsequent transformation into the triangular nanoprisms.

A seed-mediated approach was demonstrated for synthesis of silver nanoprisms by Aherne et al. [72]. Silver seeds were formed by mixing aqueous TSC (5 mL, 2.5 mM), aqueous PSSS (0.25 mL, 500 mg/L, 1000 kDa) and freshly prepared $NaBH_4$ (0.3 mL, 10 mM). This was followed by addition of $AgNO_3$ (5 mL, 0.5 mM) at a rate of 2 mL/min under continuous stirring. These seeds were utilized for growth of nanoprisms by combining varying volumes of seed solution (20–650 µL) with 5 mL distilled water, and aqueous ascorbic acid (75 µL, 10 mM). Subsequently, $AgNO_3$ (3 mL, 0.5 mM) was added at a rate of ~1 mL/min. Aqueous solution of TSC (0.5 mL, 25 mM) was later added for stabilization of the nanoprisms at the end of synthesis. The nanoprims exhibiting LSPR in the range of 538–796 nm were characterized by edge length in the range of 20–65 nm and average thickness of 5.5 nm [72]. This seed-mediated method was later modified by Haber and Sokolov [117] who eliminated PSSS from the formulation and used a peroxide-etchant for two-step nanoprism synthesis. Deionized water (39.3 mL), TSC (2 mL, 75 mM), hydrogen peroxide (256 µL, 0.6%) and $AgNO_3$ (186 µL, 10 mM) were combined. $NaBH_4$ (192 µL, 100 mM), was rapidly

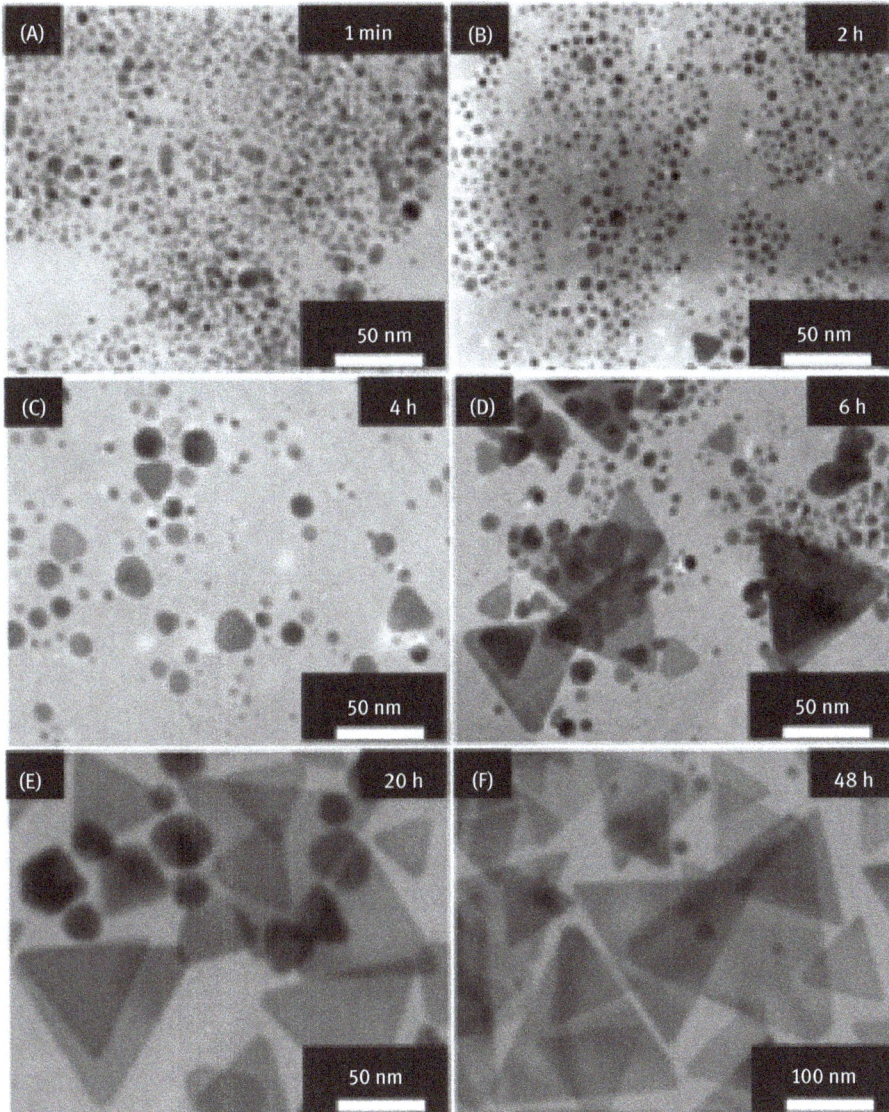

Figure 1.24: TEM image illustrating the transformation of spherical AgNPs to triangular nanoprisms. (Reprinted with permission from ref. [116]. Copyright 2010, American Chemical Society).

added to this mixture under vigorous stirring. This solution was aged overnight at room temperature. To produce silver nanoprism of ~30 nm edge length, 2.1 mL of the seed stock was mixed with L-ascorbic acid (200 μL, 5 mM) under vigorous stirring. Subsequently, $AgNO_3$ (100 μL, 10 mM) was added dropwise. To produce larger nano-plates (~100 nm edge length, 7 nm thickness), the seed solution was diluted before addition of L–ascorbic acid and $AgNO_3$. Dilution factors of 1, 2, 4, 8 and 12 were used to obtain nanoplates with absorbance peaks at ~550 nm, ~550–590 nm, ~590–630 nm, ~630–700 nm and NIR wavelength, respectively. One pot synthesis of nanoprism was also demonstrated in which the seed solution was used for nanoprism growth without the aging step.

Table 1.2 summarizes methods for size-controlled synthesis of silver nanoprisms using various synthesis protocols and reagents.

Table 1.2: Synthesis of nanoprisms through chemical method.

Reducing agent	Stabilizing agent/ Additives	AgNO$_3$	Reaction Condition	Size, Edge length and thickness (nm)	Reference
NaBH$_4$ 100–250 µL, 100 mM	TSC:1.5 mL, 30 mM; PVP: 1.5 mL, 0.7 mM; or H$_2$O$_2$ 30%, 60 µL	25 mL, 0.1 mM	RT, 30 min	31 ± 7, 7 ± 1.5 39 ± 8, 4.3 ± 1.1	Métraux and Mirkin [87]
TSC 3.5 mM; NaBH$_4$ 5 µM	—	0.1 mM	4°C 30 min, 70°C for ~48 h	100 ± 20 6.2 ± 0.7	Dong et al. [116]
Seed: NaBH$_4$ 0.3 mL,10 mM Growth: Ascorbic acid 75 µL, 10 mM	Seed: TSC 5 mL, 2.5 mM; PSSS 0.25 mL, 500 mg/L Growth: TSC 0.5 mL, 25 mM	Seed: 5 mL, 0.5 mM Growth: 3 mL, 0.5 mM	3 min (after AgNO$_3$ addition)	20–65, 5.5	Aherne et al. [72]
Seed: NaBH$_4$ 192 µL, 100 mMGrowth: L-ascorbic acid 200 µL,5 mM	Seed: TSC 2 mL, 75 mM / H$_2$O$_2$ 0.6% 256 µL	Seed:186 µL, 10 mM Growth: 100 µL, 10 mM	Seed: overnight aging	30–100[a]	Haber and Sokolov [117]

[a]AgNPs were synthesized over this size range by varying the reaction conditions

1.3.2.2.3 Synthesis of silver nanoplates

Synthesis of silver nanoplates has been extensively studied by researchers. PVP has been widely employed as a steric stabilizer or capping agent during synthesis of nanoplates. Washio et al. [118] first demonstrated that the end groups of PVP (hydroxyl groups) can behave as reducing agent for the synthesis of triangular Ag nanoplates with high yield. Triangular Ag nanoplates were synthesized by heating an aqueous solution of AgNO$_3$ (3 mL, 188 mM) and PVP at 60 °C for 21 h. Using PVP, having MW 29,000 and PVP to AgNO$_3$ molar ratio of 30, a mix of triangular Ag nanoplates (80%) and AgNPs of other shapes (20%) was obtained at the end of 21 h and the size of triangular nanoplates and other AgNPs were 350 nm and 120 nm, respectively.

By varying the reaction time, Ag triangular nanoplates with controllable edge lengths could be obtained [118]. A series of experiments were carried out by varying the molar ratio of PVP to AgNO$_3$ (5, 15, 30) as well as the molecular weight of PVP (MW 10,000, 29,000 or 55,000). The conversion percentage of AgNO$_3$ to Ag was linearly related to the molar ratio of PVP to $AgNO_3$ and the extent of conversion reduced as the MW of PVP increased. The reduction reaction was faster when PVP with lower MW was used [118].

In another study, Ag nanoplates with controlled shape were obtained by reduction of $AgNO_3$ with PVP in N-methylpyrrolidone (NMP) [119]. In a typical synthesis, 1.870 g PVP (MW ~29 kDa) was dissolved in 8 mL NMP. NMP solution containing $AgNO_3$ (3 mL, 188 mM) was then injected into it. PVP to $AgNO_3$ weight ratio was maintained at 19.5 in the total reaction mixture. The reaction was carried out at 100°C for 24 h with magnetic stirring to yield hexagonal nanoplates with average plate size 213 nm and thickness 63 nm. Upon doubling the PVP to $AgNO_3$ weight ratio to 39, large thin nanoplates with average plate size 255 nm and thickness ~29 nm were obtained. Decreasing the weight ratio of the PVP to $AgNO_3$ to 4.9 resulted in formation of small thick nanoplates with average plate size 119 nm and the thickness was more than 60 nm. The reduction rate of $AgNO_3$ was directly proportional to PVP concentration. Hence, at reduced concentration of PVP, seed particles were formed slowly resulting in small thick nanoplates. The MW of PVP was inversely proportional to the concentration of PVP end groups. Hence, reducing the MW of PVP to 10 kDa produced many NPs with multiply twinned structures.

Silver nanotriangles with edge length 94 ± 8 nm were produced by reducing $AgNo_3$ (100 mL, 0.1 mM) with hydrazine (5 mL, 2 mM) in the presence of TSC (5 mL, 34 mM) as stabilizing agent [120]. When $AgNo_3$ concentration was increased to 0.5 mM, nanotriangles were not produced. However, after aging the solution for 2 months, triangular silver NPs were formed. High concentration of hydrazine produced smaller triangles, as the nucleation step was favoured over the growth step. In another study, hydrazine hydrate ($N_2H_4 \cdot H_2O$) was used as reducing agent for synthesis of triangular nanoplates [112]. The triangular nanoplates were characterized by edge length in the range of 50–200 nm. Initially small quasi-spherical AgNPs (about 20 nm) were produced. These AgNPs were then transformed into triangular AgNPs due to Ostwald ripening. Using hydrazine hydrate, Liu et al. [121] reported synthesis of hexagonal Ag nanoplates (Ag HNPs). The precursor solution (5 mL) contained $AgNO_3$ (0.01 M), ammonium chloride (5×10^{-6} M) and PVP (0.01 M, MW ~50,000). A mixture of hydrazine hydrate (0.2 mL, 0.2 M), HNO_3 (0.2 mL, 0.1 M) and ammonium chloride (0.2 mL, 0.5 mM) prepared in 1.4 mL water was used as the reduction solution. The reduction solution (2 mL) was added at a rate 0.1 mL/min into the precursor solution maintained at a temperature of 4–10°C. By changing the concentration of reagents in the precursor or/and reduction solution, the size of Ag HNPs could be controlled. Ag HNPs with average edge length of around 40 nm were produced by reducing the concentration of $AgNO_3$ and PVP to half the typical values.

In polyol synthesis, PAM was used to serve as a stabilizer for the Ag NPs and the amino groups of PAM also formed complexes with Ag^+ ions [71]. This coordination effect could considerably reduce the reduction rate of Ag^+ ions encouraging the formation of thin nanoplates through a kinetically control process. In a typical procedure, AgNO$_3$ (0.024 g) was dissolved in 0.5 mL of EG and 0.17 mL of PAM (MW = 10,000, 50 wt.% solution in water) was mixed with 0.33 mL EG at room temperature. These solutions were introduced simultaneously into EG (5 mL, heated at 135°C for 1 h) solution. The reaction was continued at 135°C for 3 h. TEM and SEM images of the NPs synthesized revealed that 90% nanoplates and 10% multiple twinned particles were present in the sample (Figure 1.25a–b).

The nanoplates obtained were triangular in shape with edge length ~25 nm having varying degrees of truncation and an average thickness of 10 nm. High-resolution TEM images (Figure 1.25c–d) were taken perpendicular to the flat faces of an individual nanoplate along the [$\bar{1}$11] zone axis. The fringes had a separation of 2.5 A°. The {220} fringes separated by 1.4A° were seen in the perpendicular direction. The high-resolution TEM image showed six-fold rotational symmetry (inset Figure 1.25c) suggesting that the flat faces were bounded by {111} planes. According to the d-spacing, the set with a spacing of 1.4A° (triangle) was due to the {220} reflection, the outer set with a lattice spacing of 0.8A° (squared) was related to {422} Bragg reflection, whereas, the inner set with a spacing of 2.5A° (circle) was due to the 1/3{422} reflection. High-resolution TEM images taken from side face of the sample recorded in the [011] orientation confirmed that the side face was bounded by one of the {100} planes. Moreover, Figure 1.25f showed that {111} twin defects were present parallel to the flat faces in each nanoplate [71].

Shape transformation of Ag nanospheres to triangular Ag nanoplates can be achieved by hydrogen peroxide. Li el al. [122] obtained triangular Ag nanoplates by modifying Mirkin's method explained under the section on nanoprisms. The order of addition of H$_2$O$_2$ and NaBH$_4$ was changed and mild alkaline condition was maintained to produce triangular Ag nanoplates with edge length 35–95 nm and thickness 5 nm. When stirring was stopped after H$_2$O$_2$ injection, both the edge length and the thickness of triangular nanoplates became larger [122]. The effect of H$_2$O$_2$ on shape transformation of Ag nanospheres to triangular Ag nanoplates was studied under mildly alkaline conditions (pH 7–8). H$_2$O$_2$ could function as both a facet selective etchant towards metallic Ag species and a reducing agent for ionic Ag species. The reactions were as follows eqs. (1.5)–(1.7):

H$_2$O$_2$ as an oxidizing agent:

$$Ag^+ + e^- \rightarrow Ag^0 \text{ (Reduction by } NaBH_4) \tag{1.5}$$

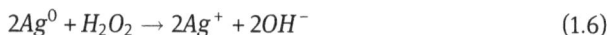

$$2Ag^0 + H_2O_2 \rightarrow 2Ag^+ + 2OH^- \tag{1.6}$$

H$_2$O$_2$ as reducing agent:

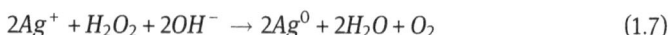

$$2Ag^+ + H_2O_2 + 2OH^- \rightarrow 2Ag^0 + 2H_2O + O_2 \tag{1.7}$$

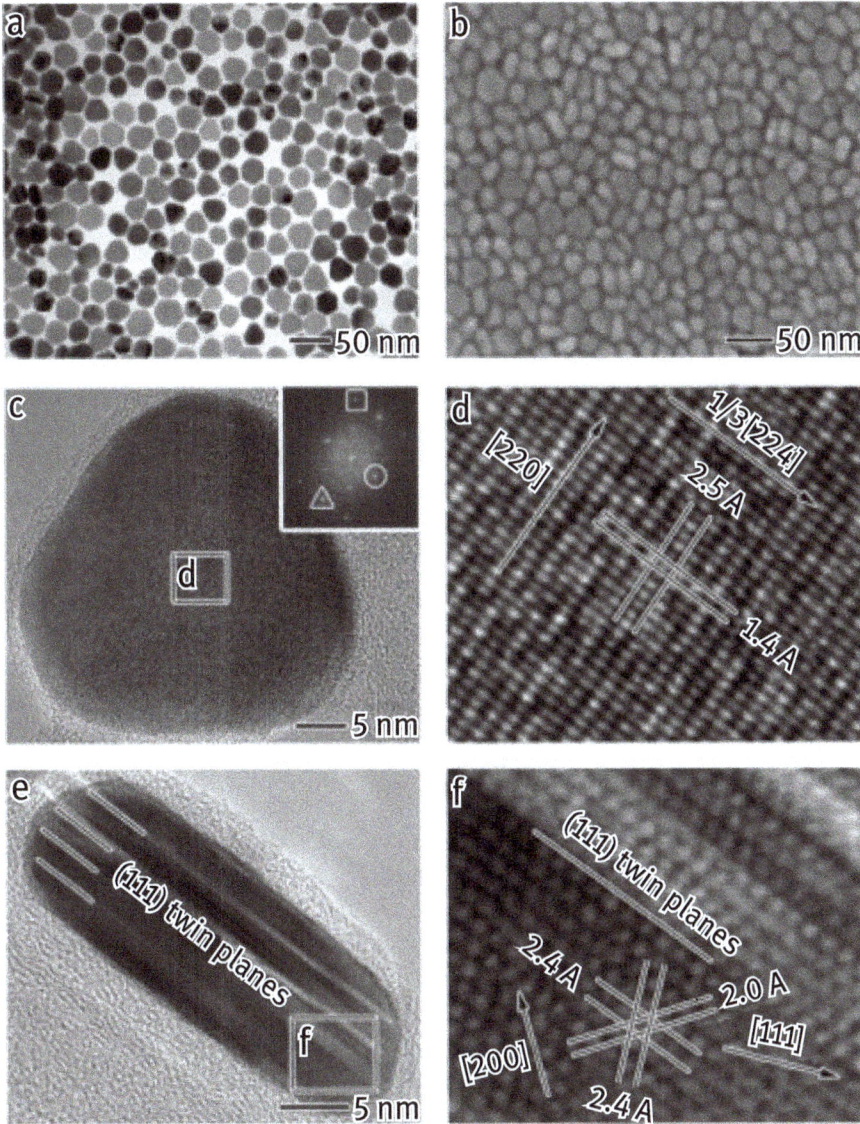

Figure 1.25: (a) TEM and (b) SEM images nanoplates produced in typical synthesis. (c, d) High-resolution TEM images taken perpendicular to the flat face of a truncated triangular nanoplate. Fast Fourier transform (FFT) of the image is shown in inset c. (e, f) High-resolution TEM images obtained from the side face of a nanoplate. (Reprinted with permission from ref. [71]. Copyright 2007, Royal Society of Chemistry).

In this process, H_2O_2 behaved as an oxidative etchant for selective dissolution of unstable facets, producing seeds with planar twinned defects or stacking faults parallel to the {111} direction. These seeds were used for the growth of AgNPs.

Further, the energetically unfavourable seeds with multiply twinned defects were etched. The dissolved Ag ions could be again reduced to Ag atoms by H_2O_2 which enabled growth of triangular AgNPs. TSC bound preferentially on the flat Ag {111} facets enabling lateral growth of the anisotropic NPs. While PVP bound onto the Ag {100} facets controlled the thickness increment of NPs.

Seed-mediated growth method was also exploited for silver nanoplate synthesis. Seeds were prepared by following a process reported by Jin et al. [123]. $AgNO_3$ (1 mL, 30 mM) and TSC (1 mL, 30 mM) were mixed in 48 mL of water. Subsequently, ice cold $NaBH_4$ (1 mL, 50 mM) and BSPP (1 mL, 5 mM) were introduced into the mixture. For nanoplate growth, CTAB (10 mL, 80 mM), $AgNO_3$ (0.25 mL, 10 mM), ascorbic acid (0.5 mL of 100 mM) and varying amounts of silver seed (30, 60, 125 and 250 µL) were mixed together. Subsequently, (1 mL of 1 mM NaOH) was added. Geometrical structures ranging from discs to triangular prisms were seen in the prepared sample. By changing the amount of silver seeds (30, 60, 125 and 250 µL) added to the growth solution, silver nanoplates of thickness 10 nm with varying diameters (d = 100, 53, 32 and 25 nm) were produced. With the help of hexadecyltrimethyl ammonium ions (CTA$^+$) as a trace additive in the seed solution, Lee et al. [124] demonstrated synthesis of Ag nanoplates. To obtain CTA$^+$ ions, $AgNO_3$ (0.2 M, 40 mL) was added to CTAB (40 mL, 0.2 M) aqueous solution. The white precipitate formed was removed by centrifugation to obtain 0.1 M CTA$^+$ additives. In seed-mediated synthesis, $AgNO_3$ (50 µL, 0.05 M) solution was added to TSC (10 mL, 2.5×10^{-4} M) solution. CTA$^+$ (25 µL of 0.1 M) was slowly added to this mixture under continuous stirring. The addition of $NaBH_4$ (25 µL, 0.02 M) produced CTA$^+$-adsorbed Ag seed solution. Later, $AgNO_3$ (100 µL, 0.05 M) was added to CTAB (20 mL, 0.1 M) aqueous solution. Subsequently, ascorbic acid (1 mL, 0.1 M) and the CTA$^+$-adsorbed Ag seed solution (100 µL) were added. Upon addition of NaOH (80 µL, 2 M) to this solution, Ag nanoplates were formed. By decreasing the amount of CTA$^+$ (100–25 µL), the mean edge length of triangular nanoplates was changed from ~78.9 nm to ~124.6 nm. Also by reducing CTA$^+$ adsorbed seed amount from 1000 to 100 µL, the edge length of triangular nanoplates increased from 70 nm to 148 nm [124].

Nanoplates sometimes need to be coated with various polymers for functionalization. Polymers such as chitosan and PSS exhibit preferential binding to Ag (111) surface which is exploited in synthesis of triangular nanoplates. In one such study, chitosan-coated silver nanoplates were obtained by a two-step approach. The seed solution was prepared by mixing an aqueous solution of $AgNO_3$ (10 mL, 2.9×10^{-4} M) with an aqueous solution of TSC (10 mL, 2.5×10^{-4} M). The mixture was cooled in an ice-bath under vigorous stirring. An aqueous solution of $NaBH_4$ (0.6 mL, 0.1 M) was added dropwise into the mixture. To facilitate growth of nanoplates, aqueous solutions of TSC (200 µL, 38.8 mM), ascorbic acid (50 µL, 0.1 M,), chitosan (10 mL, 2 mg/mL) and seeds (200 µL) were mixed at 35 ± 2°C. Immediately, $AgNO_3$ (300 µL, 0.01 M) was added dropwise under continuous stirring. The highest fraction of triangular nanoplates were observed using TSC concentration of (38.8 mM), and these nanoplates had edge lengths from 115 to 123 nm and thickness of 11 ± 2 nm (aspect ratio 11).

Similar to TSC, chitosan can bind more strongly to {111} planes. Thus, chitosan could not only increase the proportion of nanotriangles in the final product, it could also attach as a functional layer on the nanoplates [73]. In another study, Si et al. [88] reported synthesis of silver nanoplates by reducing $AgNO_3$ (50 mL, 0.1 mM,) with $NaBH_4$ (300 µL, 70 mM,) in the presence of TSC (3 mL, 30 mM), H_2O_2 (60 µL, 30 wt.%) and PSS (3 mL). Triangular nanoplates along with a few spherical and trapezoidal NPs were produced when PSS concentration was varied from 0 to 3 mg/mL. The number of spherical NPs increased when PSS concentration was above 3 mg/mL. The edge length of the triangular Ag nanoplates decreased (45 ± 12, 40 ± 10, 36 ± 9, 32 ± 9, 28 ± 8, 24 ± 9 nm) with rise in PSS concentration (0, 0.5, 1, 1.5, 2 and 3 mg/ml). It was proposed that PSS molecules preferentially bind to the surface of Ag (111) having the lowest interaction energy and inhibit their growth. Hence, the edge length was decreased and the Ag nanoplate thickness remained unchanged [88].

Table 1.3 summarizes methods for size-controlled synthesis of silver nanoplates using various synthesis protocols and reagents.

1.3.2.2.4 Synthesis of silver nanorods and nanobars

Nanorods have been synthesized via seed-mediated growth method using surfactants, polyol -based synthesis approach using additives and citrate-based reduction method using various stabilizers. In seed-mediated growth method, the seed concentration and base concentration relative to the Ag+ concentration play a vital role to make larger aspect-ratio nanomaterials [67]. Ag seeds of average diameter 4 nm were produced by chemical reduction of $AgNO_3$ (0.25 mM) by $NaBH_4$ (0.6 mL, 10 mM) in the presence of TSC (0.25 mM). The growth solution was comprised of $AgNO_3$ (0.25 mL, 10 mM), ascorbic acid (0.5 mL, 100 mM), CTAB (10 mL, 80 mM) and varying volume of the seed solution (2–0.06 mL). Later, NaOH (0.1 mL, 1 M) was added to the mixture. For nanorod synthesis, pH of the growth solution was maintained at a value higher than the pK_{a2} of ascorbic acid [67]. Silver nanorods of aspect ratio 2.5–15 and 10–15 nm width were synthesized.

Another surfactant used for nanorod synthesis is cetyltrimethylammonium tosylate (CTAT). At low surfactant concentration, CTAT forms long wormlike micelles. Ni et al. used this surfactant in seed-mediated growth method for nanorod synthesis. The seed solution prepared using silver dodecyl sulfate (AgDS, 5 mL, 0.25 mM) and borohydride (150 µL, 10 mM) was incubated at 35°C for 2 h. The aqueous growth solution comprised of CTAT (0.2–1%) doped with AgDS (0.25 mM). Subsequently, variable amount of seed solution and ascorbic acid (0.1 M) were added to the growth solution. The number of silver nanorods produced increased with increase in CTAT concentration. By decreasing the seed concentration (6.25–1.25 µM) at a fixed CTAT concentration, the aspect ratio of the rods increased, such that nanowires with aspect ratio exceeding 100 were produced. Diameter of the nanorods synthesized was in the range of 30–50 nm which was larger than the diameter of the surfactant micelles. This

Table 1.3: Synthesis of nanoplates using chemical method.

Reducing agent	Stabilizing agent	AgNO₃	Reaction conditions	Size (nm)	Reference
PVP [PVP/AgNO₃] 30	—	3 mL, 188 mM	60°C, 21 h	Triangular 350; Others 120	Washio et al. [118]
PVP in NMP 8 mL, 1.87 g; PVP/AgNO₃ 19.5	—	3 mL, 188 mM	100°C, 24 h	Hexagonal 213, T= 63	Kim et al. [119]
Hydrazine 5 mL, 2 mM	TSC 5 mL, 34 mM	100 mL, 0.1 mM	~5 min	Triangular 94 ± 8	Mansouri and Ghader [120]
Hydrazine hydrate	PVP: 10 mL, 0.375 M	10 ml, 20 mM	10 min[#]	Triangular 50–200	Li et al. [112]
Hydrazine hydrate: 0.2 mL, 0.2 M; HNO₃: 0.2 mL, 0.1 M; NH₄Cl 0.2 mL, 0.5 mM	PVP: 5 mL, 0.01 M; NH₄Cl: 5 µM	0.01 M	4–10°C, 1 h [#]	Hexagonal 60nm, T = 20	Liu et al. [121]
EG 5 mL, 135°C, 1h	PAM (50 wt.% Aq. solution) 0.17 mL in 0.33 mL EG	0.024 g, 0.5 mL	135°C, 3 h	Triangular 25, T=10	Xiong et al. [71]
NaBH₄ 300 µl, 0.1 M	TSC:3 mL, 30 mM; PVP: 3 ml, 0.7 mM; 30% H₂O₂:120 µL,	50 mL, 0.1 mM	~20 min	Triangular 35–95, T = 5	Li el al. [122]
Seed: NaBH₄ 0.6 mL, 0.1 M Growth: ascorbic acid 50 µL,0.1 M	Seed: TSC 10 mL, 0.25 mM Growth: Chitosan 10 mL, 2 g/L; TSC 200 µL, 38.8 mM	Seed: 10 mL, 0.29 mM Growth:300 µL, 10 mM	—	Triangular 115–123, T= 11 ± 2 AR: 11	Potara et al. [73]
NaBH₄ 300 µL, 70 mM	PSS 0–3 g/l, TSC 3 mL, 30 mM; 30% H₂O₂ 60 µL	50 mL, 0.1mM	—	Triangular 45 ± 12 to 24 ± 9[a]	Si et al. [88]

1 Size is specified in terms of edge length, thickness (T) and aspect ratio (AR); [#]after the injection of the reduction solution; [a]AgNPs were synthesized over this size range by varying the reaction conditions; NMP: N-methylpyrrolidone; PAM: polyacrylamide; BSPP: Bis(p-sulfonatophenyl) phenylphosphine; CTAB: cetyltrimethylammonium bromide; PSS: poly(styrene-4-sulfonate).

indicated that anisotropic growth of nanorods may have been caused by preferential attachment of CTAT to the lateral {100} planes as opposed to {111} planes. Also, TEM analysis suggested growth of nanorods from multiply twinned decahedral seeds along {110} direction. They were surrounded by five {111} facets on each end and five {001} side surfaces [125].

In polyol synthesis, addition of elevated concentration of bromide produced nanobars [125] whereas Na₂S addition resulted in rectangular silver nanorods [82]. For nanobars synthesis, AgNO₃ (48 mg dissolved in 3 mL of EG) and a mix of PVP of

MW 55,000 and NaBr (48 mg and 0.068 mg, respectively, dissolved in 3 mL of EG) were introduced dropwise into EG (5 mL) heated to 155°C in an oil bath. The reaction was stopped after 1 h. Silver nanobars having aspect ratio ~2.7 were produced as illustrated in Figure 1.26A–C [126].

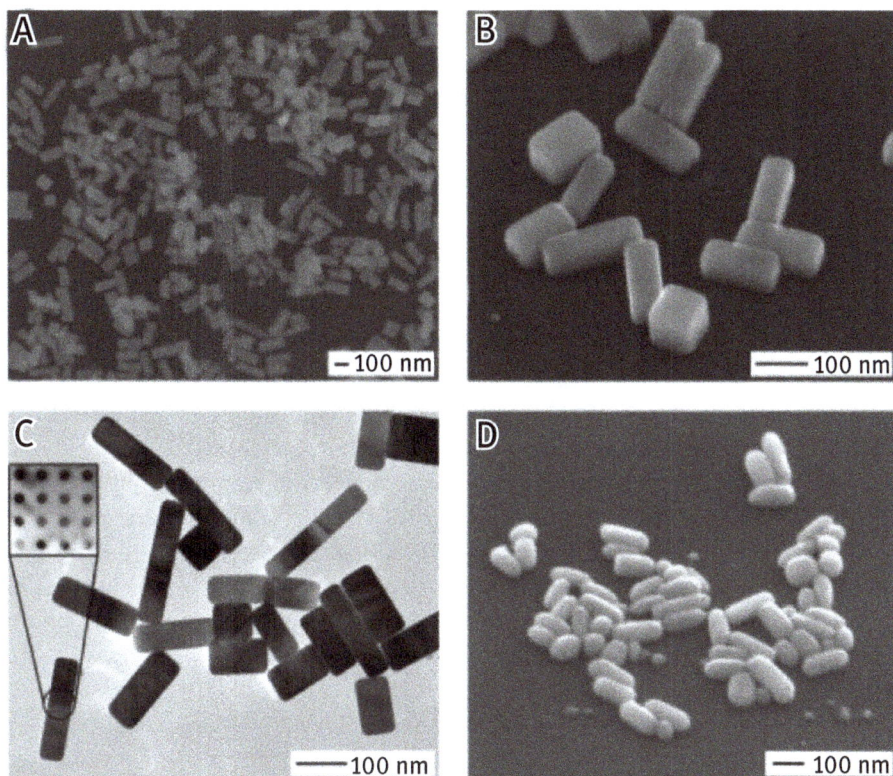

Figure 1.26: (A, B) SEM images of the Ag nanobars formed using NaBr, (c) TEM image of nanobars with inset showing a convergent beam electron diffraction pattern and (D) SEM image of nanorice formed upon aging of the nanobars. (Reprinted with permission from ref. [79]. Copyright 2007, American Chemical Society).

Convergent beam electron diffraction on several nanobars showed diffraction from the (100) planes indicating that the nanobars were single crystals bound on all sides by {100} facets like nanocubes (Figure 1.26C). Upon one week storage in 5 wt% aqueous solution of PVP, they transformed into single-crystal nanorice with rounded corners and edges (Figure 1.26D). When bromide concentration (30 µM) was reduced by half compared to that used for nanobar synthesis [79], right bipyramids having edge length ~40 nm were produced after 2.5 h. By reducing bromide concentration, the reactive multiply twinned seeds were etched away while seeds with a single twin plane remained unaffected. These seeds could produce right bipyramids after the growth phase.

Guo et al. demonstrated rectangular silver nanorods synthesis using polyol method in the presence of PVP and Na$_2$S [82]. During preparation, EG (36 mL) was heated to ~157°C for 1 h. After injection of Na$_2$S solution (30 μL, 48 mM), PVP (MW 30,000–40,000, 9 mL, 177 mM) was added rapidly to heated EG. Subsequently, AgNO$_3$ (3 mL, 284 mM) was injected dropwise. Concentration of Na$_2$S and the PVP/AgNO$_3$ molar ratio was found to alter the dimensions and morphology of the NPs produced. Large aspect ratio silver nanowires were produced using 40 μL of Na$_2$S solution (48 mM) and PVP/AgNO$_3$ molar ratio of 1.87. By reducing the molar ratio of PVP/AgNO$_3$ from 1.87 to 0.93, silver nanorods with a larger diameter (~400 nm) and cubic AgNPs were formed. Upon increasing this ratio from 1.87 to 8.4, small AgNPs of varying size were produced.

Citrate reduction approach for nanorods synthesis was reported by Rao et al. in presence of SDS as stabilizing reagent using the Turkevich method [74]. Aqueous solution of AgNO$_3$ (90 mL, 10^{-4} M) and SDS (in the range of 1–10 mM) was stirred for 30 min at 400 rpm. The solution was heated at 100°C. Later, 10 mL TSC solution (10 mM) was injected at a rate of 1 mL/min and the resultant solution was heated for 5 min. The AgNPs formed using low SDS concentration (<2 mM) were spherical and less than 50 nm in size. With increase in SDS concentration, greater variability in shape was observed At SDS concentration of 10 mM, primarily nanorods were obtained. At higher SDS concentrations, cylindrical SDS micelles were observed. Hence, AgNPs growing inside SDS micelles assumed the same shape of the micelles [74]. At pH 7, most of the particles were spherical. With increase in pH, the amount of Ag nanorods increased. At pH 9 and 9.5, almost half of the particles were Ag nanorods.

In another study, the dimension and morphology of NPs was controlled by varying TSC concentration when AgNO$_3$ was reduced using TSC in presence of sodium dodecylsulfonate (SDSN) [127]. At lower concentration of the reducing agent (TSC ≤0.2 mM), the concentration of silver monomer (produced from TSC reduction) promoted synthesis of spherical AgNPs as a result of Ostwald ripening, where smaller AgNPs were dissolved in solution and were added onto larger AgNPs (Figure 1.27A). At TSC concentration >0.2 mM, high silver monomer concentration was produced thereby causing higher growth rate on (011) facets. This resulted in formation of silver nanorods. At 0.8 mM, TSC concentration nanorods with average diameter of 20 nm and aspect ratio around 12.5 was obtained. At TSC >0.8 mM, nanocrystals could also grow on the same axis (Figure 1.27B). However, since the velocity of crystal growth was higher, nanowires with smaller diameter and higher aspect ratio were produced [127]. Using 1.2 mM TSC concentration, nanowires having diameter of ~15 nm were observed. The mean length of these nanowires was in the range of 1–1.5 μm.

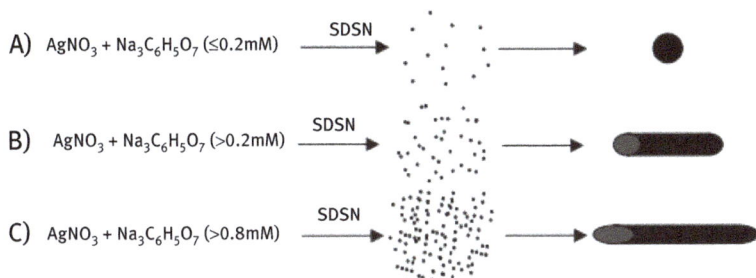

A) AgNO$_3$ + Na$_3$C$_6$H$_5$O$_7$ (≤0.2mM) $\xrightarrow{\text{SDSN}}$

B) AgNO$_3$ + Na$_3$C$_6$H$_5$O$_7$ (>0.2mM) $\xrightarrow{\text{SDSN}}$

C) AgNO$_3$ + Na$_3$C$_6$H$_5$O$_7$ (>0.8mM) $\xrightarrow{\text{SDSN}}$

Figure 1.27: Illustrating the generation of silver A) sphere B) rod C) wire. (Reprinted with permission from ref. [127]. Copyright 2004, John Wiley and Sons).

Table 1.4 summarizes methods for size-controlled synthesis of silver nanorods and nanobars using various synthesis protocols and reagents.

Table 1.4: Synthesis of nanorods via chemical method.

Reducing agent	Stabilizing agent/ additives	AgNO$_3$	Reaction conditions	Size (nm)	Reference
Seed: NaBH$_4$ 0.6 mL,10 mM Growth: Ascorbic acid 0.5 mL, 100 mM	Seed: TSC: 20 mL, 0.25 mM Growth: CTAB 10 mL, 80 mM; NaOH: 0.1 mL, 1 M	Seed: 0.25 mM Growth: 0.25 mL, 10 mM	~10 min	AR: 2.5–15 SA:10–15	Jana et al. [67]
Seed: NaBH$_4$ 150 µL, 10 mM Growth: ascorbic acid 0.1 M	Growth: CTAT 0.2–1%	*Seed: 5 mL, 0.25 mM *Growth: 0.25 mM	Seed: 35°C, ~2 h	D: 30–50	Ni et al. [125]
EG 36 mL	PVP 9 mL, 177 mM / Na$_2$S 30 µL, 48 mM	3 mL, 284 mM	157°C, 1 h	D: 200	Guo et al. [82]
EG 5 mL	PVP 48 mg / NaBr 0.068 mg, 3 mL of EG	48 mg, 3 mL of EG	155°C, 1 h	AR: 2.7	Wiley et al. [79]
TSC 10 mL, 10 mM	SDS 10 mM	90 mL, 0.1 mM	100°C	—	Rao et al. [74]

*Precursor salt used was silver dodecyl sulphate instead of AgNO$_3$; AR implies aspect ratio; D implies diameter; SA implies short axis.

1.3.2.2.5 Synthesis of silver nanowires

Silver nanowires (AgNWs) have been synthesized primarily via two methods, seed-mediated synthesis and polyol process. In seed-mediated synthesis, solution pH plays a vital role in nanowire synthesis. Moreover, certain additives, such as, halides and sulfides have been reported to facilitate controlled growth of silver nanowires. Using seed-mediated growth method, Jana et al. [67] demonstrated synthesis of nanorods as discussed in **Section** 1.3.2.2.4. Through a modification in this protocol, nanowires could be synthesized. When pH of growth solution is lower than pK_{a2} but higher than pK_{a1} of ascorbic acid ($pK_{a1} \approx 4.1$, $pK_{a2} \approx 11.8$), the monoanion of ascorbic acid is the predominant species in solution [67]. Seed solution was prepared as explained under the section on nanorods. The growth solution comprised of $AgNO_3$ (2.5 mL, 10 mM), ascorbic acid (5 mL, 100 mM), CTAB (93 mL, 80 mM) and 2.5 mL of seed solution. Subsequently, NaOH (0.5 mL, 1 M) was then added to the mixture. After 15 min, nanowires of 1–4 µm length and 12–18 nm short axes were produced [67].

Silver nanowires can also be produced using gelatin as the template. Nanocrystals of silver bromide (AgBr) were prepared by adding potassium bromide (KBr, 10 mL, 1 M) and $AgNO_3$ (10 mL, 1 M) in gelatin solution (1.5 g gelatin soaked in 10 mL DI water for 30 min and further dissolved in 20 mL water) at 50°C. To obtain silver nanowires (9 µm long and 80 nm in diameter), AgBr emulsion was developed using a developer made of metol (N-methyl-p-aminophenol sulfate), citric acid and $AgNO_3$ [128].

In a self-seeding polyol process, silver nanowires were produced by reducing $AgNO_3$ with EG in presence of PVP. In a typical experiment, EG solution of $AgNO_3$ (3 mL, 0.1 M) and EG solution of PVP (3 mL, 0.6 M, MW~55,000) were introduced into 5 mL EG refluxed at 160°C at a rate of 0.3 mL/min. After complete injection, the mixture was further refluxed for 60 min at 160°C. On purification, AgNWs demonstrated a mean diameter of ~60 ± 8 nm (Figure 1.28A). XRD pattern of AgNWs suggested fcc crystal structure of silver. TEM image (Figure 1.28D) indicated {111} twin plane parallel to longitudinal axis of nanowires.

With molar ratio of PVP to $AgNO_3$ as 1.5:1, silver nanowires with high aspect ratio (~1000) were synthesized [129]. Addition of CuCl or $CuCl_2$ during polyol synthesis of Ag° in the presence of PVP enabled the formation of Ag nanowires. EG (5 mL) was heated for 1 h at 151.5°C with continuous stirring. After 1 h, Cu solution (40 µL, 4 mM $CuCl_2$) was added to heated EG and heating was continued for 15 min. Subsequently, PVP solution in EG (1.5 mL, 0.147 M, MW~55,000) was added to the heated EG. This was followed by addition of $AgNO_3$ solution in EG (1.5 mL, 0.094 M) to the mixture. Ag nanowires having length of 10–50 µm with uniform diameters of ~100 nm were obtained. Cl^- reduced the level of free Ag^+ present initially and Cu(I) (added directly or generated in situ by reduction of Cu(II) by EG) removed the adsorbed atomic oxygen from the surface of the seeds and promoted growth of silver nanowires [83].

By adding stainless steel in polyol process, the synthesis of AgNWs with high aspect ratio was possible at a high $AgNO_3$ concentration (500 mM). For nanowire synthesis, 10 mL of $AgNO_3$ solution (0.3 M in EG), 10 mL of PVP solution (0.45 M in EG, calculated in terms of the repeating unit, MW ~55,000) and several small pieces of stainless steel grids (~0.5 g), were mixed together. The mixture was heated at 120°C in an oil bath for about 0.5 h. The temperature was increased to 140°C and the mixture was kept at this temperature until complete reduction of $AgNO_3$ (~1 h). The initial concentration of $AgNO_3$ determined the diameter of the AgNWs. With 150 mM $AgNO_3$

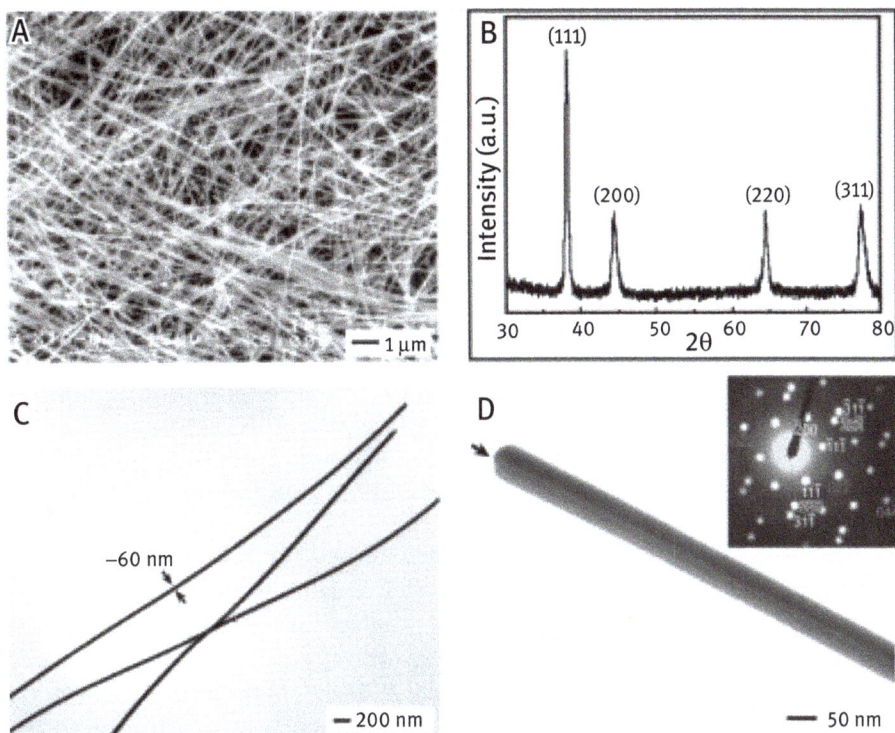

Figure 1.28: (A) SEM, (B) XRD (C) TEM image of synthesized AgNWs with PVP to AgNO$_3$ molar ratio as 1.5: 1 (D) TEM image of single AgNWs. (Reprinted with permission from ref. [129]. Copyright 2002, John Wiley and Sons).

concentration, nanowires having a mean diameter of ~120 nm, lengths of tens of micrometres and aspect ratios ~500 were formed. In the reduction process of AgNO$_3$ by EG, nitric acid was generated. Stainless steel removed the nitric acid generated *in situ*, thereby ensuring the survival of MTPs for the growth of AgNWs as presented in Figure 1.29 [130].

In a solvothermal method, 0.15 M PVP (MW 40,000) was added to 10 mL EG solution of Na$_2$S with vigorous stirring. This solution was added drop by drop to 10 mL EG solution of AgNO$_3$ (0.1 M) with continuous stirring. Subsequently, the mixture was maintained at 160°C for 2.5 h. In the initial silver seed formation stage, Ag$^+$ ions were reduced by silver sulfide (Ag$_2$S). Subsequently, Ag$^+$ ions from Ag$_2$S colloids were released into solution enabling the synthesis of silver nanowires. The diameter of silver nanowires increased from ~170 to 310 nm with increase in Na$_2$S concentration from 0.1 to 0.3 mM [80].

In two-step polyol process, the diameter of nanowires can also be controlled using KCl [84]. In a typical synthesis reaction, PVP was dissolved in 110 mL EG PVP (MW 360,000) to silver molar ratio was maintained as 1.5. EG was heated at 160°C in

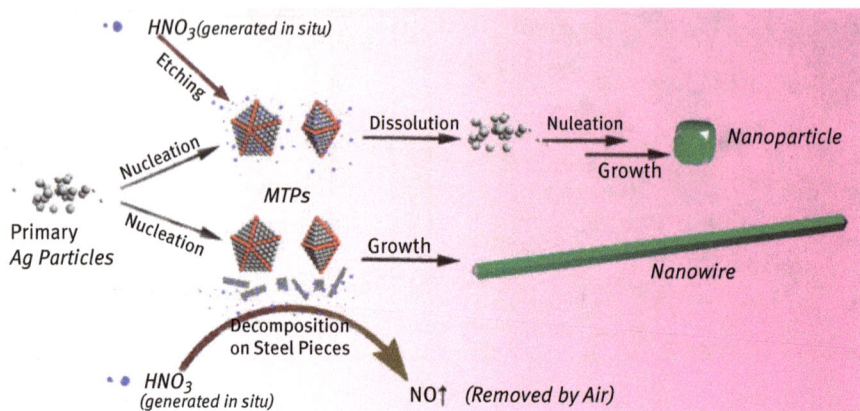

Figure 1.29: Schematic of Ag nanowire synthesis using polyol mechanism in presence of stainless steel grids. (Reprinted with permission from ref. [130]. Copyright 2008, American Chemical Society).

an oil bath under continuous stirring. AgNO$_3$ (0.6 g) and KCl (2.25 mg) were dissolved in 10 mL EG and the mixture was added to the PVP solution. In a subsequent step, AgNO$_3$ (0.93 g) and varying amounts of KCl were mixed in 80 mL EG and this solution was introduced into the seed solution for growth of nanowires. The diameter of nanowire formed was 100 nm. Introduction of Cl$^-$ ions into the growth formulation prepared in the 2nd step at 0.05 times the concentration of AgNO$_3$ reduced the diameter of Ag nanowires to 55 nm It was proposed that chloride ion preferentially adsorb on to (100) face of silver and thus controls the diameter of the nanowires.

Long silver nanowires (average length ~28 μm, and aspect ratio ~130) with uniform diameter were synthesized via polyol process using PVP as preferential growth agent [131]. Preferential growth of nanowires could be achieved by lowering the reaction rate by controlling the rate of addition of PVP and AgNO$_3$ solution. In the synthesis process, EG (5 ml) was heated at 160°C with stirring for 30 min. AgNO$_3$ (3 mL, 0.085 M) and PVP (3 mL, 0.1275 M) prepared in EG as solvent with molar ratio of AgNO$_3$ to PVP as 1:1.5 were mixed by dropwise addition over 8 min. Upon heating the reaction mixture for 70 min at 160°C, nanowires (28 μm long, aspect ratio ~130) were produced. By reducing the rate of addition of AgNO$_3$, the Ag content in the polyol mixture and consequently the rate of reduction of Ag$^+$ ions were reduced. Hence, the reduced Ag atoms could deposit along the PVP chains yielding silver nanowires. Moreover, since no additives were added, the etching of Ag atoms was decreased. This slower etching rate protected MTPs in the seed formation stage such that a high yield of Ag nanowires could be achieved. When AgNO$_3$ concentration was increased from 0.07 M to 0.085 M, formation of nanowires with higher aspect ratio and diameter was observed. Increase in AgNO$_3$ concentration to 0.1 M resulted in formation of shorter nanowires. The effect of temperature on nanowire formation was also studied. At 150°C, the MTP content in the initial seed formed was lowered, leading to lower yield of Ag nanowires. A higher temperature of 160°C resulted in higher surface diffusion causing greater extent of Ostwald ripening. Thus, longer nanowires with higher aspect ratio were synthesized. At 170°C, a higher yield of silver nanowires was obtained without any change in aspect ratio.

Recently, Zhang el al. [85] studied the role of bromide ion in thin silver nanowires synthesis using the polyol method. For synthesis, 30 mL of pure EG, 15 mL of AgNO$_3$ (0.1 M EG solution), 15 mL of PVP (0.2 M EG solution) and varying concentrations of sodium chloride (NaCl) and KBr were mixed by stirring at room temperature. The solution was heated at 75°C for 30 min. This was followed by heating at 120°C with stirring for 8 h. The amount of halide ions was kept constant whereas the ratio of chloride to bromide ion was changed. During the preheating stage, AgCl and AgBr nanocubes were formed. The AgCl cubes were larger than the AgBr cubes. At high temperature (120°C), silver ions were reduced to Ag atoms and MTPs as well as NPs were produced on the surface of the nanocubes. During the growth stage, the MTPs were grown into silver nanorods and they later formed nanowires in presence of PVP as PVP interacted more strongly with the {100} facets than with the {111} facets. Presence of bromide ions could effectively passivate the {100} facets and limit lateral growth, owing to their smaller size. Since AgBr cubes are much smaller than AgCl cubes, the formation of AgBr cubes resulted in thinner Ag nanowires. Hence, the average diameter decreased steadily when KBr concentration was increased from 0 to 0.08 mM and the number of Ag nanowires with diameters less than 30 nm increased steadily. On increasing the ratio of Br$^-$ to AgNO$_3$, both the {100} facets and {111} facets were passivated by bromide ions. Under this condition, bromide ions could no longer limit the lateral growth, thus, nanowires with larger diameters were synthesized. Therefore, for KBr concentration higher than 0.08 mM, nanowire diameter was increased to 40 nm. The addition of bromide ions produced more stable AgBr complexes that were less soluble than AgCl. Hence, the rate of Ag$^+$ release was significantly reduced.

Silver nanowire synthesis using TSC and DMF as reducing agent has also been reported. In one such study, Caswell et al. [69] demonstrated nanowire synthesis without addition of any surfactant or externally added seeds. Two silver solutions were mixed to produce silver nanowires having average width 35 ± 6 nm and length in the range of 166 nm–12 μm. Solution-A was comprised of 100 mL DI water, 1.5–2 μL of 1 M NaOH and 40 μL of 0.1 M AgNO$_3$. This solution was heated up to boiling temperature with stirring. Later, 5 mL of 0.01 M TSC was added and the solution was boiled for 10 min. Solution-B was prepared by mixing 150 mL DI water and 1.5–2 μL of 1 M NaOH with 20 μL of 0.1 M AgNO$_3$ and the solution was heated up to boiling temperature. The two solutions were mixed and the solution was boiled for 60 min. The role of citrate was crucial in this synthesis. It strongly complexed the silver ion to form Ag-citrate complex and promoted reduction of silver. The complex (with pK$_1$ = 7.1) could also act as a capping agent. Mdluli and Revaprasadu [132] reported synthesis of silver nanowires using DMF as reducing agent in the presence of PVP without addition of seed particles through a facile aqueous process. PVP (0.005–0.02 g/mL, MW = 30,000) and $AgNO_3$ (0.1 mol/dm^3) solutions (250 mL each) were separately prepared using DI water. Equal volume (10 mL) of each solution was mixed to obtain a homogeneous mixture. DMF (10 mL) was then added and the resultant solution was left unstirred in the dark for 30 days.

Novel branched silver nanowires such as Y-shaped, K-shaped, as well as other multi-branched nanowire structures were synthesized by Cong et al. [132]. For synthesis, 0.1275 g $AgNO_3$ was dissolved in an aqueous solution of PVP (MW 30 kDa, 2.5 mL, 1 M). Subsequently the whole solution was added into 25 mL PEG (MW 600 Da) under stirring. The mixture was heated at 160°C for ~30 min until a dark cyan solution was obtained. Subsequently, the temperature was reduced to 100°C and held constant for 4 h. This was followed by cooling to room temperature. For a PVP/$AgNO_3$ molar ratio of 3.3:1, branched Ag nanowires were produced. The thickness of PVP coating and the location of PVP chains on the surface of a seed could be modified by adjusting the molar ratio of PVP/$AgNO_3$, which in turn, changed the growth of branched nanowires. However, no branched structures were formed when the reaction was carried out at a higher temperature of 160°C for ~4 h. Similarly, using PVP with a higher degree of polymerization (MW >30 kDa), resulted in NPs instead of branched structures.

Table 1.5 summarizes methods for size-controlled synthesis of silver nanowires using various synthesis protocols and reagents.

1.3.2.2.6 Synthesis of silver nanocubes

Nanocubes can be synthesized by increasing $AgNO_3$ concentration in polyol process established for nanowire synthesis. Certain additives can enable the growth of nanocubes by selectively etching the twinned seeds or by increasing the reduction rate of Ag^+ ions. In a self-seeding polyol process reported by Sun and Xia [68], anhydrous EG (5 ml) was heated at 160°C for 1 h. EG solution of $AgNO_3$ (3 mL, 0.25 M) and PVP (0.375 M, 3 mL, MW 55,000) were injected simultaneously into heated EG at a rate of 0.375 ml/min and heating at 160°C was continued for additional 45 min. The silver nanocubes formed were recovered by centrifugation and were dispersed in water. In this synthesis, the concentration of $AgNO_3$ was 3 times higher than that used for nanowire synthesis (discussed in Section 1.3.2.2.5) and the molar ratio between PVP and $AgNO_3$ was maintained at 1.5. Slightly truncated single-crystalline silver nanocubes bounded by {100}, {110} and {111} facets with mean edge length of 175 nm, and standard deviation of 13 nm were obtained. When the molar ratio between PVP and $AgNO_3$ was increased from 1.5 to 3, MTPs were produced. By controlling the growth time, silver nanocubes with different dimensions could be synthesized. Monodispersed silver nanocubes can be produced by adding a small amount of HCl in the conventional polyol synthesis protocol discussed earlier.

In a typical reaction reported by Im et al. [134], EG (5 mL) was heated in an oil bath at 140°C for 1 h with stirring. HCl (1 mL, 3 mM) solution in EG was quickly added

Table 1.5: Synthesis of nanowires using chemical method.

Reducing agent	Stabilizing agent/ Additives	AgNO$_3$	Reaction condi- tions	Size	Reference
Seed: NaBH$_4$ 0.6 mL,10 mM Growth: Ascorbic acid 5 mL, 100 mM	Seed: TSC 20 mL, 0.25 mM Growth: CTAB 93 mL, 80 mM/NaOH 0.5 mL, 1 M	Seed: 0.25 mM Growth: 2.5 mL, 10 mM	~15 min	L 1–4 µm, SA 12–18 nm	Jana et al. [67]
EG 5 mL 160°C	PVP 3 mL, 0.6 M [PVP:AgNO$_3$] = 1.5:1	3 mL, 0.1 M	160°C, 1 h	AR ~1000D~60 ±8 nm	Sun and Xia [129]
EG 5 mL 151.5°C for 1 h	PVP (1.5 mL, 0.147 M)/CuCl$_2$ (40 µL, 4 mM)	1.5 mL, 0.094 M	151.5°C, 15 min	L = 10–50 µm, D ~ 100 nm	Korte et al. [83]
EG	PVP 10 mL, 0.45 M/ Stainless steel grid small pieces (~0.5 g)	10 mL, 0.3 M	120 °C, 30 min; 140 °C, ~1 h	D ~ 120 nm, L = tens of µm AR~500	Zhang et al. [130]
EG	PVP 0.15 M; Na$_2$S 10 mL, 0.15 mM	10 mL, 0.1 M	160°C, 2.5 h	D ~ 170–310 nm[a]	Chen et al. [80]
EG 5 mL 160°C, 30 min	PVP 3 mL, 0.1275 M [PVP/AgNO$_3$] = 1.5	3 mL, 0.085 M	160°C, 70 min	L = 28 µm, AR 130	Nekahi et al. [131]
EG 30 mL	PVP (15 mL, 0.2 M)/ NaCl, KBr (0.08 mM)	15 mL, 0.1 M	75°C,30 min 120°C, 8 h	D < 30 nm	Zhang el al. [85]
PEG 25 mL	PVP (2.5 mL, 1 M) [PVP:AgNO$_3$] = 3.3:1	0.1275 g	160°C, 30 min 100°C, 4 h	Branched NW: L = 5–15 µm, branched part >2 µm (L) D ~200 nm	Cong et al. [133]

The size is specified in terms of edge length; aAgNPs were synthesized over this size range by varying the reaction conditions; CTAC: Cetyltrimethylammonium chloride; PVE: Poly(methyl vinyl ether).

to this solution. After 10 min, EG solution of AgNO$_3$ (3 mL, 94 mM) and PVP (3 mL, 147 mM, MW ~55,000) were injected simultaneously into this solution at a rate of 45 mL per hour. The mixture was heated at 140°C. At the end of 26 h, nanocubes with 130 nm edge length were synthesized. On addition of AgNO$_3$ and PVP into the hot EG solution, both twinned and single-crystal seeds of silver is produced through nuclea-tion. Since the reactivity of twinned particles is higher, they are more susceptible towards etching by nitric acid formed *in situ*, thus, high yield of single-crystal nanocubes can be achieved through this method. HCl plays a dual role. It causes *in situ* generation of nitric acid and the Cl$^-$ ions originating from HCl adsorb on the surfaces of silver seeds to prevent agglomeration through electrostatic stabilization.

Reduction of the reaction temperature from 160 to 140°C reduced the net reaction rate and increased the efficiency of selective etching by nitric acid.

Further, the authors scaled up the synthesis process and the synthesized nano-cubes exhibited sharp corners having average edge length of 125 nm (Figure 1.30A). XRD peak corresponding to (200) indicated that the sample was comprised of nanocubes and were oriented with their (100) planes parallel to the supporting substrate (Figure 1.30B). The synthesized NPs were single crystal as indicated by TEM image and an electron diffraction pattern (Figure 1.30C) [134].

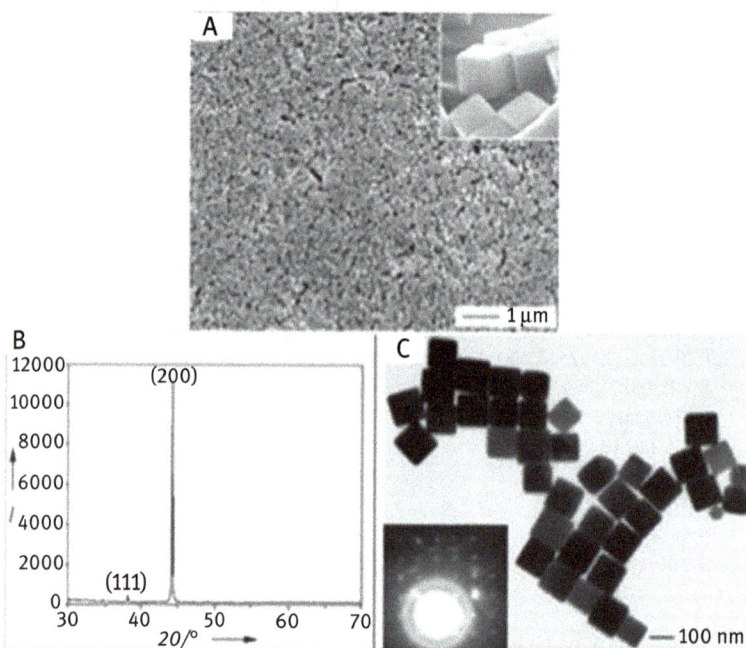

Figure 1.30: (A) SEM, (B) XRD and (C) TEM images of nanocubes in a scale-up synthesis. (Reprinted with permission from ref. [134]. Copyright 2005, John Wiley and Sons).

The addition of trace amount of Na_2S or NaHS in the conventional polyol synthesis process can reduce the reaction time from several hours to a few minutes. In a typical synthesis described by Siekkinen et al. [81], EG (6 mL) was heated at 150–155°C for 1 h with stirring. A sulfide solution in EG (3 mM Na_2S or NaHS) was made 45 min before injection. This solution (80 μL) and EG solution of PVP (1.5 mL, 20 mg/mL, MW ~55,000) and $AgNO_3$ (0.5 mL, 48 mg/mL) were added to heated EG successively. The concentration of the sulfide ion is critical for nanocube formation. To produce high-quality, monodispersed silver nanocubes, sulfide ion concentration in the range 28–30 μM is required for fast reduction of Ag^+ at elevated temperature. Within 3–8 min, monodispersed silver nanocubes with edge length in the range of 25–45 nm were synthesized by adjusting the reaction time.

Poly(methyl vinyl ether) (PVE) has also been employed as stabilizing and shape-modifying agent [113]. AgNO$_3$ (5 mM) solution was prepared in 12.5 wt% PVE and this solution was subsequently reduced using NaOH (25 mM). Homogeneous small-sized cubes with average length of 29 nm (55% yield) were produced, however, the yield was only 55%. Silver nanostructures, including nanocubes can be produced by reducing AgNO$_3$ by EG using PVP as an adsorption agent in the presence of varying concentration of NaCl [135]. While at low NaCl concentration, silver nanowires were obtained, at higher NaCl concentration, both silver nanocubes and bipyramids were obtained. Nanocube synthesis through the polyol process necessitates the use of organic solvent and relatively high temperature.

To overcome these issues, Zhou et al. [136] developed an aqueous method for nano-cube synthesis. In a typical experiment, aqueous solution of Cetyltrimethylammonium chloride (CTAC, 5 mL, 20 mM) and ascorbic acid (0.5 mL, 100 mM) were mixed and heated at 60°C for 10 min. Aqueous solutions of silver trifluoroacetate (CF$_3$COOAg) (50 µL, 10 mM) and iron(III) chloride (FeCl$_3$, 80 µL, 4.29 µM) were then injected simultaneously. The solution pH was maintained at 3.1. Silver nanocubes with average edge length 35–95 nm was synthesized by this aqueous method [136]. AgCl octahedra were produced within the first few minutes upon addition of CF$_3$COOAg into an aqueous CTAC solution. Small Ag$_n$ nuclei (e. g., Ag$_2$, Ag$_3^+$, and Ag$_4$) were generated both on the surface and in the interior of AgCl octahedra due to the reducing agent and light exposure. Once the Ag$_n$ nuclei on the surface started to grow, the underlying AgCl octahedron disappeared because of reduction and dissolution. Thus, AgCl served as a precursor to elemental Ag during the evolution of Ag$_n$ nuclei into single-crystal seeds and finally nanocubes were formed.

Table 1.6 summarizes methods for size-controlled synthesis of silver nanocubes using various synthesis protocols and reagents.

1.3.2.2.7 Synthesis of nanoflowers

Flower-like silver NPs can be synthesized using citric acid as a reducing as well as shape-directing agent in presence of CTAB [137]. Yang et al. [138] presented high yield and controlled synthesis of flowerlike silver nanostructures by reducing aqueous solution of AgNO$_3$ (10 mL, 0.1 M) with p-phenylenediamine (0.1 g, 1 mM) in the presence of ethyl alcohol solution of PVP (10 mL, 40g/L) at room temperature [138].

1.3.2.3 Photochemical method

Photochemical method has been widely used for size- and shape-controlled synthesis of AgNPs. The advantages of photochemical method are high spatial resolution, controlled *in situ* generation of reducing agents and versatility in fabricating NPs on media such as polymeric films, glass, cells and surfactant micelles [89, 139]. As

Table 1.6: Synthesis of nanocubes using chemical method.

Reducing agent	Stabilizing agent/Additives	AgNO₃	Reaction condition	Size (nm)	Reference
EG 5 mL160°C, 1 h	PVP 3 mL, 0.375 M; [PVP: AgNO₃] = 1.5:1	3 mL, 0.25 M	160°C, 45 min	175 ± 13	Sun and Xia [68]
EG 5 mL140°C, 1 h	PVP 3 mL, 147 mM HCl 1 mL, 3 mM	3 mL, 94 mM	140°C, 26 h	130	Im et al. [133]
EG 6 mL150- 155°C, 1 h	PVP 1.5 mL, 20 g/mL/Na₂S or NaHS 80 µL, 3 mM [PVP/ AgNO₃] = 1.9	0.5 mL, 48 mg/mL	3–8 min	25–45[a]	Siekkinen et al. [134]
NaOH:25 mM	PVE (12.5 wt%)	5 mM	RT, 30 min	29	Shervani et al. [112]
Ascorbic acid, 0.5 mL, 100 mM	CTAC 5 mL, 20 mM/FeCl₃ (80 µL, 4.29 µM)	CF₃COOAg 50 µL, 10 mM	60°C for 10 min, 6 h	35–95	Zhou et al. [136]

The size is specified in terms of edge length; [a]AgNPs were synthesized over this size range by varying the reaction conditions; CTAC: Cetyltrimethylammonium chloride; PVE: Poly(methyl vinyl ether).

discussed in Section 1.3.2.1.2, irradiation can induce the reduction of Ag^+ to Ag^0 in the presence of plasmonic seeds, either directly or in the presence of a reducing agent. However, harmful and strong reducing agents are typically not required and the reactions can often proceed at room temperature. Turning off irradiation can stop the reaction from proceeding further [140]. This process can be utilized for synthesizing AgNPs of varying shapes, such as, Ag HNPs, penta-twinned Ag nanorods and tetra-hedral, bipyramidal and decahedral AgNPs. Various protocols for size- and shape-controlled synthesis of AgNPs are discussed in the following sections.

1.3.2.3.1 Synthesis of silver nanospheres

Spherical AgNPs are usually synthesized via photoreduction under ultraviolet (UV) light exposure. Sometimes photosensitive dye or catalyst is also employed. AgNPs with average particle size in the range 15.2–22.4 nm were synthesized by photoreduction of 3 ml aqueous $AgNO_3$ (0.23–2.3 mM) in the presence of PVP (0.25–1 wt%, MW 40,000 Da) using 254 nm UV light [64]. PVP played an important role in the reduction of silver. It was demonstrated that the $>C = O$ functional group of PVP could absorb 254 nm light and the excited species generated ($>C = O^*$) could also reduce Ag^+. Carboxymethylated chitosan has also been employed as a redu-cing agent as well as a stabilizing agent for synthesis of AgNPs using UV light irradiation [141]. AgNPs with a narrow size distribution of 2–8 nm were obtained with alkaline $AgNO_3$/carboxymethylated chitosan aqueous solution at pH 12.4. AgNPs could be formed by exposing an aqueous solution containing the sodium salt of N-cholyl amino acid and $AgNO_3$ to sunlight [142]. In alkaline aqueous medium, N-cholyl amino acid worked as a stabilizing cum reducing agent.

Recently, Pu et al. [143] proposed a mechanism for photochemical synthesis of AgNPs in the presence of proteins, such as, bovine serum albumin, lysozyme, trypsin, protease K (Pro K), glucose oxidase and glutathione reductase. In this synthesis route, Pro K solution (or other protein solution, 0.3–17.5 µM) and AgNO$_3$ (0.9 mM) were well mixed. Subsequently, NaCl (2 mM) was added and the mixture was then exposed to UV light (365 nm) for 20 sec. It was found that proteins could act as template to regulate the shape, size and surface properties of the AgNPs synthesized.

In addition to direct synthesis, AgNP synthesis has also been performed using photosensitive dye or a photocatalyst. In a process proposed by Sudeep and Kamat [144], thionine dye (10 µM) and AgNO$_3$ (50 µM) in toluene/ethanol mixture (1:1) was exposed to visible light. The reduction of Ag$^+$ ions was induced due to excitation of the dye by visible light. After around 90 min of photolysis, AgNPs of 20 nm diameter was produced. Irradiation period was important for controlling the size. With the help of the photocatalyst Sn(IV) tetra(N-methyl-4-pyridyl) porphyrin tetratosylate chloride (SnP), Quaresma et al. demonstrated synthesis of spherical AgNPs in water at pH 7 [89]. This method exploited the advantages of photochemical reaction and catalytic reaction. Spherical AgNPs having 5–10 nm diameter were produced using an optimized molar ratio of [PVP]/ [AgNO$_3$] of 107. The concentration of other reagents was as follows: [AgNO$_3$] = 0.56 mM; triethanolamine [TEA] = 40 mM; [SnP] = 16 mM; and [PVP] = 60 mM. The mixture was irradiated with white light produced with a halogen bulb having 50W power.

1.3.2.3.2 Synthesis of silver nanoprisms

Seed-mediated photochemical approach was used for the synthesis of nanoprisms. Spherical AgNPs can be transformed into triangular prisms using photochemical method. Mirkin and co-workers [70] were the first to report such transformation. In a typical synthesis, spherical AgNPs were synthesized by injecting NaBH$_4$ (1 mL, 50 mM) into a solution containing AgNO$_3$ (100 mL, 0.1 mM,) and TSC (0.3 mM). Later BSPP (2 mL, 5 mM), the stabilizing agent, was added dropwise into the mixture. This solution was irradiated with a conventional 40 W fluorescent light. After 70 h mostly all of the nanospheres were converted to nanoprisms (edge length = 100 nm, standard deviation 15%). With 0.3:1 molar ratio of BSPP to citrate optimum transformation was obtained [70]. This process was modified by using dual-beam illumination of the NPs [93]. It was observed that, the prism edge length was controlled by the wavelength of incident light. The nanoprism size (edge length) was controlled by the primary beam whereas unimodal growth was achieved using the second beam. Nanoprisms of desired edge lengths were obtained using a primary light source with wavelength range 450–700 nm combined with a fixed secondary beam (340 nm). Average edge lengths of 38 ± 7 nm, 50 ± 7 nm, 62 ± 9 nm, 72 ± 8 nm, 95 ± 11 nm and 120 ± 14 nm were achieved using primary excitation wavelength 450 ± 20, 490 ± 20, 520 ± 20, 550 ± 20, 650 ± 20 and 750 ± 20 nm, respectively.

The same group later studied photo-mediated growth of triangular silver nanopr-isms. A seed solution was prepared by mixing nanopure water (95 mL), AgNO₃ (0.5 mL, 20 mM) and TSC (1 mL, 30 mM) in a 250 mL three-neck flask, cooled in an ice bath. Nitrogen gas was bubbled through it for 20 min under vigorous stirring and later freshly prepared aqueous NaBH₄ (1 mL, 50 mM) was added rapidly. Subsequently, 5 drops of NaBH₄ was added at regular intervals of 2 min over the next 15 min. This was followed by dropwise addition of BSPP (1 mL, 5 mM) and NaBH₄ (1 mL). The resulting solution stirred for 5 h in an ice bath and aged overnight at ~4°C in absence of light. This seed solution (20 mL) at pH 9.6 was irradiated with a halogen lamp (150 W) using an optical bandpass filter centered at 550 ± 20 nm for 5 h. Large prisms with edge length 130 nm were formed along with small ones of 65 nm having average thickness 9.1 ± 0.5 nm. This photochemical process involved reduction of Ag⁺ by citrate on the surface of the AgNPs and oxidative dissolution of small Ag seeds in presence of O₂. BSPP formed complexes with silver. The small silver seeds and BSPP were important for maintaining a constant concentration of Ag⁺ ions (20 μM). Under plasmon excitation, large silver particles functioned as photocatalyst and aided Ag⁺ reduction by citrate [92].

Subsequently, the same group synthesized silver right triangular bipyramids by irradiating a mixture of nanopure water (17.0 mL), AgNO₃ (0.60 mL, 10 mM), BSPP (10 mM), TSC (0.30 mL, 0.10 M) and 0.50 mL of Ag seeds (prepared as discussed above) with halogen lamp (150 W) using an optical bandpass filter centred at 550 ± 20 nm [145]. The pH of the solution was altered by adding either HNO₃ (0.5 M) or NaOH (0.1 M). This synthesis displayed three distinct stages as shown in Figure 1.31.

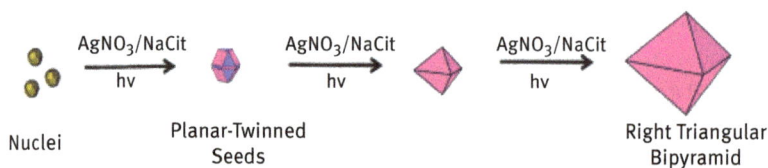

Figure 1.31: Schematics indicating growth of Ag right triangular bipyramids from AgNO₃ and spherical Ag nuclei. (Reprinted with permission from ref. [145]. Copyright 2010, American Chemical Society).

In the nucleation stage, spherical planar-twinned seeds were produced. In second stage, these seed were rapidly transformed into small bipyramids. In the last stage, bipyramids increased in size until all the available Ag⁺ ions were exhausted. As in the case of nanoprism synthesis discussed earlier, oxidation of citrate and reduction of Ag⁺ governed the process. The reaction was pH sensitive since oxidation of citrate evolved H⁺ and BSPP concentration controlled the reduction of Ag⁺. Thus, by varying the pH and the [BSPP]/[Ag⁺] molar ratio, both triangular bipyramids and nanoprisms could be synthesized. A fast reaction rate was achieved at high pH (10 or 11) and at a relatively low [BSPP]/[Ag⁺] ratio close to 1.0 which enabled the growth of (100)-faceted right triangular bipyramids by preferential Ag deposition on (111) facets compared to (100) facets. Hence, the planar-twinned seeds were converted into small right bipyramids bound with (100) facets which eventually formed right triangular bipyramids.

The reaction rate was reduced by increasing the concentration of [BSPP]/[Ag+] ratio or by decreasing the pH of the reaction. At slower reaction rate, Ag deposition on (100) facets of the planar twinned seeds was achieved, producing NPs with higher aspect ratios, such as truncated triangular bipyramids and prisms having a larger surface area defined by (111) facets (Figure 1.32). At pH 11, with rise in [BSPP]/[Ag+] ratio to 2.25, the two transverse vertices of the bipyramids became truncated. By increasing BSPP concentrations, NP morphologies changed from bipyramids to truncated bipyramids at pH 10. At pH 9, truncated bipyramids transformed to nanoprisms and as pH was lowered from 8 to 6, progressively larger nanoprisms were formed [145].

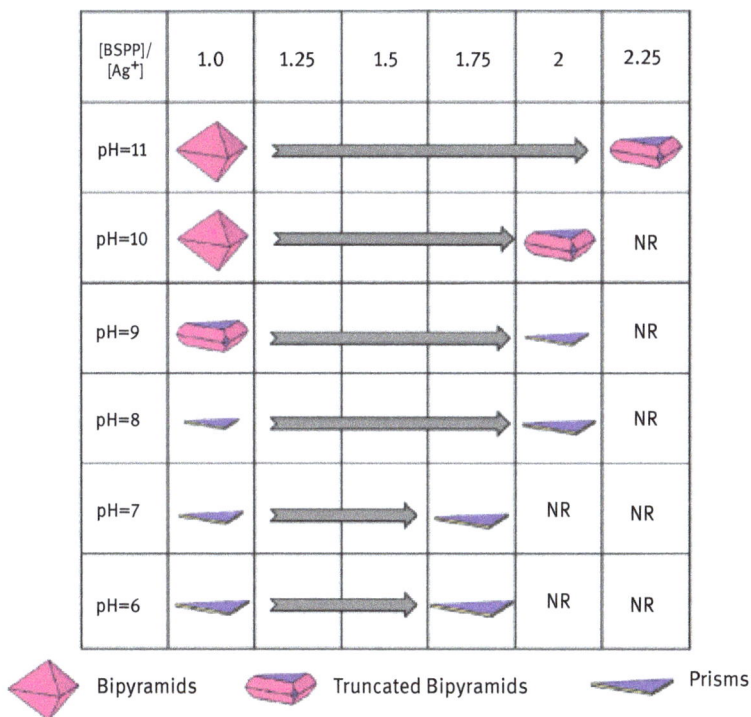

[BSPP]/[Ag+]	1.0	1.25	1.5	1.75	2	2.25
pH=11	Bipyramids	→				Truncated Bipyramids
pH=10	Bipyramids	→			Truncated Bipyramids	NR
pH=9	Truncated Bipyramids	→			Prisms	NR
pH=8	Prisms	→			Prisms	NR
pH=7	Prisms	→		Prisms	NR	NR
pH=6	Prisms	→		Prisms	NR	NR

Bipyramids   Truncated Bipyramids   Prisms

Figure 1.32: Nanoparticle morphologies obtained at various pH and [BSPP]/[Ag+] ratio during photochemical synthesis. (Reprinted with permission from ref. [145]. Copyright 2010, American Chemical Society).

Illumination with narrow band, low-cost light emission diodes (LED) was utilized for photochemical synthesis of Ag nanoprisms by Saade and Araújo [146]. Seed solution of silver nanospheres having diameters 3–5 nm was prepared by addition of 10 μL of

NaBH$_4$ (20 mM) at room temperature into an aqueous solution of AgNO$_3$ and TSC (30 mL, 0.25 and 1 mM, respectively). The solution was stirred vigorously before illumination was turned on. This seed solution was later exposed to LEDs having emission wavelength centred at 467, 475, 520 and 630 nm for ~8 h yielding nanoprisms with average size 32.4, 54.6, 98.6 and 151.6 nm with standard deviation 8.6, 16.3, 19.1 and 23.5, respectively.

By controlling the reaction temperature at 40°C, silver nanoprisms with various edge lengths were synthesized [147]. The seed solution was prepared by mixing ultrapure water (88 mL), AgNO$_3$ (1 mL, 10 mM) and TSC (10 mL, 10 mM) in a flask cooled in an ice bath. This solution was stirred for 0.5 h, subsequently freshly prepared ice-cold NaBH$_4$ (0.8 mL, 10mM) was injected dropwise over 4 min, and stirring was continued for 2 min. Later 0.1 mL NaOH solution (1 M) was introduced and the seed solution and exposed to 478, 503, 523 and 590 nm LED while maintaining the temperature at 40°C. Silver nanoprisms having average edge lengths of 38 ± 4, 46 ± 6, 56 ± 6, and 71 ± 7 nm, respectively, were obtained at wavelength of 478, 503, 523 and 590 nm.

Further, seed-mediated sunlight-driven formation of shape-controlled AgNPs was demonstrated by Tang et al. [148]. Silver seeds were made by dropwise addition of NaBH$_4$ (1 mL, 8.0 mM) into an aqueous solution of AgNO$_3$ (100 mL, 0.1 mM) in the presence of TSC under vigorous stirring. The seeds were exposed to simulated sunlight to obtain silver nanoprisms and nanodecahedrons by varying the TSC concentration. Silver nanoprisms were obtained with low TSC concentration (0.5 mM) and decahedral NPs were produced at high TSC concentration (5 mM) after 5 h.

In all preceding silver nanoprism synthesis methods, citrate stabilized spherical NPs were used as seeds. Tanimoto et al. [149] demonstrated synthesis of hexagonal silver nanoprisms by irradiating silver citrate using visible light irradiation. TSC (100 mL, 0.1 M) solution was added dropwise into AgNO$_3$ (100 mL, 0.3 M) solution under stirring at the rate of about two or three drops per second. The precipitate of silver citrate formed by stirring was separated and the steps were repeated multiple times after resuspension in ultrapure water. Finally, the precipitate was resuspended in ethanol instead of ultrapure water and recovered precipitate was air dried. Silver citrate solution (6.6 mM) was prepared by dissolving the precipitate (169 mg) in ultrapure water (50 mL) containing ammonia solution (0.5 mL). Later this solution was irradiated with visible light for production of hexagonal silver nanoprisms (distance between two parallel sides was ~40–120 nm).

1.3.2.3.3 Synthesis of silver nanodiscs

Silver nanodiscs could also be synthesized by seed-mediated photochemical transformation as discussed for silver nanoprisms in the preceding subsection. The only difference was that, nanodisc synthesis needed preparation of growth solution. In a typical synthesis [91], an aqueous colloidal suspension of AgNP seed was prepared by adding ice-cold NaBH$_4$ (0.5 mL of 25mM), TSC (20 mL, 10 mM) and AgNO$_3$ (0.5 mL,

5 mM) under constant stirring followed by aging over 24 h. The growth solution was preapred using the seed suspension (100 µL), TSC (50 µL, 10 mM), AgNO₃ (50 µL of 5 mM) and ultrapure water (800 µL). Irradiating the growth solution at 457 nm with a linearly polarized argon laser beam for 90 min produced silver nanodisc having average diameter 38 nm and height 10.7 nm. The final size of the particles was determined by the concentration of silver ions in solution. The aspect ratio was dependent on the wavelength, such that longer wavelength resulted in larger aspect ratio.

The size and shape of AgNPs could be controlled by employing monochromatic light of specific laser wavelengths to irradiate the seed crystals. High-power monochromatic laser lines were employed to produce monodispersed AgNPs. When solution containing AgNO₃ (100 mL, 0.1 mM,), TSC (0.5 mM) and NaBH₄ (0.5 mL, 10 mM) was exposed to argon ion laser lines 514.5 nm (14.4 W/cm²), 501 nm (5.6 W/cm²) and 488 nm (6.4 W/cm²), Ag nanodiscs and Ag nanotriangles were produced whereas exposure of 476.5 nm (3.2 W/cm²) and 457.9 nm (1.3 W/cm²) lines produced pyramidal and pentagonal NPs. In a subsequent study, monodispersed silver triangular nanoplates were synthesized by mixed irradiation with 514.5, 496.5 and 488 nm laser lines whereas decahedral nanoplates were observed using 476.5 and 457.9 nm laser lines [150, 165].

Sun and Xia [151] introduced PVP (0.5 mL, 5 mg/mL, MW 55,000) in a mixture of TSC (0.5 mL, 30 mM), AgNO₃ (1 mL, 5 m M), NaBH₄ (0.5 mL, 50 mM) and 47.5 mL of deionized water. The mixture was irradiated for 40 h with halogen lamp to obtain triangular silver nanoplates having edge length ~30 to ~90 nm [151]. Recently, photochemical synthesis of silver nanodiscs using a conventional metal-halide lamp, similar to that used for nanoprism synthesis, was developed [152]. Highly monodispersed silver nanodiscs were synthesized by exposing a solution comprised of PVP (0.015 mL, 0.05 M, MW 40,000), AgNO₃ (0.2 mL, 0.005 M), TSC (0.5 mL, 0.05 M), NaBH₄ (0.08 mL, 0.1 M) and pure water (7 mL) to 400 W metal-halide lamp. After 5 h exposure, nanodiscs having average diameter 30.7 nm were formed. Silver nanodiscs size ranged from 15 nm to over 60 nm and the size was controlled by dilution. The final reaction mixture was diluted with the growth solution made of AgNO₃ (0.128 mM), TSC (3.207 mM) and PVP (10.757 mg/L) at an early stage of the synthesis (i. e., 30 min after the injection of NaBH₄) and subsequently this solution was illuminated for 4.5 h. In this synthesis process, when AgNO₃ was replaced by silver salt of trifluoroacetate larger nanodiscs were produced. The dissociation of Ag⁺ from CF₃COO⁻ was much slower than that from NO₃⁻, causing lower initial Ag⁺ concentration compared to the formulation containing AgNO₃. This resulted in production of fewer seeds and larger nanodiscs.

Nicotinamide adenine dinucleotide model compound, N-benzyl-1,4-dihydronicotinamide (BNAH) was employed for synthesis of triangular Ag nanoplates [153]. In a typical synthesis, AgNO₃ (90 µL, 20 mM), BNAH (2.86 mL, 0.32 mM) and TSC (50 µL, 40 mM) aqueous solutions were mixed and exposed to 125 W sunlamp (λ > 350 nm) for 30 min. Upon irradiation, the spherical AgNPs first formed were converted to hexagonal nanoplates (~50 nm) and finally to triangular Ag nanoplates (~40 nm) over 30 min.

1.3.2.3.4 Synthesis of silver nanorods

Silver nanorods have also been synthesized using plasmon excitation of Ag seeds in an appropriate growth solution. Ag seeds with planar twin defects were produced earlier for nanoprism synthesis; however, spherical seed particles with pentagonal twin structures were synthesized by Mirkin and co-workers [154] for nanorod synthesis. For seed preparation, nanopure water (19 mL), $AgNO_3$ (0.4 mL, 10 mM), BSPP (0.4 mL, 10 mM), TSC (0.2 mL, 0.1 M) and NaOH (1 mL, 0.1 M) were mixed. This mixture was then irradiated with a UV lamp (254 nm) for 30 min. The seed solution was purified by centrifuging at 15,000 rpm for 60 min followed by dispersal of the Ag seeds in 20 mL water. To synthesize Ag nanorods, the seeds (0.5 mL) were added to an aqueous solution containing $AgNO_3$ (0.2 mM) and TSC (1.0 mM). This solution was irradiated for 24 h using a 150 W halogen lamp with a bandpass filter selecting excitation wavelength (λ_{ex}) from 600 to 750 nm. Nanorods with uniform diameter and length of 67 ± 5 nm and 330 ± 85 nm, respectively, were produced for λ_{ex} centred at 600 ± 20 nm. The aspect ratio of the silver nanorods increased as λ_{ex} increased from 600 to 750 nm.

1.3.2.3.5 Synthesis of decahedral silver NPs

Decahedral silver NPs can be synthesized by illuminating precursor/seed solution using light of appropriate wavelength. To produce size-controlled nanodecahedra, multiple growth steps/regrowth steps were carried out. PVP was used as the stabilizer and L-arginine was used as the photochemical promoter [94]. In a typical synthesis process, the precursor solution was prepared by adding a mixture of TSC (0.5 mL, 0.05 M), PVP (0.015 mL, 0.05 M, MW 40,000), L-arginine (0.05 mL, 0.005 M) and $AgNO_3$ (0.2 mL, 0.005 M) to DI water (7 mL). Later, $NaBH_4$ (0.08 mL, 0.1 M) was injected into this mixture. This pale yellow solution was magnetically stirred for several minutes to get a bright yellow solution. To obtain smaller decahedra with diameter of 35–45 nm, the above solution was exposed to the blue light for 2–15 h. This decahedral AgNP seed solution (4 mL) was mixed with the precursor solution (8 mL) and exposed to white light of a metal halide lamp to get decahedra from 40 nm up to 120 nm. To achieve complete conversion of the AgNPs decahedral NPs, oxidative etching was carried out using hydrogen peroxide [155]. In one study, these decahedral AgNPs were used for production of nanorods using citrate-based reduction method [156]. Instead of using metal halide lamp, Jakab et al. [157] synthesized nanodecahedra by illuminating the precursor solution with blue diode array lamp with λ_{max} of ~460 nm for 10 h. Lu et al. [158] adjusted the ratio of seed and precursor in the regrowth process for obtaining highly mono-dispersed decahedral AgNPs with LSPR in the range of 499–590 nm. In a typical regrowth experiment, decahedral AgNP seed solution (2 mL) was mixed with precursor solution (2–4 mL) as discussed earlier and this was followed by LED exposure for 6–9 h. Plasmon-driven seed-mediated regrowth occurred as λ_{max} of the LED was matched with the LSPR wavelength of the decahedral AgNP seeds. This phenomenon reduced the formation of silver nanoprisms and nanoplates in the solution and resulted in production of highly monodispersed decahedral AgNPs. Decahedral AgNPs with edge length (mean ± SD) of 28.3 ± 3.6, 38.9 ± 4.0, 41.7 ± 4.5, 53.5 ± 5.1, 61.4 ± 7.6, 67.4 ± 4.7 nm were reported by varying the volumetric ratio of seed to precursor solution and illumination wavelength [158].

L-arginine can act as photochemical promoter and is reported to control the size and morphology of decahedral silver NPs. Cardoso-Ávila et al. [159] studied the role of other aminoacids such as L-lysine and L-histidine in photochemical synthesis of decahedral silver NPs. Precursor solution was prepared by combining DI water (4 mL), TSC (0.285 mL), PVP (0.009 mL, MW 10,000 Da), $AgNO_3$ (0.114 mL) and $NaBH_4$ (0.045 mL) under continuous stirring. Subsequently, amino acid (0.01, 0.03, 0.05 and 0.070 mL) was added and the solution was kept at rest for 30 min before illumination using blue LED (80, 50 and 15 mW/cm^2) for 135 nm [159]. The yield of the decahedral NPs was higher for L-arginine and L-lysine. Also, for the highest irradiating power (80 mW/cm), better yield of decahedral NPs was achieved. Comparing Triton X-100 and PVP as surfactants for decahedral NP production showed that yield was 75% (for mean size 55 nm) with PVP and L-lysine (0.03 mL) at 46°C in contrast to yield of 56% (for mean size of 54 nm) with Triton X-100 and L-lysine (0.03 mL) at 46°C [159]. Later other researchers studied the effect of green and blue irradiation on AgNP synthesis using protoporphyrin IX as a photopromoter. Spherical AgNPs (diameter 3 nm) were transformed into flat rounded NPs (21 nm) and decahedral NPs (78 nm) via green and blue irradiation, respectively [160]. AgNPs with varying morphology and size were produced by changing the irradiation wavelength and exposure time. Successive blue and green irradiation for exposure duration less than 30 min, generated morphologies, such as, twine rounded (42 nm), flat elongated AgNPs (peanuts, 17 nm) and flat rounded AgNPs (11 and 24 nm). Decahedral AgNPs with rounded/sharp corners (size 71–78 nm) were produced with longer exposure time (>45 min).

Synthesis of monodispersed decahedral silver NPs without PVP and L-arginine was demonstrated by Huang and co-workers [161]. To produce decahedral silver NPs, TSC (1 mL, 3.0×10^{-1} M) and $AgNO_3$ (1 mL, 10 mM) were mixed with DI water (97.8 mL) with rapid stirring. Subsequently, $NaBH_4$ (0.2 mL, 10 mM) was added dropwise under vigorous stirring which was continued for 0.5 h. The spherical Ag seed formed was irradiated with the green LEDs ($\lambda_{max} = 520 \pm 18$ nm) with average power of ~25 mW/cm^2 at 0°C for 100 h. Although decahedral AgNPs (edge length ~65 nm) were formed, they were not stable under ambient conditions [161]. Later, decahedral AgNPs were synthesized by controlling the temperature of the seed solution at 20°C during illumination with 478 nm LEDs [147]. Seed solution was prepared in ultrapure water (88 mL) by mixing $AgNO_3$ (1 mL, 10 mM) and TSC (10 mL, 10 mM). The mixture was stirred for 30 min. Subsequently, aqueous ice cold $NaBH_4$ (0.8 mL, 10 mM) was injected dropwise into the mixture over the next 4 min. This Ag colloid was rapidly stirred for 2 min in an ice bath. Decahedral AgNPs with various edge lengths were obtained by repeating the regrowth steps. In the first regrowth step, 5 mL of the bright-yellow silver seed solution was added to 5 mL of the decahedral silver NPs solution and exposed to 478 nm LEDs at 20°C. These regrowth steps were repeated by using the decahedral AgNPs obtained in the previous step as seed. The sizes of the decahedra thus produced were 37 ± 5, 46 ± 5, 54 ± 6 and 63 ± 7 nm, in successive regrowth steps.

A one pot decahedral AgNP synthesis protocol was developed by the Huang and co-workers [32] based on photo-assisted TSC reduction. During synthesis, TSC (1 mL, 4.5×10^{-1} M) and AgNO$_3$ (1 mL, 10 mM) was mixed with pure water (98.0 mL) and irradiation of the solution was irradiated was performed with 24 blue LEDs (λ_{max} = 460 ± 12 nm, average power ~100 mW/cm^2). Colloidal AgNPs with more than 85% nanodecahedra (43.5 ± 5.2 nm) were obtained after 90 min of irradiation. The XRD spectra revealed that the grain size of the Ag nanodecahedra synthesized were larger than those made by the plasmon-mediated photochemical processes and the Ag nanodecahedra formed were highly stability.

A new photochemical synthesis method using Irgacure 2959 (I-2959) in absence of PVP, L-arginine and NaBH$_4$ was developed. On excitation, I-2959 produced ketyl radicals which could reduce Ag$^+$ to Ag^0 [140]. AgNP seed (3 nm) solution was prepared by photolyzing an aqueous solution containing 0.2 mM AgNO$_3$, 0.2 mM I-2959 and 1 mM TSC using UVA (355 nm). This seeds solution was irradiated at various wavelength, 405, 455, 590/627 and 720 nm to obtain larger spherical NPs, dodecahedra, nanoplates and nanorods, respectively, as shown in Figure 1.33 [140]. In another study [162], an aqueous solution containing 0.2 mM AgNO$_3$, 0.2 mM I-2959 and 1 mM TSC was purged with N$_2$ for 30 min. Subsequently, the mixture was irradiated using UV LED (355 nm) for 50 min for generating seeds. This seed solution was later irradiated using blue LED (wavelength 455–465 nm) for 30 min to obtain Ag nanodecahedra with edge length ~35 nm.

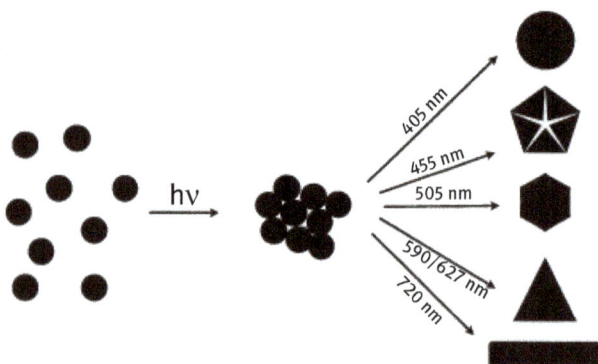

Figure 1.33: Diagram illustrating plasmon-mediated transformation of AgNP seeds into various shapes in presence of Irgacure 2959 (Reprinted with permission from ref. [140]. Copyright 2010, American Chemical Society).

1.3.2.3.6 Synthesis of icosahedral silver NPs

Recently, Keunen et al. [163] synthesized icosahedral silver NPs photochemically. For synthesis of precursor solution, PSSS was used instead of PVP. $CuSO_4$ was added as Cu^{2+} promotes formation of icosahedral silver NPs by producing icosahedral nuclei from the precursor. H_2O_2 was added as an oxidizing agent. The precursor solution was exposed to violet LED (395–420 nm). Later, photo-assisted tartrate reduction method was developed to obtain Ag icosahedra with diameter in range of 80–150 nm [33]. The mixture of sodium tartrate (1 mL, 0.45 M) and $AgNO_3$ (1 mL, 10 mM) were irradiated for 48 h using 24 UV lamps (λ_{max} 310 ± 12 nm, average power 25 mWcm^{-2}).

Protocols for size- and shape-controlled synthesis of AgNPs through photochemical route are summarized in Table 1.7 and Table 1.8.

This This is moved to table footnote.

1.3.2.4 γ-ray irradiation method

Various reducing agents are commonly employed for AgNP synthesis via chemical and photochemical methods. Although these methods are effective, several of the chemicals used in the formulations are toxic and hazardous and may have adverse effect on various organisms in the ecosystem. Traces of these toxic and hazardous chemicals may remain associated with AgNPs synthesized using chemical and photochemical methods. The γ-ray irradiation method, relying on hydrated electrons produced during irradiation, can reduce Ag^+ ions to neutral Ag^0 atom while avoiding the use of toxic reducing agent eqs. (1.8)–(1.9). The neutral Ag^0 atoms can react with Ag^+ ions to form Ag clusters which aggregate to form AgNPs [164].

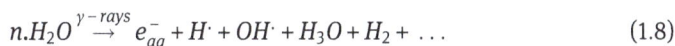

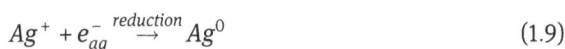

$$n.H_2O \overset{\gamma-rays}{\rightarrow} e^-_{aq} + H^\cdot + OH^\cdot + H_3O + H_2 + \ldots \tag{1.8}$$

$$Ag^+ + e^-_{aq} \overset{reduction}{\rightarrow} Ag^0 \tag{1.9}$$

Chitosan, an amino polysaccharide derived from chitin, can be used for forming size-controlled AgNPs through γ-ray irradiation. When irradiated with γ-rays, chitosan degrades into fragments which have either amine or carboxylic groups associated with them. The chitosan fragments help in binding the silver clusters through ionic interaction between Ag^+ and carboxyl groups, through Ag–O bonds or through a combination of both these interactions.

Protonated amino groups ($-NH_3^+$) derived from chitosan adsorb on the surface of the AgNPs and provide stability via electrostatic repulsion [65, 166]. In typical synthesis by Chen et al. [164], dry chitosan (3 g) was dispersed a solvent comprised of acetic acid and distilled water (1:100; 150 mL) under stirring to obtain a

Table 1.7: Synthesis of Ag nanospheres and nanoprisms using photochemical route.

Reducing agent	Stabilizing agent/Additives	AgNO₃	Reaction conditions	Size (nm)	Reference
Nanospheres					
PVP 0.25–1 wt%	—	3 mL, 0.23–2.3 mM	254 nm UV light, 120 min	15.2–22.4[a]	Huang et al. [64]
Carboxy-methylated chitosan 0.5%	—	0.5 mM	UV	2–8	Huang et al. [141]
Nanoprisms					
NaBH₄ 1 mL, 50 mM	TSC 0.3 mM BSPP 2 ml, 5 mM [BSPP:citrate] 0.3:1	100 mL, 0.1mM	Fluorescent light 40 W, 70 h	L100 ± 15%	Jin et al. [70]
NaBH₄ 2 mL, 50 mM	TSC:1 mL, 30 mM BSPP:1 mL, 5 mM	0.5 mL, 20 mM	Halogen lamp 550 ± 20 nm, 5 h	L 65–130 T 9.1 ± 0.5	Xue et al. [92]
Seed: as above	Growth: BSPP 10 mM, TSC 0.3 mL,0.1M	0.60 mL, 10 mM	Halogen lamp 550 ± 20 nm	Right triangular bipyramids	Zhang et al. [145]
NaBH₄ 10 μL, 20 mM	TSC 30 mL, 1 mM	0.25 mM	LED 467, 475, 520 and 630 nm, 8 h	32.4 ± 8.6 54.6 ± 16.3 98.6 ± 19.1 151.6 ± 23.5	Saade and Araújo [146]
NaBH₄ 0.8 mL, 10 mM	TSC: 10 mL, 10 mM); NaOH 0.1mL, 1M	1 mL, 10 mM	LED: 478, 503, 523 and 590 nm, 40°C	38 ± 4, 46 ± 6 56 ± 6 71 ± 7	Wang et al. [147]

1 Reaction conditions listed include type of source and power, irradiation wavelength, irradiation time and temperature (only when conducted at controlled temperature); size is specified as diameter for nanospheres and edge length (L) and thickness (T, listed only when provided) for nanoprisms; [a]AgNPs were synthesized over this size range by varying the reaction conditions.

Table 1.8: Synthesis of Ag nanodisc and decahedral silver nanoparticles using photochemical route.

Reducing agent	Stabilizing agent/ Additives	AgNO₃	Reaction conditions	Size (nm)	Reference
Nanodiscs					
Seed: NaBH₄ 0.5 mL, 25 mM	Seed: TSC: 20 mL,10 mM Growth: TSC: 50 µL, 10 mM	Seed: 0.5 mL, 5 mM Growth: 50 µL, 5 mM	457 nm Argon laser, 90 min	D 38 H 10.7	Maillard et al. [91]
NaBH₄: 0.5 mL, 10 mM	TSC:100 mL, 0.5 mM	0.1 mM	Laser 514.5 nm (14.4 W/cm²)	36 ± 4	Zheng et al. [150, 165]
NaBH₄ 0.5 mL, 50 mM	TSC:0.5 mL, 30 mM; PVP:0.5 mL, 5 mg/mL	1 mL, 5 mM	Halogen lamp, 40 h	~30– 90	Sun and Xia [151]
NaBH₄ 0.08 mL, 0.1 M	PVP 0.015 mL, 0.05 M TSC 0.5 mL, 0.05 M	0.2 mL, 0.005 M	400 W Metal- halide lamp, 5 h	30.7	Kim and Lee [152]
Nanodecahedral AgNPs					
NaBH₄ 0.08 mL, 0.1 M	TSC:0.5 mL, 0.05 M; PVP: 0.015, 0.05 M/L-arginine 0.05 mL, 0.005 M	0.2 mL, 0.005 M	Blue light, 2–15 h	D 35– 45	Pietrobon and Kitaev [94]
NaBH₄ 0.2 mL, 10 mM	TSC: 1 mL, 0.3 M	1 mL, 10 mM	Green LEDs λmax = 520 ± 18 nm at 0°C, 100 h	L 65	Lee et al. [161]
TSC 1 mL, 0.45 M		1 mL, 10 mM	Blue LEDs λmax = 460 ± 12 nm, 90 min	L 43.5 ± 5.2	Yang et al. [32]

1 Reaction conditions listed include type of source and power, irradiation wavelength, irradiation time and temperature (only when conducted at controlled temperature); size is specified as diameter (D) and height (H, listed only where available) for nanodiscs and either as diameter (D) or as edge length (L) for nanodecahedral AgNPs.

transparent solution. An aqueous AgNO₃ solution (20 mL, 30mg/mL) was added to the chitosan solution dropwise with continuous stirring and subsequently isopropanol (23 g) was added as free radical scavenger. The mixture was stirred in N₂ atmosphere for more than 2 h to ensure complete mixing of the viscous solution. The solution was irradiated under ⁶⁰Co γ ray source. The rate of irradiation was fixed at 40.9 Gy/min. AgNPs with average diameter of 4–5 nm were obtained through this route. In a similar experiment carried out by Long et al. [65] using oligochitosan, AgNPs having size in the range 5–15 nm were produced.

Later, Yoksan and Chirachanchai [166] synthesized size-controlled AgNPs in the size range 7–30 nm in chitosan–aqueous acetic acid solution exposed to atmospheric condition. The diameter of the particles increased with increase in γ-ray dose and with initial AgNO₃ content. For 0.1 mmole AgNO₃ and 0.1% (w/v) chitosan, the size of AgNPs synthesized were 20, 20, 25 and 30 nm for γ -ray dose of 10, 15, 20 and 25 kGy, respectively, while for 0.5% (w/v) chitosan, the size of AgNPs produced decreased to 14, 14, 17 and 17 nm for γ-ray dose was 10, 15, 20 and 25 kGy, respectively.

When 0.1% (w/v) chitosan solution with varying $AgNO_3$ content was subjected to 25 kGy γ-ray irradiation, the size of AgNPs produced were 9, 12, 17, 19 and 23 nm for $AgNO_3$ content 0.02, 0.04, 0.06, 0.08 and 0.10 mmole, respectively. At higher chitosan concentration, i. e., 0.5% (w/v) and $AgNO_3$ contents 0.02, 0.04, 0.06, 0.08 and 0.10 mmole, the average diameter of silver NPs formed were 7, 9, 11, 12 and 14 nm, respectively. Thus, at a higher concentration of chitosan, smaller diameter AgNPs with greater monodispersity was obtained. Attempts have also been made to synthesize AgNPs by γ-ray irradiation in presence of sodium alginate [167]. Alginate could bind to AgNPs through Ag–O bonding and AgNPs in the size range 6–30 nm were produced.

1.3.2.5 Sonochemical method

Sonochemical reduction is another prospective method for fabrication of AgNPs of small size and large surface area. Ultrasound irradiation produces acoustic cavitation in a liquid involving the formation of bubbles, followed by their growth and sudden collapse. The implosive collapse of bubbles creates localized hotspots through adiabatic compression of the gas phase within the collapsing bubbles. This generates very high temperature (5000 K), pressure (1800 atm), microjets with high speed (e. g. 400 km/h) and high cooling rates exceeding 10^{11} K/s. AgNPs can be synthesized under these extreme conditions produced by acoustic cavitation [168, 169].

Amorphous AgNPs ~20 nm size were synthesized by sonochemical reduction of $AgNO_3$ (1 g/100 ml of DI water) under argon–hydrogen (95:5) atmosphere at 10°C [169]. Silver nanorods (100 nm diameter, 4–7 μm long) were synthesized in an aqueous formulation containing silver nitrate (0.085 g), methenamine (HMTA) (0.04 g) and PVP (0.02 g) in 50 mL distilled water using a probe sonicater over 60 min.[175]. HMTA and PVP served as reducing agent and as a surfactant/soft template, respectively. The morphology of silver nanorods was determined by the ultrasonication time, the concentration of PVP and by the reaction temperature. After 15 min of irradiation, numerous small spherical AgNPs were seen. Continued ultrasonication, resulted in aggregated AgNPs having a linear arrangement, however, isolated AgNPs were also observed. After 60 min irradiation, dense, compact and regular-shaped Ag nanorods with 4–7 μm length were formed with mean diameter of ~100 nm. At optimum PVP concentration (0.40 g/L), the compact and regular-shaped silver nanorods with length 4–7 μm were formed. At lower PVP concentration (0.20 g/L, with 0.01 mol/L $AgNO_3$ and 0.80 g/L HMTA) needle-shaped structures with length more than 10 μm were produced whereas at a higher PVP concentration (0.60 g/L), the length of the silver nanorods formed was only ~2 μm. With further increase in PVP concentration (0.80 g/L), only silver nanospheres were produced. As the temperature was increased from 10°C to 20°C silver nanospheres gave way to short nanorods

(~400 nm), and at 30°C length of the nanorods were increased to 4–7 μm. Further rise in temperature to 50°C produced a hexagonal morphology due to over-aggregation.

Recently, Kumar et al. demonstrated AgNP synthesis using starch through soni-cation [170]. Spherical, polydispersed, amorphous AgNPs with size ranging from 23 to 97 nm (mean size ~45.6 nm) were prepared using this method. Pulsed sonoelectro-chemical technique was adopted to synthesize AgNPs of diverse shapes from an aqueous solution containing $AgNO_3$ and nitrilotriacetate (NTA, $N(CH_2 COOH)_3$) [171]. The concentration of $AgNO_3$ (0.6–20 g/L) and NTA tailored the size and shape of the NPs, such that rods of varying size and dendrites could also be achieved in addition to spherical AgNPs.

Silver nanostructures, i. e., dendritic rod, dendritic sheet and flower-like den-drites and spherical and oval AgNPs were prepared through ultrasonication of a solution containing silver perchlorate ($AgClO_4$) and PVP [172]. Spherical AgNPs (of size range 7–10 nm) were formed at molar ratio of PVP monomer to Ag^+ (R) of 100, while 15 nm NPs were prepared at R of 50 when a low current density (0.5 mA cm^{-2}) and $AgClO_4$ concentration 3 mM was used. For R of 5, oval-shaped NPs were obtained. With 3 mM $AgClO_4$, current density 1.25 mA cm^{-2} and R of 50, dendritic nanostruc-tures were obtained. Upon increasing the current density (2 mA cm^{-2}), flower-like dendrites consisting of staggered nanosheets were observed.

1.3.2.6 Electrochemical method

Electrochemical reduction of metal ions in solution can also be used for synthesis of AgNPs through appropriate choice of electrode-type, current density, supporting electrolyte and solvent [173]. NP size can be controlled by adjusting the current density. AgNPs of size range 2–7 nm could be synthesized electrochemically in acetonitrile containing tetrabutylammonium salts [173]. The anode and cathode consisted of a silver sheet and platinum sheet, respectively. A typical synthesis was carried out in acetonitrile using 0.1 M tetrabutylammonium acetate at 25 ± 0.1°C. The size of NPs formed decreased (6 ± 0.7–1.7 ± 0.4) with increase in current density. At high current densities (7 mA cm^{-2}), irregular aggregates were observed. No AgNPs were formed when the platinum cathode was replaced with aluminium.

Using electrochemical mechanism, polyphenylpyrrole-coated silver NPs have been synthesized at the liquid/liquid interface [174]. The experiment employed aqu-eous and non-aqueous phase solutions. The aqueous phase contained lithium sulfate and Ag_2SO_4 while the organic phase contained N-phenylpyrrole and tetraphenylar-sonium tetrakis (pentafluorophenyl) borate in 1,2-dichloroethane. N-phenylpyrrole served as reducing agent and provided electrons for reducing the Ag^+ ions and thus facilitated their transfer from the aqueous phase to the organic phase. AgNPs were subsequently formed through polymerization and growth of the metal cluster [174].

Spherical chitosan-coated AgNPs (size range 2–16 nm) were reported by Reicha et al. [175] using electrochemical process followed by UV irradiation. The electrolyte solution was prepared by dissolving chitosan (1 wt%) into acetic acid (1% v/v) solution. The electrochemical cell was comprised of silver (anode) and platinum (cathode) plates placed 5 cm apart. The electrochemical process was carried out at constant potential of 2 V and at 25°C, using time period of 6, 12, 18 and 24 h. The solution obtained was exposed to UV radiation (254 nm) at 25°C for 1, 3 and 6 h in order to increase NPs concentration [175]. Chitosan-coated AgNPs having average diameter 10–16 nm were obtained after 12 h followed by 1 h UV irradiation. The size of AgNPs was reduced to 2–8 nm when electrochemical process was carried out for 18 h followed by 1 h UV irradiation.

A two electrode set-up, with Ag sheets as both anode and cathode was used by Khaydarov et al. [63]. The electrodes were immersed in an electrochemical cell filled with 500 mL of distilled water. AgNPs were synthesized electrochemically over the temperature range 20–95°C at a constant voltage of 20 V. For reaction time 50–70 min in the temperature range 50–80°C, AgNP concentration in the range of 20–40 mg/L was obtained. The AgNPs synthesized were nearly spherical with average size 7 nm and size range 2–20 nm after filtration and H_2O_2 (up to 0.005%) treatment [63]. Using a similar set-up, DMF and thioglycolic acid (TGA)-capped AgNPs were prepared by Rabinal et al. [176]. The organic molecules DMF and TGA were used as capping agents. The TGA-capped AgNPs did not show any LSPR peak, however, they demonstrated a sharp absorption curve similar to a direct band gap semiconductor.

Well-dispersed spherical AgNPs were synthesized by Yin et al. [177] through electrochemical method, without the use of sacrificial anode. The synthesis was carried out in an aqueous electrolyte solution containing KNO_3 (0.1 M), $AgNO_3$ (5.0 mM) and PVP (MW 40,000). Platinum sheet and a coiled platinum wire were used as cathode and anode, respectively. The experiment was conducted in a potentiostat at room temperature under mechanical stirring/ultrasonication. PVP not only stabilized the silver clusters but also improved the formation rate of AgNPs and reduced silver deposition rate on the cathode. Size of AgNPs formed could be controlled by adjusting the electrolysis parameters. Silver clusters produced in the vicinity of the cathode was distributed throughout the solution through mechanical stirring. AgNPs having mean diameter 16.6 ± 6.22 nm were obtained. Ma et al. [178] further tested the effect of addition of two surfactants, i. e. the anionic surfactant SDBS and non-ionic surfactant PEG on AgNP synthesis in presence of PVP in the electrolyte. Addition of SDBS was favourable for the synthesis of AgNPs with smaller diameter and narrower particle size distribution (~10.1 ± 1.89 nm). In presence of PEG (MW 6000), the AgNPs formed were not well dispersed. During synthesis, the platinum cathode was rotated to accelerate the transfer of AgNPs away from the cathode to the bulk solution.

Electrochemical-assisted polyol process in presence of PVP and KNO_3 was demonstrated by Lim et al. [179]. Silver nanospheres with average size ~11 nm were obtained. The amount of Ag nanospheres produced was increased by increasing the $AgNO_3$ concentration (1–20 mM) [179].

1.3.2.7 Microwave-based synthesis

Over the past few years, there has been increasing interest in exploring microwave
-based synthesis of AgNPs. Although most studies are currently limited to lab scale
synthesis, this method may offer good possibilities for scale up. Heat-assisted reduction of silver ions and complexes through chemical method is one of the preferred
methods for AgNP synthesis as discussed in preceding sections. However, since
conventional heating requires a longer induction period [98], rapid heating through
microwave irradiation may be utilized for synthesis of AgNPs. Apart from dielectric
constant of the medium, factors affecting the yield and characteristics of AgNPs
synthesized using microwave irradiation include: power input and time of irradiation; dielectric constant and refractive index of the medium; and type and concentration of precursor salt, capping agents and other additives; and reaction conditions.

As discussed in preceding sections, AgNP synthesis through the polyol method,
based on EG and/or PEG requires heating up to high temperatures in an oil bath.
Since these polyol solvents have relatively high dipole moment, AgNP synthesis
through polyol process may also be performed using microwave irradiation instead
of conventional heating [25, 95]. The high dielectric constant for EG (41.4 at 25 °C) and
PEG causes high dielectric loss, such that rapid heating occurs under microwave
irradiation. Moreover, other ions present in solution may also cause rapid heating.
This process works in two stages. In the first stage, EG/PEG serves as a reducing agent
and metal hydroxide, i. e., $M(OH)_2$ crystallizes. Subsequently, an intermediate phase
precipitates as the starting hydroxide dissolves and water molecules are generated.
Dissolution of the intermediate phase leads to reduction of metal ions in solution
followed by nucleation and growth of AgNPs [95]. The general mechanism of metal
reduction can be represented by the following reactions eq. (1.10)–(1.11):

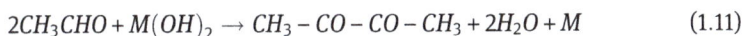

$$2CH_2OH - CH_2OH \rightarrow 2CH_3CHO + 2H_2 \qquad (1.10)$$

$$2CH_3CHO + M(OH)_2 \rightarrow CH_3 - CO - CO - CH_3 + 2H_2O + M \qquad (1.11)$$

Komarneni et al. [95] synthesized AgNPs using the microwave -based polyol
approach. For AgNP synthesis, AgNO$_3$ (15 mg) was mixed with EG (100 ml), NaOH
and PVP (257 mg). PVP of varying MW was used in combination with NaOH, i. e.,
8000, 10,000 and 40,000 and in one of the formulations PVP with MW 40,000 was
used in absence of NaOH. These formulations were treated at 150 °C for 15 min using a
microwave digestion system employing a frequency of 2.45 GHz. The double-walled
vessel used had inner liner and cover made of Teflon PFA and the outer shell was
made of Ultem polyetherimide. Strings of AgNPs were obtained with PVP of MW
8000, 10,000 and 40,000. The formation of strings reduced as the MW of PVP
increased and at the highest MW discrete AgNPs of size ~100 nm were obtained.
For the formulation with no NaOH and PVP of MW 40,000, discrete AgNPs of diverse

shapes, including rods (50 nm), were observed. In this microwave-based polyol method, synthesis of AgNPs could be achieved within 15 min instead of 2–6 h required for the conventional polyol process [95].

Navaladian et al. [98] demonstrated microwave-assisted synthesis of anisotropic AgNPs using silver oxalate in a glycol medium (EG) in presence of PVP as capping agent. Reduction of silver was obtained by microwave irradiation (2.45 GHz, on: off cycle of 10:5 sec) for 60 sec as evident from the greenish-yellow colour development, with λ_{max} at 367 nm due to LSPR of the AgNPs synthesized. On increasing the irradiation time to 75 sec, a red-coloured colloidal AgNP suspension was observed whereas, irradiation time of 90 sec led to the formation of a black colloidal suspension with λ_{max} due to LSPR at 422 nm and 462 nm, respectively. This red shift in λ_{max} with increasing irradiation time is indicative of increase in size of the AgNPs formed. An additional peak at 569 nm observed for the AgNPs synthesized using 75 sec microwave irradiation also indicated presence of anisotropic AgNPs. From HRTEM images, the AgNPs formed after 60 sec microwave irradiation were found to be spherical with size range 3–9 nm and average size between 5 and 6 nm. The TEM images of AgNPs formed over 75 sec confirmed the presence of anisotropic AgNPs, including short rods, triangles and other non-symmetric morphologies with diameter range 15–45 nm and length in the range 20–70 nm. The AgNPs formed after 90 sec microwave irradiation, exhibited both spherical or multi-edged non-spherical shapes and their size ranged from 50 to 120 nm. When EG was replaced with diethylene glycol, a yellow-colour development was observed after 75 sec of microwave irradiation and with continued irradiation the colour turned black, thereby, illustrating the effect of the media on change in characteristics of AgNPs formed.

Kundu et al. [38] established a method for size-controlled synthesis of AgNPs using microwave irradiation for 1 min. The surfactant, poly oxyethylene isooctyl phenyl ether (TX-100) was used together with the reducing agent, 2,7- dihydroxy naphthalene (2,7-DHN) in an alkaline solution (NaOH) for reducing Ag^+ ions released by the precursor salt, $AgNO_3$. AgNPs of variable size were synthesized by varying the Ag^+/TX-100 molar ratio by varying the concentration of $AgNO_3$ (0.363–0.380 mM), concentration of 2,7-DHN and pH of the reaction medium. For synthesis of size-controlled AgNPs, 200 μL of $AgNO_3$ (at concentration 0.363, 0.37, 0.373, 0.377, 0.379 and 0.380 mM) was added to TX-100 (4 mL, 10^{-2} M) solution. Later, 2,7-DHN (1 mL, 10^{-2} M) and NaOH (50 μL, 1 M) were added to the solution containing $AgNO_3$ and TX-100 and well mixed before exposing it to microwave irradiation for 1 min in cyclic on and off mode (off after every 10 sec). The TEM images of AgNPs formed are shown in Figure 1.34.

The size of spherical AgNPs was 4 ± 0.6 nm, 12.5 ± 3 nm and 32 ± 3.5 nm for final Ag^+ ion concentration of 0.38 mM, 0.733 mM and 1.06 mM, respectively. The AgNP colloidal suspension with mean sizes 4, 12 and 32 nm exhibited λ_{max} in UV-vis absorption spectra at 414, 418 and 426 nm, respectively. The method was also explored for synthesis of self-assembled silver nanochains using a strongly basic pH (>11.5) and by lowering the concentration of TX-100. To synthesize nanochains, the pH of the reaction mixture was increased by increasing the concentration of NaOH to 9.09 mM. The concentration of

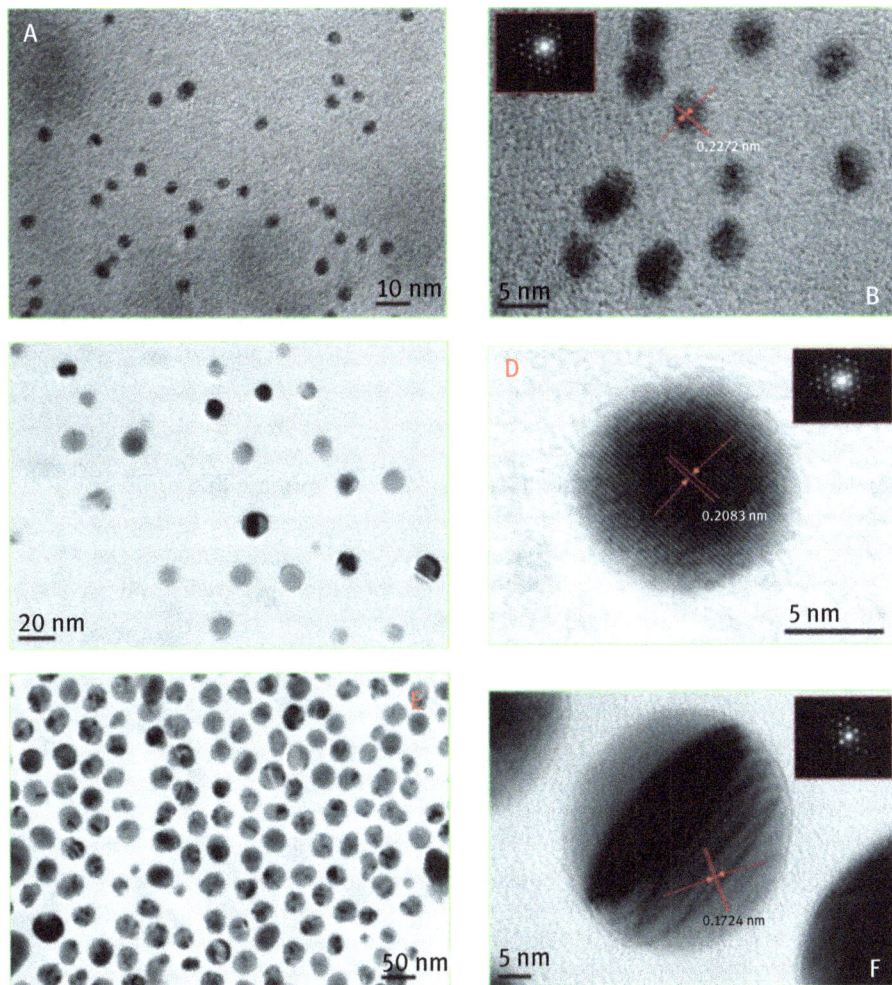

Figure 1.34: TEM images of AgNPs formed after 60 s of microwave irradiation with varying AgNO₃ concentrations. (A, C, E) and (B, D, F) represents low- and high-magnification images of the AgNPs (inset showing SAED pattern) (Reprinted with permission from ref. [38]. Copyright 2009, American Chemical Society).

TX-100 was decreased such that its final concentration in the solution was 7.27 mM. The reaction mixture was subjected to microwave irradiation for 1 min. The TEM images of AgNPs formed are shown in Figure 1.35.

The mechanism of formation of size-controlled AgNPs and nanochain synthesis was also proposed for this method. It is reported that phenolic compounds can undergo cleavage to phenoxy radicals in the presence of UV-light or at high temperatures. Thus, some radical species or solvated electrons were possibly generated from 2,7-DHN upon microwave irradiation of the reaction mixture and these radicals/

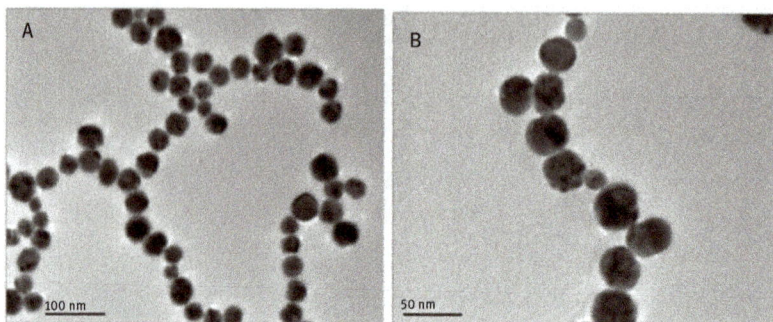

Figure 1.35: TEM images of the Ag nanochain structure (A) at low magnification and (B) at high magnification (Reprinted with permission from ref. [38]. Copyright 2009, American Chemical Society).

solvated electrons caused reduction of Ag^+ to Ag^0 forming spherical Ag nuclei. Subsequently, these small nuclei grow to form Ag clusters and eventually formed crystalline AgNPs. It was observed that more Ag nuclei were formed as the Ag^+ ion concentration was increased, this promoted formation of larger-sized AgNPs at higher $AgNO_3$ concentration (as shown in schematics in (Figure 1.36). The non-

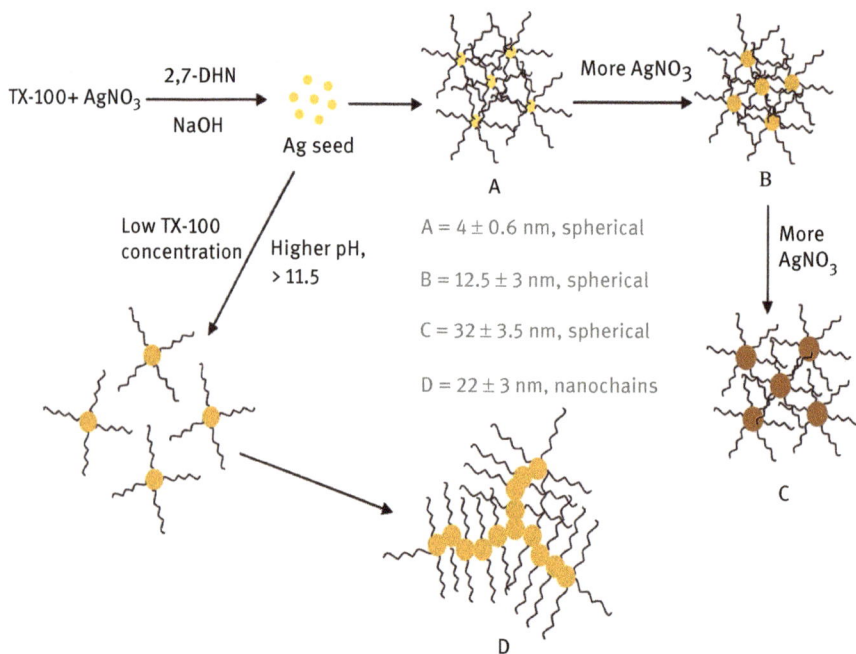

Figure 1.36: Schematic representation for the formation of size-controlled AgNPs and the nanochains using 2,7-DHN as reducing agent and microwave irradiation (Reprinted with permission from ref. [38]. Copyright 2009, American Chemical Society).

ionic surfactant TX-100 molecules acted as stabilizer and prevented unwanted growth and aggregation due to their adsorption on particular facets of AgNPs thereby facilitating 3D growth of the AgNPs. The formation of nanochain structure at lower concentrations of TX-100 and higher pH was attributed to availability of more unoccupied surface for adsorption of surfactants on the Ag nuclei. The AgNPs present in close proximity, thus generated nanochain structures, whereas at higher concentrations of TX-100, well-distributed AgNPs were formed [38].

Zhu et al. [97] synthesized AgNPs of varying shapes ranging from nanospheres, nanocubes, nanowires, nanorods, and nanoplate using the polyol process based on PEG and PVP. In a typical experiment for AgNP synthesis, aqueous solution of $AgNO_3$ (0.5 mL, 1 M) and PVP (1.25 mL, 1 M), and 12.5 mL of PEG were mixed in a 100 mL round bottom flask and heated under magnetic stirring in a microwave oven. Effect of various factors such as, reaction temperature (100°C, 140°C and 170°C, microwave power (300 W, 500 W, and 700 W) and molar ratio of PVP to $AgNO_3$ was explored (as shown in Table 1.9) and the microwave-based method was also compared with the conventional oil heating method.

Table 1.9: Summary of peak position and FWHM for AgNPs synthesized by Zhu et al. [97].

Experiment	Temperature (°C)	Peak position (nm)/FWHM (nm)	
		30 min	60 min
Microwave method	100	410/148	412/123
	140	430/60	442/77
	170	432/132	440/150
Oil heating	100	400/150	400/145
	140	434/118	440/110
	170	464/207	460/190

The AgNPs formed were separated by centrifugation at 8,000 rpm and dispersed in ethanol for subsequent characterization. The UV-Vis peaks and full width at half maximum (FWHM) values shown in Table 1.9 indicated a red shift in λ_{max} and wider size distribution for increase in irradiation time from 30 to 60 min. A red shift in LSPR peak was also observed as the reaction temperature was increased, thereby indicating a higher rate of Ag^+ reduction and faster particle growth at higher temperature. Typically, at any temperature, FWHM for AgNPs synthesized by microwave irradiation depicted a narrower size distribution (~50 nm) and showed regular spherical or polygon shaped AgNPs. In contrast, AgNPs formed by oil heating exhibited a wide size distribution ranging from 20 to 160 nm and AgNPs of irregular shapes were observed at 140°C.

Since PVP works as capping agent for stabilizing the AgNPs, the molar ratio of PVP to $AgNO_3$ (1:1 to 15:1) was also studied for AgNP synthesis by carrying out the reaction at 140° for 30 min under 700 W microwave heating. The peak position and corresponding FWHM ($\lambda_{max}/\lambda_{FWHM}$) were 448/154, 430/60, 438/62, and 438/66 nm for PVP/AgNO$_3$ ratio 1:1, 5:1, 10:1 and 15:1, respectively. The widest extinction band was obtained for 1:1 molar ratio, indicating higher polydispersivity in the AgNPs formed under these conditions. A molar ratio of 5:1 or more exhibited greater homogeneity in size, however, increasing the molar ratio beyond 5:1 had little effect on mean size and size distribution of the AgNPs.

In another study by Yin et al. [180], the effect of varying conditions on synthesis of AgNPs using AgNO$_3$, formaldehyde and TSC through microwave irradiation was tested and compared with the conventional oil heating method. A total of 8 formulations designated as A1–A8 were used. AgNPs were formed using: AgNO$_3$ solution as precursor salt (0.002 M in A2; and 0.1 M in A1, A3–A6); formaldehyde as reducing agent (1 M in A1–A3; and 1.5 M in A4–A6) and TSC as stabilizer (1.5 M in A1–A3; 1 M in A4; and 0.1 M in A5–A6). Apart from A3 and A6, all other formulations were subjected to continuous microwave irradiation (650W, 2.45 GHz) over 1 min under stirred conditions, while A3 and A6 were subjected to heating in an oil batch at 150°C. Two control formulations, i. e., A7 comprised of only AgNO$_3$ (0.1 M) and TSC (1.5 M) and A8 comprised of only AgNO$_3$ (0.1 M) and formaldehyde (1 M) were also prepared and subjected to microwave irradiation. When TSC at the highest concentration (1.5 M) was mixed with AgNO$_3$ (A1–A2), a precipitate of silver citrate (Ag$_3$C$_6$H$_5$O$_7$) (identified by powder XRD) was immediately formed; however, the silver citrate precipitate completely dissolved upon stirring for 10 min. Thus, at high concentration of citrate anions, the silver cations were effectively stabilized by formation of solvated silver citrate complexes. The solution remained clear and colourless even after formaldehyde addition and AgNPs were formed after microwave irradiation for 1 min. While in A1, A2, A4 and A5 synthesis could be completed in 1 min by microwave irradiation, the total time for AgNP synthesis required in A3 and A6 subjected to conventional heating in oil bath (150°C) was 5 and 8 min, respectively. The time required for temperature of the reaction mixture to reach from room temperature to boiling temperature was 0.6 min and 4 min for microwave irradiation and conventional heating, respectively, under vigorous stirred conditions. Only trace levels of AgNPs were synthesized in the controls (A7–A8) even under microwave irradiation, demonstrating the requirement of both TSC and formaldehyde for synthesis of AgNPs. The coordination of citrate to Ag$^+$ ions facilitated the reduction of the solvated Ag$^+$ ions by formaldehyde.

The AgNPs synthesized by both microwave irradiation and conventional heating exhibited cubic crystal structure of Ag as determined by XRD. The size range (and average size) of AgNPs in samples A1–A6 were 60–260 (132), 40–120 (70), 20–480 (66), 20–260 (65), 10–90 (24) and 20–360 (71), respectively. Thus, a narrower particle size distribution of AgNPs was obtained under microwave irradiation for a minute as compared to conventional heating. The highest yields were obtained for AgNPs in samples A5 and A6 having identical composition (98% and 94%, respectively) whereas the yield obtained in A1 and A4 was only 53% and 74%, respectively. Thus, microwave irradiation resulted in formation of high density of silver nuclei by homogeneous and rapid heating of the silver citrate colloids in presence of formaldehyde and small-sized AgNPs were produced in the subsequent process of crystal growth. Temperature gradients formed during conventional heating may have adversely affected the nucleation and crystal growth processes to result in a broader particle size distribution.

Researchers from the same group as Yin et al. [180], later explored the synthesis of shape-controlled AgNPs. Yamamoto et al. [181] performed microwave -based synthesis of AgNPs from aqueous $AgNO_3$ using either TSC with formaldehyde or PVP as additive. While the former resulted in formation of nanospheres, the latter resulted in formation of nanoprisms. The authors demonstrated that during synthesis of AgNPs from aqueous $AgNO_3$, TSC and formaldehyde using microwave irradiation, initially $[Ag(NH_3)_2]^+$ (similar to Tollen's reagent) is formed and this species is later reduced by formaldehyde. In this process, TSC concentration significantly affected the size and size distributions of spherical AgNPs. Four formulations designated as A1–A4 was used. Maintaining $AgNO_3$ at a constant concentration of 0.1 M, TSC was varied from 0.1 M to 1.5 M (0.1 M in A1 and A2, 1 M in A3 and 1.5 M in A4) while the formaldehyde concentration used was either 1.5 M (A1–A3) or 1 M (A4). When TSC was present at low concentration (A1), citrate acted as an effective stabilizer and spherical AgNPs (~24 nm) with relatively narrow size distribution (10–90 nm) were formed upon microwave irradiation for 1 min. When a formulation with identical composition was subjected to heating in an oil bath (A2), AgNPs of spherical shape having fcc crystal structure was obtained; however, the average size of AgNPs was 71 nm and a wide size distribution (20–360 nm) was observed. Thus, homogeneous and fast nucleation of silver from the reduction of silver ion generated by microwave irradiation resulted in AgNPs of smaller size and narrower size distribution compared to conventional heating. However, at a higher TSC concentration of 1 M and 1.5 M, the average size of AgNPs formed with 1 min microwave irradiation was 65 and 132 nm, respectively. Thus, TSC appears to promote coalescence of AgNPs at higher concentration. A possible reason for this may be the dielectric loss of in reaction medium due to rapid heating caused by microwave irradiation and subsequently, particle growth due to localized super heating [181]. Moreover, inter-molecular hydrogen bonding between the citrate moieties and AgNPs may have increased the probability of collision between AgNPs and this may have promoted coalescence of AgNPs. Thus, AgNPs of larger size and wider particle size distribution were obtained at higher concentration of TSC. From these results, it appears that controlling the heating rate in microwave-based synthesis may ensure a narrower size distribution of spherical AgNPs.

Yamamoto et al. [181] also synthesized prism-shaped AgNPs through microwave irradiation using aqueous solution of $AgNO_3$ with PVP (MW 40,000) as additive. The PVP concentration was held constant (3 M). At low concentration of $AgNO_3$ (0.01 M), spherical AgNPs with average size of 25 nm (range: 10–35 nm) were formed upon microwave irradiation for 30 min. However, on extending the time of microwave irradiation to 90 min, both spherical AgNP (average size: 50 nm, range: 40–60 nm) and two-dimensional triangular AgNPs (edge length: 70–250 nm) were obtained. At higher $AgNO_3$ concentration (0.1 M), a combination of spherical AgNPs (average size 70 nm) and triangular AgNPs (200–380 nm edge length) were obtained on microwave irradiation for 20 min. With increase in microwave irradiation time to 75 min, primarily prisms (230–420 nm edge length) were observed. The prisms included some truncated triangular prisms also along with perfectly triangular prisms. The triangular prisms were characterized by fcc structures as revealed through SAED and XRD spectra. At low concentrations of $AgNO_3$ (0.01 M), no prism-shaped AgNPs were obtained even after 120 min of microwave irradiation. Thus, high concentration of $AgNO_3$ was favourable for shape transformation and formation of prism-shaped AgNPs. The homopolymer PVP having individual units of N-methyl pyrollidone (NMP) may have played a key role in shape transformation. The oxygen atoms in the polar imide group in NMP may have a strong affinity to Ag^+. Electron transfer from this group may have facilitated thermal reduction of Ag^+. The resulting AgNPs were also stabilized due to surface adsorption on the PVP chain. NMP has a large dipole moment (32.2) at room temperature, hence, a large dielectric loss (8.855) may be expected under microwave irradiation at 2.45 GHz. This may have accelerated the coherent heating of PVP and PVP-stabilized AgNPs, resulting in shape transformation from spherical to relatively large prism-shaped Ag morphologies [181].

Kundu et al. [182] reported synthesis of silver nanocubes within 2 min using microwave irradiation. Silver nanocubes were synthesized in a 1000 W microwave oven employing gold NPs (AuNP) as seed and PSS as stabilizing agent. As reported earlier, PSS has the ability to support growth of AgNPs in specific crystallographic directions leading to synthesis of shape-controlled AgNPs. The AuNP seeds were synthesized by addition of ice-cold $NaBH_4$ (0.6 mL, 0.1 M) to an aqueous reaction mixture (20 mL) containing 0.25 mM each of $HAuCl_4$ and TSC under stirred condition. The reduction of gold ions by $NaBH_4$ immediately turned the mixture to pink, indicating synthesis of AuNPs (size: 4 ± 0.7 nm). These AuNPs were used as seeds within 2–5 h for synthesis of Ag nanocubes under microwave irradiation. For synthesis of Ag nanocubes, $AgNO_3$ (200 µL 0.1 M) was added to PSS (4 mL, 0.1%) and the mixture was stirred for 2–3 min. Subsequently, 50 µl of AuNP seed solution was added to the mixture with stirring and it was exposed to microwave irradiation for 30–120 sec in a cyclic manner with pause after every 10 sec. AgNP formation was evident within 20 sec of microwave irradiation through development of yellow colour in the solution and this was also revealed by UV-visible spectrometry. When synthesis was done in the presence of PSS but in absence of AuNP seed, hexagonal single-crystal NPs were synthesized at the polymer/bulk-solution interface.

Both microwave irradiation and PSS were essential for synthesis of silver nanocubes. Rapid heating of $AgNO_3$ solution due to microwave irradiation in presence of

AuNPs as seed led to thermal reduction of Ag^+ to Ag^0. In contrast, in the absence of PSS, only large Ag clusters were reported. PSS also formed PSS-Ag salt in solution through interaction between the Ag^+ ions and the $-(SO_3)^-$ groups of PSS. Nucleation and growth of AgNPs occurred at the polymer/electrolyte interface on microwave irradiation using the Ag^+ ions from the PSS–Ag salt rather than from the bulk solution. The adsorption to PSS may have been mediated by Ag^+, i. e., the Ag^+ ions deposited on the negatively charged AuNP seed may have exhibited growth in the crystallographic direction guided by PSS. When PSS was replaced with SDS, slower growth of AgNPs and AgNP morphology devoid of distinct facets were observed.

Darmanin et al. [183] employed a modified polyol process and demonstrated the feasibility for synthesizing PVP-capped silver nanoplates/nanoprisms using microwave irradiation. They proposed use of a polyol with only one hydroxyl group as reducing agent, such as, diethylene glycol monomethyl ether (DME), diethylene glycol monoethyl ether (DEE), diethylene glycol monopropyl ether (DPE), diethylene glycol monobutyl ether (DBE) and diethylene glycol monohexyl ether (DHE). For synthesis of nanoprisms, 120 mg of PVP (MW 10,000 kDa) was added to 6 mL of DEE in a 10 mL glass vessel. The mixture was homogenized by stirring and microwave heating was performed at 120°C for 5 min. After cooling under stirred condition, $AgNO_3$ (8 mg) was added to the homogenized solution. This solution was again heated at 140°C in a microwave oven at an input power of 25 W for 30 min using the dynamic heating mode. This ensured the desired initial heating rate of ~15°C/min. The solution turned dark pink due to formation of Ag nanoplates. The reaction mixture was cooled using compressed air at a cooling rate of ~25°C/min. The Ag nanoplates were precipitated from the mixture using acetone and collected by centrifugation at 3000 rpm for 15 min. The pellet was washed with acetone and resuspended in ethanol or other solvents, such as, ethanol, methanol or water, to obtain a clear, pink-purple solution. The colour of the nanoplates colloidal suspension was observed to vary from pink to purple, depending on the solvent used for dispersion. The λ_{max} due to LSPR was also affected by the solvent used for redispersion of AgNPs due to differences in refractive index of the solvents. For example, a red shift in SPR peak was observed when the spectra was taken from nanoplates dispersed in ethanol (higher refractive index) as compared to that obtained for nanoplates in methanol or water [183].

Darmanin et al. [183] also studied the effect of chain length of the alkyl group in the glycol derivatives and found that the chain length of the alkyl group had a prominent effect on the height of the Ag nanoprisms. It was evident from the change in UV-Vis optical absorption spectra. For synthesis using DHE, the AgNP suspension exhibited orange colour due to the presence of very small Ag nanoprisms. With decreasing alkyl chain length from DBE to DME, the NP suspension exhibited colour change from red to blue. Thus, decrease in alkyl chain length, increased the height of the Ag nanoplates as revealed from TEM micrographs illustrated in Figure 1.37. Alkyl chain length may have impacted nanoprism stabilization. However, the shape dispersity was not affected by

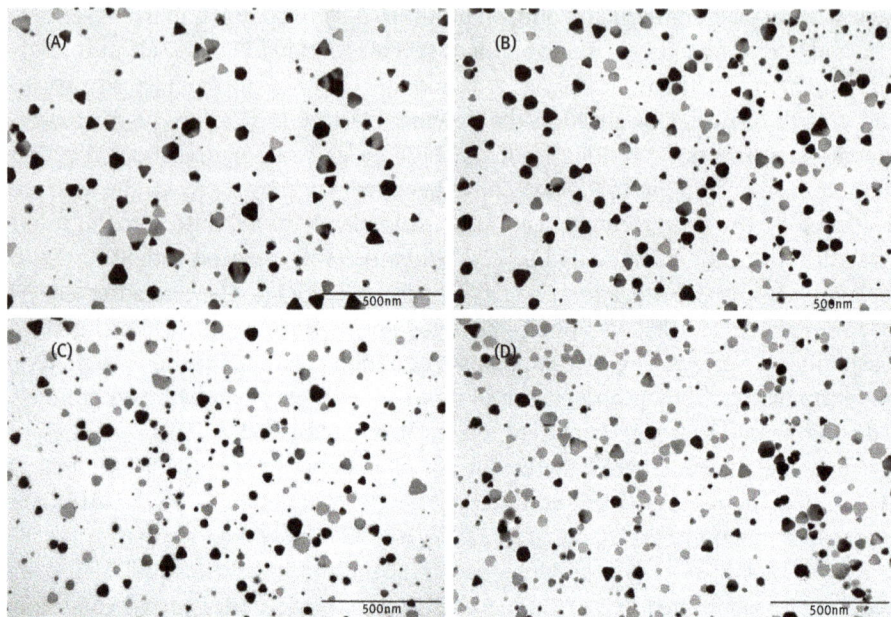

Figure 1.37: TEM images of AgNPs formed using microwave irradiation with various solvents: (A) DME, (B) DEE, (C) DPE and (D) DBE (Reprinted with permission from ref. [183]. Copyright 2012, Elsevier).

the alkyl chain length of the solvent. Darmanin et al. [183] also studied the effect of heating temperature (120, 140, 160, 180 and 200°C) and the ratio of PVP (10 kDa) and AgNO$_3$ on silver nanoprisms synthesis using DEE as reducing agent. They reported that a temperature of less than 140°C was not sufficient to generate nanoprisms, however, higher temperature (140–180°C) led to increased truncation of edges of the nanoprisms. At 200°C, the nanoprisms were found to transform to quasi-spherical AgNPs. By maintaining the mass of AgNO$_3$ unchanged, the effect of PVP to AgNO$_3$ ratio (R) on nanoprism synthesis was also studied. For low R-values (R = 4 or less), very large polydispersed nanoprisms were formed whereas at very high R-values (46), nanoprisms were not synthesized. For R-values ranging from 6 to 32, nanoprisms were synthesized as a result of complete conversion. The optimum temperature and PVP to AgNO$_3$ ratio for nanoprisms synthesis was found to be 140°C and 12. The mechanism of modified polyol method reported by Darmanin et al. [183] may be explained by the fact that during the synthesis of nanoprisms employing monohydroxy species such as, DEE, the reaction is kinetically controlled and occur at sufficiently slow rate so as to ensure completion of the reaction. Therefore, nanoprism synthesis was analysed at various time using TEM (Figure 1.38) and UV-Vis spectroscopy (Figure 1.39).

It was observed that in 0.5 min, only spherical AgNPs with SPR peak at 413 nm were formed and the AgNP suspension appeared yellow. At 2 min, the AgNPs were observed to be predominantly spherical in shape, although a small amount of

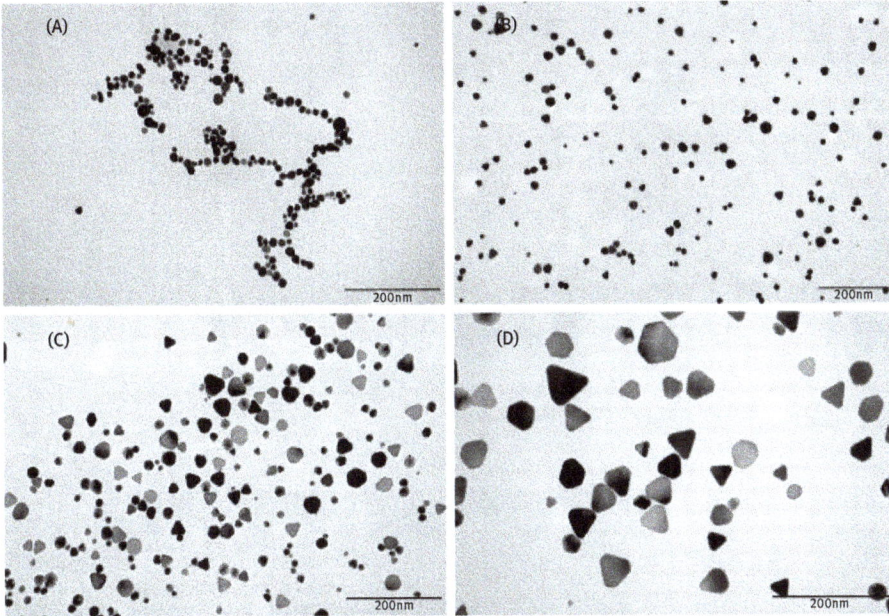

Figure 1.38: TEM images of the AgNPs in solutions as a function of the synthesis time (A) 2 min, (B) 5 min, (C) 10 min and (D) 20 min; (power = 25 W); reducing agent: EEE. (Reprinted with permission from ref. [183]. Copyright 2012, Elsevier).

nanoprisms were also present. The LSPR peak was obtained at 431 nm along and an additional peak at 378 nm, demonstrating the growth of nanoprisms. After 5 min microwave irradiation, the colour of solution changed to orange. An increase in nanoprism formation (mean height 15 nm) was observed, and LSPR peaks were observed at 349, 378 and 448 nm. Beyond 5 min, with further increase in nanoprism formation, the colour changed from orange to red, pink and purple (at the end of 20 min). The LSPR peak in nanoprism suspension exhibited red shift to 496 nm (where mean height of nanoprisms was 20 nm) in red-coloured solution formed after 10 min microwave irradiation. However, with increasing microwave irradiation for 15 min and 20 min (pink and purple solution, respectively), LSPR peak at 523 and 536 nm (mean height of 45 nm), respectively, were observed. The colour stabilized beyond 20 min and thus represented completion of the reaction, as shown in Figure 1.39.

In another study, Dzido et al. [184] reported studies for continuous synthesis of metallic NPs in a reactor with uniform microwave irradiation, as shown in Figure 1.40. They proposed a methodology for AgNP synthesis based on an understanding of the synthesis mechanisms. They also studied the effects of various variables: precursor type and its concentration, temperature, residence time, and density of microwave power, on formation of AgNPs. Characteristics of the AgNPs formed were explained using the classical LaMer–Dinegar (LMD) mechanism of NPs formation applied to a non-isothermal, continuous-flow perspective.

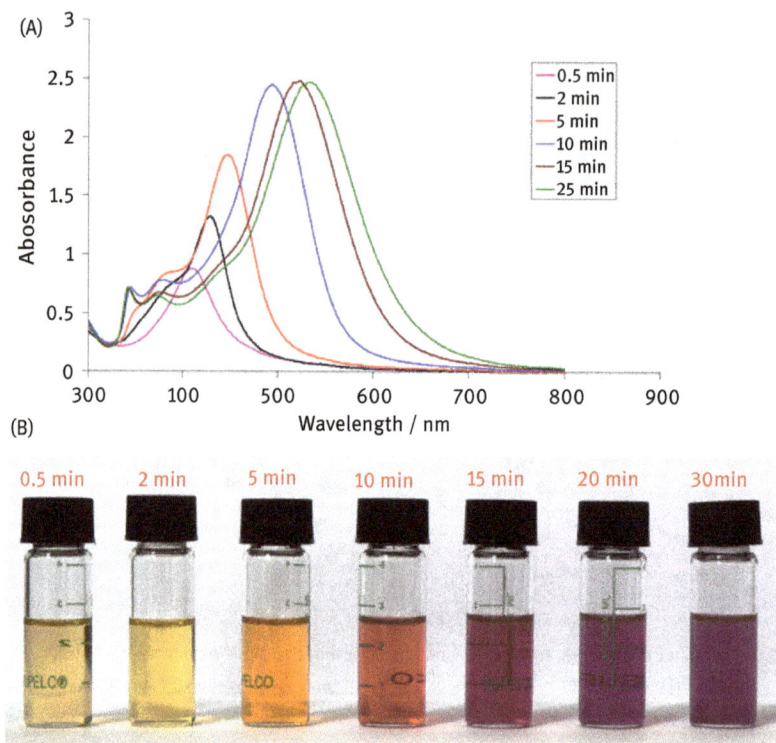

Figure 1.39: (A) UV spectra and (B) solution colour as a function of the synthesis time for microwave -based synthesis with DEE as reducing agent (Reprinted with permission from ref. [183]. Copyright 2012, Elsevier).

Various combinations of variables were adopted for optimized synthesis of AgNP. Most of them exhibited AgNPs with LSPR peak at ~419 nm in the UV-Vis spectra. However, the peak intensities depended on the applied substrate. The peaks were narrow and clearly distinct in all silver acetate-derived AgNPs, irrespective of the inlet concentration and reaction time. Whereas for $AgNO_3$-derived AgNPs, the scenario appeared to be more complex. They were narrow and pronounced for low (10 mM) inlet concentration; however, the peaks were wide and red-shifted (up to a maximum of 422–424 nm), at $AgNO_3$ concentration of 50 mM. The spectra of AgNPs synthesized using higher concentration exhibited the presence of both small and large AgNPs. However, negligible effect on LSPR spectra was observed when the reaction time was extended fourfold (from 6 to 24 sec), indicating no significant change in the mechanism of NP formation and particle size distribution. Moreover, in case of silver acetate-derived AgNPs, minor changes in LSPR peaks were obtained with changes in reaction time (6–24 s) and the UV-vis spectra got further narrower and red-shifted from 415 to 417 nm and from 418 to 420 nm, for 20 and 50 mM $AgNO_3$ concentration, respectively.

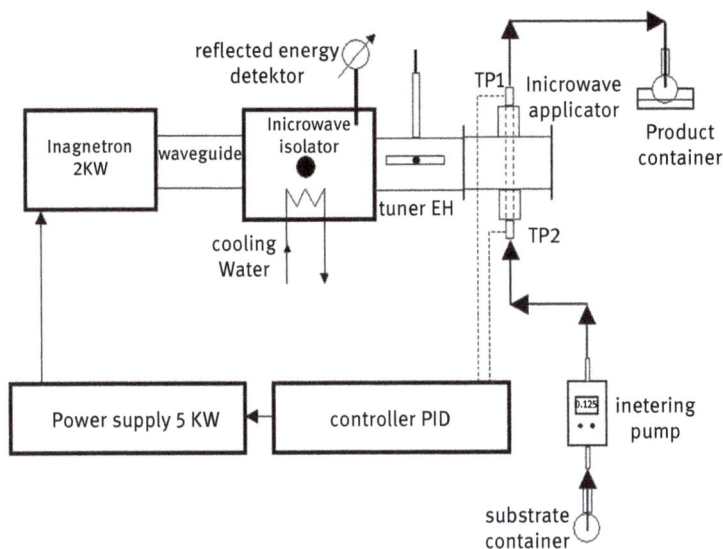

Figure 1.40: Experimental set-up of the microwave apparatus applied for synthesis of AgNPs in continuous flow through mode (Reprinted from ref. [184]. Published by Springer).

1.3.2.8 Biological method

Biological methods of synthesis can overcome the limitations of AgNP synthesis associated with chemical and physical methods, i. e., it can avoid the use of toxic and hazardous chemicals and can perform synthesis of AgNPs without the use of high temperature and pressure during synthesis. Most biological processes occur under atmospheric pressure and ambient temperature, resulting in vast energy savings [18]. Thus, eco-friendly and cost-effective, biomimetic approaches for NP synthesis have gained significant attention. Since the last few decades, a variety of microorganism- and plant-based approaches have successfully been exploited for biosynthesis of AgNPs. Microorganisms including bacteria, fungi, yeast, algae and various plant extracts that have been reported for the biosynthesis of AgNPs as listed in Table 1.10, Table 1.11, Table 1.12 and Table 1.13. Although the interest in biological methods for synthesis of AgNPs is more recent, formation of NPs due to interaction between microorganisms and metals have been reported since long [185]. Metal NPs can be produced intracellularly by specific microorganisms possessing metalloregulatory molecules, particularly proteins and peptides involved in metal detoxification processes. Under certain conditions, addition of metal cations in the cell free culture medium may also lead to extracellular production of NPs [186].

There are two predominant methods for biological synthesis of NPs using microorganisms. These are intracellular and extracellular synthesis methods. All microorganisms cannot be used for intracellular synthesis of AgNPs since silver

Table 1.10: Bacteria used for biosynthesis of silver nanoparticles.

Bacterial strains	size (nm)	References
Pseudomonas stutzeri AG259	200	Klaus et al. [185]
Bacillus licheniformis	50	Kalimiuthu et al. [99]
E. coli	2–100	Gurunathan et al. [187]
Lactobacillus fermentum	11.2	Sintubin et al. [188]
Bacillus cereus	4–5	Ganesh Babu and Gunasekaran [189]
Bacillus subtilis	5–60	Saifuddin et al. [190]
Bacillus Koriensis	9	Sharma et al. [191]
Rhodococcus species	5–15	Ahmad et al. [192]
Streptomyces albogriseolus	16	Samundeeswari et al. [193]
P. putida NCIM 2650	70	Thamilselvi and Radha [194]
Streptomyces sp. MBRC-91	10–60	Manivasagan et al. [195]
Streptomyces viridochromogenes	2.2–7.3	El-Naggar and Abdelwahed [196]
actinobacteria Streptomyces spp. 211A	7–16	Tsibakhashvili et al. [197]
Bacillus licheniformis Dahb1	18.69–63.42	Shanthi et al. [198]
Klebsiella pneumonia	50	Shahverdi et al. [199]
Bacillus licheniformis	50	Kalimuthu et al. [99]
Bacillus subtilis	5–60	Saifuddin et al. [190]
Streptomyces sp. ERI-3	10–100	Zonooz and Salouti [200]
Streptomyces sp. SS2	67.95 ± 18.52	Mohanta and Behera [201]
Bacillus amyloliquefaciens	14	Wei et al. [202]

Table 1.11: Fungus-mediated AgNP synthesis.

Fungal strain	Size (nm)	References
Fusarium oxysporum	5–50	Ahmad et al. [207]
Aspergillus fumigatus	5–25	Bhainsa and D'Souza [208]
Aspergillus niger	20	Abd El-Aziz et al. [210]
Trichoderma viride	5–40	Fayaz et al. [211]
Fusarium semitectum	10–60	Basavaraja et al. [212]
Pycnoporus sanguineus	7–14	Chan and Don [213]
Penicillium citrinum (MTCC 9999)	20–30	Goswami et al. [214]
Cladosporium sphaerospermum F16	15	Abdel-Hafez et al. [215]
Trichoderma Reesei	5–50	Vahabi et al. [209]
MKY3 (yeast)	2–5	Kowshik et al. [216]
Phaenerochaete chrysosporium	50–200	Vigneshwaran et al. [217]
Trichoderma asperellum	13–18	Mukherjee et al. [218]
Verticillium	25 ± 12	Mukherjee et al. [219]
Phoma sp. 3.2883	71–74	Chen et al. [220]

ions are highly toxic to microorganisms. A requirement for intracellular biological synthesis of AgNPs is that the microorganisms should be resistant to silver ions. Some microorganisms contain a unique silver resistance machinery that can

Table 1.12: Synthesis of silver nanoparticles using algae (Reprinted with permission from ref. [221]. Copyright 2015, Springer Nature).

Algae	Size (nm)	Shape	References
Microalgae			
Chlamydomonas reinhardtii	5–15	Spherical/rectangular	Barwal et al. [225]
Chlorella vulgaris	28	Nanoplates	Sharma et al. [221]
Chlorococcum humicola	2–16	Spherical	Jena et al. [222]
Chlorella pyrenoidosa	5–10	Spherical	Aziz et al. [224]
Macro algae			
Caulerpa racemosa	5–25	Spherical/triangular	Aziz et al. [224]
Ulva compressa	40–50	Spherical	Dhanlakshmi et al. [227]
Sargassum cinereum	45–76	Spherical	Mohandass et al. [228]
Sargassum muticum	5–15	Spherical	Azizi et al. [229]
Turbinaria conoides	2–19	-	Vijayan et al. [230]
Gelidiella acerosa	22	Spherical	Vivek et al.[231]
Gracilaria dura	6	Spherical	Shukla et al. [232]
Cyanobacteria			
Plectonema boryanum	<10	Spherical	Lengke et al. [233]
Oscillatoria willei	100–200	Spherical	Ali et al. [234]
Arthrospira platensis	11.6	spherical	Mahdieh et al. [235]
Microcoleus spp.	44–79	Spherical	Sudha et al. [236]

Table 1.13: Plant-mediated silver nanoparticle synthesis.

Plants	Size (nm)	References
Aloe Vera	30.5	Ashraf et al. [247]
Acalypha indica	20–30	Krishnaraj et al. [238]
Euphorbia prostrata	10–15	Singh et al. [248]
Ginkgo biloba, Pinus desiflora	~32	Song and Kim [246]
Panax ginseng, Platanus orientalis	~32	Song and Kim [246]
Azadirachta indica	41–60	Panigrahi [241]
Ananas comosus	5–30	Ahmad and Sharma [240]
Musa balbisiana, A. indica & *O. tenuiflorum*	Upto 200	Banerjee et al. [249]
Lotus	23–45	Talib and Hui-Fen-Wu [245]
Ricinus communis	20–30	Mani et al. [242]
Olive	24–36	Khalil et al. [250]
Tridax procumbens Leaf Extract	5–10	Shaik et al. [251]
Dracena Leaf extract	87.1	Swamy and Prasad [252]

convert reactive silver ions into stable silver atoms. This is possible only up to a certain Ag$^+$ concentration threshold. For instance, *Bacillus licheniformis* can survive, grow and synthesize AgNPs when silver ion concentration in the surrounding environment is low (<1 mM); however, when the concentration exceeds the threshold level (10 mM) it is reported to cause cytotoxicity [99].

The major mechanism of antibacterial action of AgNPs includes: binding of Ag^+ to DNA through electrostatic interaction thereby disintegrating the structure of DNA; inhibition in DNA replication; and interaction of Ag^+ ions with S-containing proteins/ enzymes, such as, NADH dehydrogenase II in the respiratory electron transport chain. Such binding hinders the function of enzymes and promote synthesis of reactive oxygen species (ROS) and highly reactive radicals within the cell. Catalase enzyme helps in survival of the cells by scavenging the free radicals generated. However, in the presence of free radicals beyond the minimum inhibitory concentration (MIC), catalase cannot destroy all the free radicals, thus, the cell ultimately undergoes apoptosis [99]. Some bacterial cells can have high MIC, and can synthesize AgNPs by reducing silver ions using their defence mechanism as long as the Ag^+ concentration is lower than the MIC.

A large variety of microorganisms can facilitate synthesis of AgNPs through the extracellular synthesis method. In this method, the harvested culture supernatant (often referred as the cell-free extract) is utilized for AgNP synthesis. Various factors, such as, temperature, pH, concentration of metal salts and time of incubation are important factors that affect the yield and characteristics of the AgNPs formed, such as, shape, size and stability. In one study, the average size of AgNPs formed was 50 nm at room temperature while at a higher temperature of 60°C, the size reduced to 15 nm [100]. AgNP synthesis is typically favoured under alkaline conditions. For a study on extracellular AgNP synthesis conducted over the pH range 8–12 using 5 mM $AgNO_3$ and reaction temperature of 60°C, Gurunathan et al. [187] demonstrated maximum yield of AgNPs (10–15 nm) at pH 10. High pH also reduced the synthesis time significantly. Increase in pH beyond 10 exhibited decrease in absorbance at 420 nm due to LSPR, indicating lower yield AgNPs. Increase in pH up to 10 exhibited decrease in AgNP size, however, increase in pH beyond 10 caused increase in size of AgNPs. In controls (without the cell free extract) where $AgNO_3$ solution was incubated over the pH range 8–11, no AgNP synthesis was observed. Thus, an excess of OH^- ions facilitated the reduction of Ag^+ ions during extracellular biosynthesis. At alkaline pH, reducing power of proteins involved may have increased significantly. This may have led to fast conversion of Ag^+ ions to AgNPs, such that complete conversion was obtained in less than 30 min. Thus, a change in alkalinity beyond pH 10 may lead to aggregation of AgNPs.

1.3.2.8.1 Synthesis of silver NPs using bacteria

A bacterial strain isolated from a silver mine, *Pseudomonas stutzeri* AG259, was the first bacterial strain reported to synthesize AgNPs [185]. Since then, AgNP synthesis has been reported using a variety of bacteria (Table 1.10). $AgNO_3$ concentration is one of the key factors that affect the synthesis of AgNPs by bacteria. While Ag^+ at low concentration may induce AgNP synthesis in organisms, higher concentration may induce cell death.

Kalishwaralal et al. [203] reported extracellular synthesis of AgNPs (10–50 nm) by reduction of aqueous Ag^+ ions using the culture supernatant of *Brevibacterium casei*. Synthesis of spherical- and triangular-shaped AgNPs (12–65 nm) was reported using the culture supernatant of *Bacillus flexus* [204]. The AgNPs synthesized were stable in aqueous solution for a period of five months at room temperature in the dark. Wei et al. [202] reported the synthesis of spherical and triangular AgNPs (~14.6 nm) by solar irradiation of the mixture containing $AgNO_3$ and cell-free extracts of *Bacillus amyloliquefaciens*. Light intensity, NaCl addition and concentration of cell-free extract affected the synthesis of AgNPs. However, no synthesis was observed under dark conditions. Under optimized conditions, i. e., 3 mg/mL extract concentration, 2 mM NaCl and solar intensity at 70,000 lux, 98.23% of the Ag^+ (1 mM) was converted to AgNPs within 80 min of incubation. AgNP synthesis was also observed in the heat-inactivated extract, thereby demonstrating that enzymatic reactions were not the only mechanism involved in AgNP synthesis. Although the exact mechanism of light-driven synthesis of AgNPs by the cell-free extracts was not elucidated, hydroxyl groups of tyrosine residues, carboxylic acid-containing peptides, and carboxyl groups of aspartic acid and glutamic acid residues were hypothesized to play a role in reducing Ag^+ and causing anisotropic growth of AgNPs [205].

1.3.2.8.2 Synthesis of silver NP using fungi

Fungi-mediated synthesis offer various advantages over bacteria-mediated synthesis of AgNPs, such as, economic feasibility, ease of scaling up and downstream processing and the increased surface area due to presence of mycelia [60]. Mukherjee et al. [206] also suggested that fungi produce significantly higher amount of proteins, such that higher productivity of AgNPs may be expected [206]. The fungus, *Fusarium oxysporum*, has been used for synthesis of metallic NPs, including AgNPs. AgNPs in the size range 5–15 nm were successfully synthesised using the fungal extract. These AgNPs were capped and stabilized by biomolecules, such as, proteins, released into the media by the fungus. The most important consideration in *F. oxysporum* mediated synthesis of AgNPs is that the strain was capable of extracellular synthesis. Most previous research on fungal-mediated synthesis was on intracellular synthesis [207]. Other fungi employed for AgNP synthesis are listed in Table 1.11. In a later study, Bhainsa and D'Souza [208] reported use of *Aspergillus fumigatus* for extracellular synthesis of AgNPs in the size range of 5–25 nm. The size of NPs synthesized was comparable to that reported for chemical and physical processes although the size range was larger than that reported for *F. oxysporum* [207]. Biosynthesis of AgNPs by *A. fumigatus* occurred rapidly within 10 min of exposure [208], while other strains such as, *F. oxysporum* required about an hour.

Vahabi et al. [209] reported extracellular biosynthesis of AgNPs by exposing the mycelium of *Trichoderma reesei* to $AgNO_3$ solution where AgNPs were obtained after 72 h. The rate of production of AgNPs was much lower, the AgNP size range was broader (5–50 nm) and the composition less homogenous compared to AgNPs synthesized using either *A. fumigatus* or *F. oxysporum*. However, since *T. reesei* is an organism that has been studied extensively, it may be possible to induce it to produce high quantities of enzymes, up to 100 g/L, for enabling large-scale production AgNPs in the future.

Similar to bacteria, fungi also have some disadvantages in terms of safety. Various well-studied fungi, such as *F. oxysporum* are known to be pathogenic, such that their use may pose a health risk. Two fungi *Trichoderma asperellum* and *Trichoderma reesei* are known to produce AgNPs on exposure to silver precursor salts and have been proven to be non-pathogenic. These organisms may be preferred for AgNP synthesis at a commercial scale. Hence, *T. reesei* has been used and explored in various sectors including food, pharmaceuticals, paper and textile industries [209].

1.3.2.8.3 Synthesis of silver NP using algae

Algae are grouped as microalgae (microscopic) and macroalgae (macroscopic). Microalgae can exist as prokaryotic or eukaryotic organisms growing individually or in colonies. Various types of algal species belonging to diverse classes have been explored for their ability to synthesize metallic NPs. There is much interest in algal biosynthesis of AgNPs due to the high metal accumulation ability of various species of algae. Moreover, algal culture is relatively easy to handle and offers other advantages similar to that of bacteria- and fungi-based synthesis [221]. Algal-based biosynthesis of NPs can be achieved by incubation of the algal culture filtrate with metal precursor salt under controlled conditions with or without stirring for a certain duration. The characteristic of NP can be affected by the concentration of metal precursor solution and the type of algal culture used for synthesis. Synthesis of NPs may be achieved by reduction of metals through various biomolecules such as, polysaccharides, peptides, and various pigments. Stabilization and capping of the metal NPs can be done by proteins, amino acids or cysteine residues and sulphated polysaccharides. It has been reported that algae-based biosynthesis of NPs requires relatively less time than the other biosynthesis methods [18, 221, 223].

Aziz et al. [224] biosynthesized AgNPs employing *Chlorella pyrenoidosa*. The algal extract was extracted by collecting the pellets of mid-exponential phase (22 days) cultures by centrifugation at 5000 rpm for 5 min and 4 °C. The washed and pelleted algal biomass (1.5 g) was resuspended in 50 mL of deionized water and incubated at 100°C for 5 min. Subsequently, the mixture was allowed to cool to room temperature and the supernatant was filtered and used for biosynthesis of AgNPs. Biosynthesis of AgNPs was done by adding 10 mL of algal extract of *C. pyrenoidosa* to 90 mL of $AgNO_3$ (1 mM) solution, followed by incubation at 28 ± 2 °C for 24 h. The biosynthesis of AgNP was monitored by colour change and by measuring UV-Vis spectra over a period of 72 days until no further colour change was observed, as shown in Figure 1.41. The AgNPs obtained were predominantly in the size range 2–15 nm with only a few exceeding 20 nm. Various algae used for AgNP synthesis is summarized in the Table 1.12.

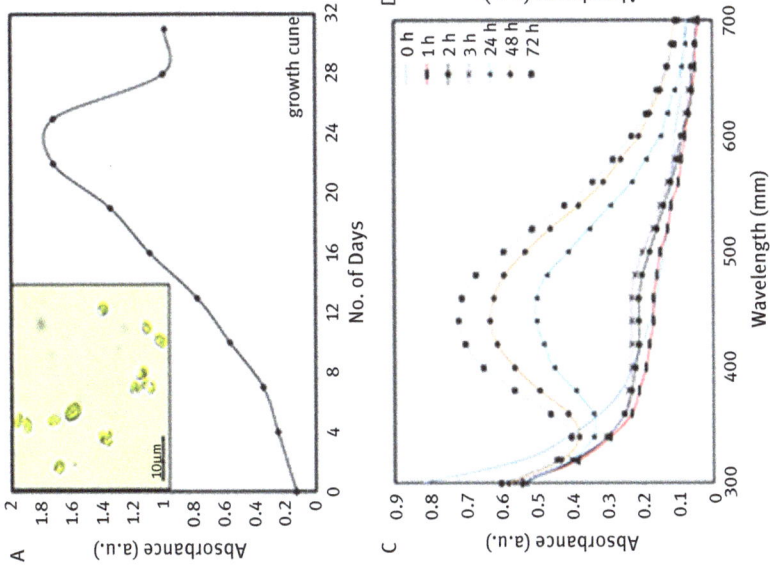

Figure 1.41: (A) Growth kinetics of *Chlorella pyrenoidosa* (inset figure shows a microscopic image with a scale bar of 10 μm) (B) Optical images of AgNPs showing colour change from 0 h to 72 h (C) UV-Vis spectra of AgNPs at various reaction times (D) Reaction saturation curve indicating the evolution of the SPR peak as a function of time (Reprinted with permission from ref. [224]. Copyright 2015, American Chemical Society).

1.3.2.8.4 Synthesis of silver NP using plants

A major advantage of using plants for NP synthesis over microbial biosynthesis is the lack of concerns over pathogenicity [60]. The plants used for AgNP synthesis includes a wide variety of plants, from algae to angiosperms; however, there are limited reports for AgNP synthesis using lower plants. The most preferred choice among plants are angiosperms, where various plant parts such as, bark, leaf, root and stem have been widely used for AgNP synthesis. A wide variety of medicinally important plants have been used, such as, *Jatropha curcas* [237], *Acalypha indica* [238], *Ocimum tenuiflorum* [239], *Ananas comosus* [240], *Azadirachta indica* [241], *Ricinus communis* [242], *Tinospora cordifolia* [243], *Aloe vera* [244] and lotus leaf [245]. An interesting study by Bar et al. [237] illustrated synthesis of AgNPs from $AgNO_3$ using the extract from *Jatropha curcas*. It resulted in synthesis of homogenous AgNPs (10–20 nm) within 4 h.

Krishnaraj et al. [238] also synthesized AgNP of size 20–30 nm using *Acalypha indica* leaf extracts. Reduction of Ag^+ started immediately at 30°C and AgNP formation was observed within 1 min of incubation. More than 90% of the Ag^+ reduction was obtained in less than 50 min. The shape of NPs produced was found to be spherical, flower-shaped and/or triangular with a broad size range of 5–108 nm. In contrast, the purified AgNPs from *Musa sativa* were found to be heterogenous in size [238].

Another plant that demonstrated a potential to reduce Ag^+ is *Ocimum sanctum*. AgNPs of size 3–20 nm were produced by reduction of 1 mM $AgNO_3$ using the leaf extract of this plant. The NPs synthesized were spherical in shape and it was demonstrated that they were stabilized by biomolecules present in the leaf extract. An extensive study was done by Song et al. [246] for AgNP synthesis from diverse plant leaf extracts, such as, Pine, Persimmon, Ginkgo, Magnolia and *Platanus*. Magnolia leaf broth exhibited the best reduction ability leading to 90% reduction of Ag^+ within 11 min. Extracellular synthesis of AgNPs by reduction of aqueous Ag^+ ions by the extract of Geranium leaves (*Pelargonium graveolens*) was also reported.

In comparison with the previous studies on biosynthesis of AgNPs using microorganisms, faster reduction of the Ag^+ ions was observed in case of plant extract-mediated synthesis. The time taken for complete reduction of the metal ions using microorganism typically ranged from 24 to 124 h, while more than 90% reduction of Ag^+ was achieved using extract of geranium leaves within 9 h [60]. Some of plants that have been used for AgNP synthesis are listed in Table 1.13.

1.3.2.8.5 Microwave-assisted biological synthesis of AgNPs

Since microwave -based synthesis is quick and yields homogenous size distribution of NPs, various researchers have combined biological reduction methods with microwave irradiation. Saifuddin et al. [190] employed a green synthesis approach for AgNPs by subjecting the culture supernatant remaining after growth of *Bacillus subtilis* to microwave irradiation. AgNPs were synthesized by mixing equal volumes of the culture supernatant (adjusted to pH 9) and aqueous $AgNO_3$ (1 mM) and sub-jecting the solution to microwave irradiation (frequency 2.45 GHz, power output 100 W) in cyclic mode (on: off 15:15 sec). The formation of AgNPs was observed by sampling aliquot (2 mL) at various intervals up to 15 cycles and measuring their UV-Vis spectra. UV-Vis spectra exhibited an LSPR peak at ~410 nm and its intensity increased over time. No change in λ_{max} was observed throughout the reaction, suggesting presence of well-dispersed stable AgNPs. The AgNPs were reported to be stable for more than 6 months and this stability was attributed to presence of capping agents, such as, proteins, secreted by the *B. subtilis*. The AgNPs synthesized were predominantly spherical with size range 5–50 nm; however, a few triangular AgNPs were also seen in TEM micrographs. The reduction of Ag^+ did not occur in the absence of cell-free extract indicating that biomolecules released by cell, such as, NADH -dependent reductase enzyme were involved in the reduction process and other biomolecules were involved in stabilization of the AgNPs. In this combined approach, AgNP synthesis was initiated after the 5 cycle and AgNP synthesis was completed within the 12th cycle, i. e., within 180 sec of microwave irradiation whereas without the use of microwave synthesis occurred over 72 h 40°C.

Manikprabhu et al. [253] synthesized AgNPs by reducing $AgNO_3$ (10^{-3} M and 10^{-2} M) using 1 mL of pigment produced by *Streptomyces coelicolor* klmp33. Maintaining the pigment volume constant, $AgNO_3$ solution and the pigment were combined in varying ratios (v/v) ranging from 1:1 to 20:1. These mixtures were exposed to microwave irradiation at a fixed frequency of 2.45 GHz for a period of 90 s. The solution colour changed from colourless to brown and λ_{max} due to LSPR was observed in the range 400–450 nm. No colour change due to AgNP synthesis was observed when $AgNO_3$ solution in the absence of cell free extract was exposed to microwave irradiation. FTIR spectra of the purified NPs exhibited absorption peaks at wavenumbers 1149, 1616, 1645, and 3333 cm^{-1} due to cyclic C–O–C, C = O and OH functional groups, respectively, thereby demonstrating the role of biomolecules in stabilization of AgNPs.

Banerjee et al. [249] employed leaf extract of the medicinal plants, *Musa balbisi-ana* (banana), *Azadirachta indica* (neem) and *Ocimum tenuiflorum* (black tulsi) for microwave -assisted AgNP synthesis. The leaves were cleaned, dried, shredded and weighed (~10 g). After boiling in 100 mL DI water for 20 min, the extracts were filtered to obtain clear solutions and were stored at 4°C. For AgNP synthesis, the extract was mixed with aqueous solution of $AgNO_3$ (1 mM) and the solution was exposed to microwave irradiation (power 300 W) for 4 min in cyclic operation. Colour change from yellowish brown to reddish brown indicated completion of AgNP synthesis. Synthesis of AgNPs was monitored periodically over 30 min by measuring the UV-Vis spectra. The absorption maxima were observed between 425 and 475 nm. In all the AgNP colloidal suspensions synthesized using banana, neem and tulsi, various bands of absorbance in FTIR spectra were observed at 1,025, 1,074, 1,320, 1,381, 1,610 and 2,263 cm^{-1} which corresponded to -C-O-C, ether linkages, -C-O-, terminal methyl, -C = C- groups of aromatic rings and alkyne bonds, respectively. These bands represented stretching vibrational bands due to compounds, such as, flavonoids and terpenoids which possibly played a role in capping and stabilization of the AgNPs. AgNPs of various shapes such as, triangles, pentagons and hexagons were obtained with size up to 200 nm. EDS exhibited a strong silver peak together with oxygen and carbon peaks, due to presence of biomolecules on the surface of AgNPs.

Roy et al. [254] synthesized shape-controlled AgNPs through microwave-assisted green synthesis using pomegranate juice. Anisotropic AgNPs of four different shapes i. e., spherical, oval, rod and flower shaped AgNPs were synthesized using freshly prepared pomegranate extract as shown in Figure 1.42. Pomegranate seeds were collected and grinded to extract the juice. The extract diluted with DI water was boiled for 20 min and filtered before use. Anisotropic AgNPs were synthesized by varying the concentration of AgNO$_3$. CTAB was used as stabilizer and pomegranate juice served as the reducing agent.

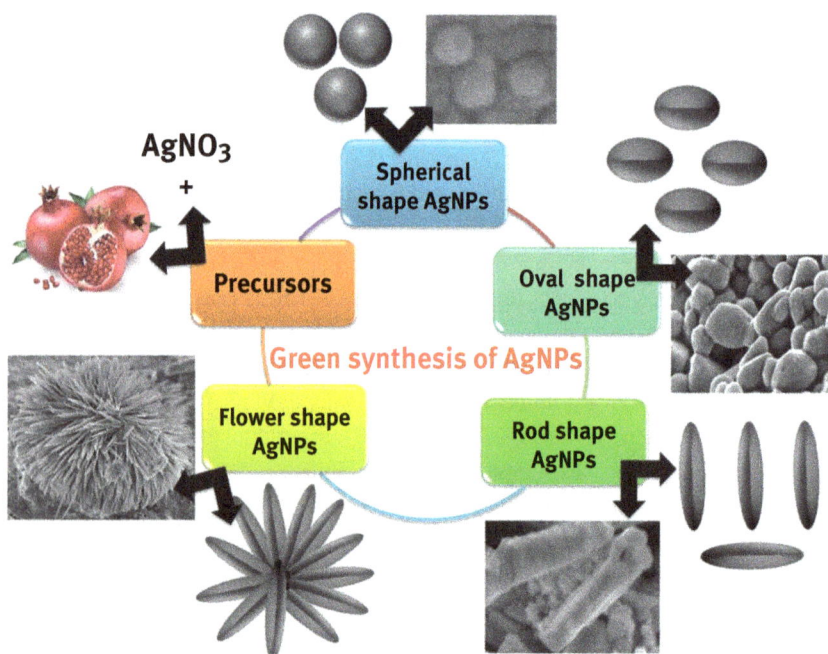

Figure 1.42: Synthesis of shape-controlled AgNPs using pomegranate extract in a microwave -assisted synthesis process (Reprinted with permission from ref. [254]. Copyright 2015, Royal Society of Chemistry).

For synthesis of spherical AgNPs, CTAB (1.82 g) followed by the pomegranate extract (3 mL) was added to an aqueous solution of AgNO$_3$ (50 mL, 0.01% w/v) and the mixture was exposed to microwave irradiation (300 W) for 5 min. Appearance of brown colour indicated the formation of AgNPs. Oval-shaped silver NPs (O-AgNPs) were synthesized through a seed-mediated approach. A seed solution for O-AgNPs was prepared by mixing 1.0 mL chilled pomegranate seed extract into a mixture containing AgNO$_3$ (10 mL, 0.25 mM) and TSC (10 mL, 0.25 mM) and well mixed for 2 h. The growth solution consisted of CTAB (7.0 mL, 0.020 M), AgNO$_3$ (0.50 mL, 10.0 mM) and pomegranate juice (0.50 mL). The seed solution (0.50 mL) was added to the growth solution and the solution was incubated at 27°C for 12 h to allow formation of O-AgNPs. Rod-shaped and flower-shaped AgNPs (R-AgNPs and F-AgNPs, respectively) were also synthesized using pomegranate juice. The growth solution for R-AgNP synthesis was prepared by mixing CTAB (10 mL, 0.2 M), AgNO$_3$ (1.2 mL, 1 mM) and distilled water (25 mL). Later, pomegranate juice (0.5 mL) and O-AgNPs (20 mL) were added to the growth solution and the solution was incubated in a water bath at 30°C for 24 h. The growth solution for F-AgNP was prepared by adding CTAB (5 mL, 0.2 M) and AgNO$_3$ (3 mL, 4 mM) to distilled water (33.0 mL). Synthesis of F-AgNPs was initiated by adding pomegranate juice (8.0 mL) and R-AgNPs (10 mL) to the growth solution followed by incubation in water bath at 30°C for 24 h.

In the spherical-and oval-shaped AgNPs synthesized, ~85–90% of the AgNPs were spherical and oval, respectively (Figure 1.43A–B). TEM image of the rod shaped AgNPs indicated diameter of the rods as ~100 nm (Figure 1.43C).

A large flower-shaped AgNP with average size 550 nm is illustrated in Figure 1.43E. The flower-shaped AgNPs demonstrated a very dark contrast image indicating presence of rich three dimensional (3D) structures.

1.3.3 Safety considerations

Synthesis of nanomaterials requires various tools and techniques, which needs to be handled carefully. For instance, synthesis using physical approaches uses laser beams and high temperature processes. For each of these techniques, special guidelines need to be followed during their use for AgNP synthesis. Exposure to laser beam may cause injury. The extent of damage would depend on wavelength of the laser beam, beam divergence and duration and frequency of exposure. Risks associated with direct laser beam exposure primarily include possible eye damage and tissue burns. Exposure of a laser beam may cause eye damage due to thermal burn which may permanently damage the retina. Upon exposure to laser beam, the laser light may get absorbed by the eye lens and this may cause immediate thermal burns depending on the level of exposure. The secondary effects of laser, i. e., non-beam hazards include laser generated airborne contaminants, electrical damage and system failures. In case of exposure to light with wavelength below 400 nm, the retina is not affected. For UV laser light, a major portion of the light is absorbed by the lens in the eye. This may cause development of cataracts depending on the level and frequency of exposure over a period of time. Other parts of the eye, such as, cornea and the conjunctive tissue can also be impaired by laser light [255]. Hence, extra precaution should be taken while handling the laser beams. The operator must wear protective masks for eyes and skin.

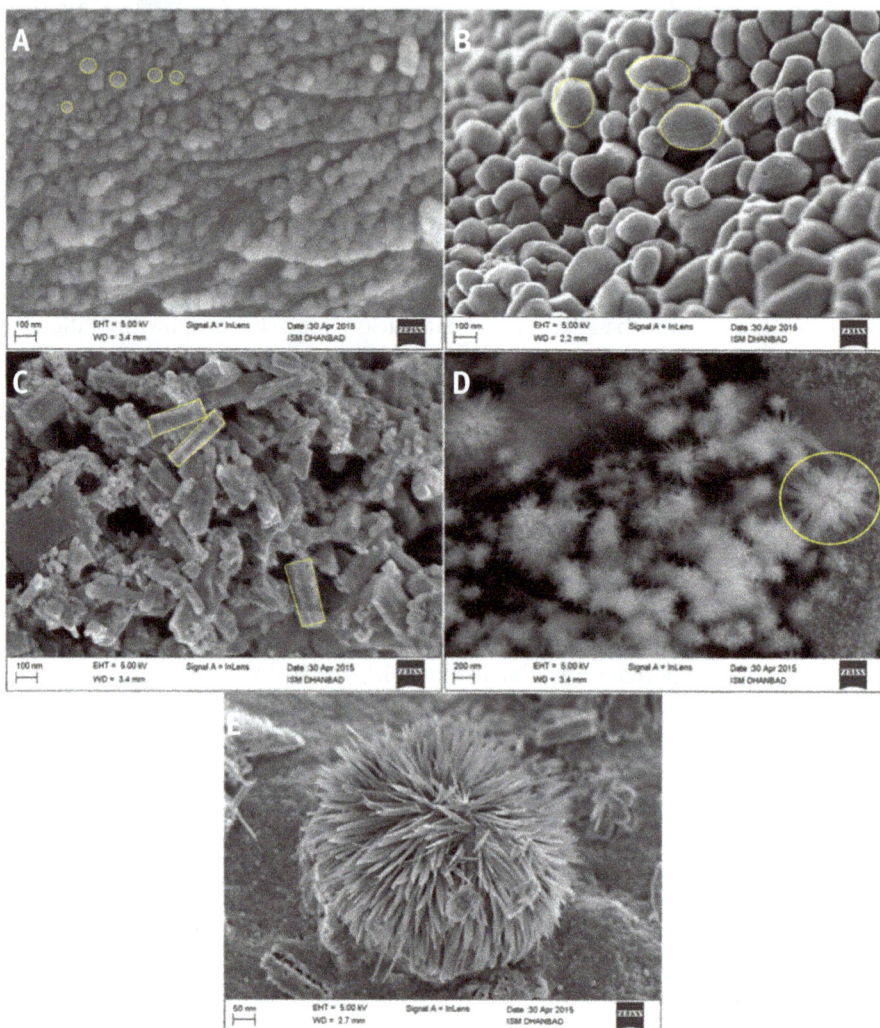

Figure 1.43: SEM images of the (A) spherical-, (B) oval-, (C) rod- and (D, E) flower-shaped AgNPs formed using pomegranate juice extract (Reprinted with permission from ref. [254]. Copyright 2015, Royal Society of Chemistry).

Several of the chemicals used for synthesis of size- and shape-controlled AgNPs needs to be handled with care. The metal precursor salt commonly used for AgNP synthesis, $AgNO_3$, is toxic, corrosive and a strong oxidizer. Moreover, $AgNO_3$ is light sensitive and should be stored in dark or in amber-coloured glass bottles. Exposure to $AgNO_3$ may cause eye burns and its absorption through the skin may give rise to adverse effects. Ingestion of silver salts is also associated with the adverse effect on gastrointestinal tract. Ingestion of silver salts may lead to argyria [256], which is characterized by permanent blue-greyish pigmentation of the skin, mucous membranes, and eyes. Exposure to silver salts through inhalation may lead to severe irritation of upper respiratory tract with coughing, burns, breathing difficulty, and may also cause coma. Chronic effect of silver salts includes methemoglobinemia, which is associated with symptoms, such as, headache, weakness, dizziness, shortness of breath, cyanosis and rapid heart rate. It may lead to unconsciousness and may even cause death. Chronic exposure to silver salts over long periods may also cause argyria. Argyria caused by accumulation of silver within the body is characterized by a permanent blue-grey discoloration of the skin, eyes, mucous membranes, and even internal organs.

NaBH$_4$ commonly used as reducing agent is a corrosive and flammable solid that may release hydrogen gas. Solutions of NaBH$_4$ should not be kept in tightly sealed vials to avoid possible explosions. TSC is slightly hazardous and may cause irritation to the skin, eyes, and respiratory tract. Hydrogen peroxide (H$_2$O$_2$) used in some processes for AgNP synthesis is corrosive and needs to be handled with care. It is an eye, skin and lung irritant and is known to cause burns. Exposure through liquid or spray mist may cause tissue damage to mucous membranes, particularly in the eyes, mouth and respiratory tract. Direct contact may cause skin burns. Inhalation of the spray mist may produce severe irritation that may manifest through coughing, choking, or shortness of breath. Prolonged exposure may result in skin burns and ulcerations. Symptoms may include redness, watering, and itching of eyes due to inflammation; itching, scaling, reddening and blistering due to skin inflammation. Organ damage due to repeated or prolonged exposure to H$_2$O$_2$ is also reported.

During preparation of silver nanoprisms, the vials should not be capped tightly due to chances of release of gas from unreacted borohydride and hydrogen peroxide. However, typically most experiments involve use of dilute solutions, hence the risks are negligible or very less. However, in case of chemical spills on the skin, clothing or entry into the eyes, the exposed body parts or area should be thoroughly washed with water immediately for at least 15 min. Subsequently, medical care must be administered, if required. Splash-proof safety goggles should be used during the experiments involving hazardous chemicals and the experiments should be conducted within a fume hood. All waste, unused or contaminated solutions should be discarded following appropriate disposal norms as recommended. Moreover, the AgNPs formed should be stored and handled with adequate care and should be properly disposed after use since these NPs may have unique toxicological characteristics of their own. Leaching of AgNPs and direct discharge of the residual Ag containing solution into waterways may adversely affect aquatic life. This may cause damage to beneficial microorganism in the environment and may also lead to severe damage to various higher organisms due to their immunotoxic potential [257, 258].

1.4 Conclusions and future perspective

Over the last few decades, AgNPs have been used in diverse applications areas such as surface plasmonics, SERS, chemical sensing, biological sensing, pollutant removal through catalysis and antimicrobial applications. AgNPs have been studied extensively due to their unique chemical, physical, electronic, optical, magnetic and catalytic properties. These properties are strongly dependent on size and shape of AgNPs. AgNPs with various shapes and sizes can be obtained using physical, chemical, photochemical, biological and microwave -assisted synthesis routes. The shape of AgNPs may range from spherical to various anisotropic shapes, i. e., nanorods, nanowires, nanoprisms, nanoplates, nanocubes and nanodecahedra. This review discusses various synthesis routes for synthesizing shape- and size-controlled AgNPs. The role of each chemical in a formulation, i. e., precursor salt, reducing agent, stabilizing agent, other additives and their concentration, the order of addition and the reaction parameters, such as, reaction temperature, pH, time and energy source needs to be understood before initiating the synthesis of AgNPs with controlled size and shape. The perspective provided in this review will not only help in determining the options available for synthesis of AgNPs with the desired size and shape, it will also help in understanding how the AgNPs generated can be characterized. Various characterization techniques that can be employed for determining the size, shape, morphology and stability has also been discussed in this review.

Physical synthesis of AgNPs predominantly produces spherical AgNPs. Moreover, only some specific methods, such as, laser ablation allows synthesis of size-controlled AgNPs. The major limitation associated with physical synthesis approaches is the high-energy consumption. Using chemical methods, both size- and shape-controlled AgNPs can be produced through proper selection of reducing and stabilizing agent and appropriate concentration of these reagents with respect to the Ag precursor salt. Moreover, certain specific additives and/or etchants can also be added to synthesize anisotropic AgNPs. Seed-mediated AgNP synthesis also exhibits better potential for synthesis of size- and shape-controlled AgNPs. Photochemical method and microwave -assisted synthesis also offers good potential for synthesis of shape- and size-controlled NPs. For synthesis of anisotropic NPs, multiple synthesis steps may be required. Spherical AgNPs achieved in one step, may later be transformed to the desired step. For instance, Ag nanoprisms could be synthesized by irradiating Ag seed solution with light at a selected wavelength. Commonly, bipyramids, nanodiscs, nanorods and nanodecahedron synthesis involve two-step synthesis process. Ag seeds prepared in the first step are grown in the second step using an appropriate growth solution and by selecting appropriate wavelength of light for irradiation or by selecting appropriate duration of microwave irradiation. Although chemical, photochemical or microwave-based synthesis routes have been reported to produce shape- and size-controlled AgNPs efficiently, there is potential for chemical contamination in the AgNPs formed. Residual toxic/hazardous chemicals used during synthesis may remain associated with the AgNPs and this may restrict their use in some applications. Physical synthesis approach may overcome such limitations and generate pure AgNP suspension free of chemical contamination.

Other synthesis methods such as gamma irradiation, sonochemical, electrochemical method can also produce spherical AgNPs with different sizes. However, these methods are less suitable for shape-controlled synthesis of AgNPs. Although biological approach/green synthesis approach for AgNP synthesis is cheaper, environment friendly and easy to use, in most cases the AgNPs synthesized are spherical while in other cases random variation in shape is observed. Moreover, some of the microbial strains used for AgNP synthesis are reported to be pathogenic. In combination with other method, such as, microwave irradiation, biological methods may be utilized for shape-controlled synthesis. Further studies are needed to achieve better control over shape and size during AgNP synthesis using biological methods.

Abbreviations

2,7-DHN	2,7-DHN: 2,7- dihydroxy naphthalene
Ag HNPs	Hexagonal Ag Nanoplates
Ag_2S	Silver sulfide
AgBr	Silver bromide
AgCl	Sodium chloride
$AgNO_3$	Silver nitrate
AgNPs	Silver nanoparticles
AgNWs	Silver nanowires
BNAH	N-benzyl-1,4-dihydronicotinamide
BSPP	Bis(p-sulfonatophenyl) phenylphosphine dihydrate dipotassium
CF_3COOAg	trifluoroacetate
CTA^+	Hexadecyltrimethyl ammonium ions
CTAB	Cetyltrimethylammonium bromide
CTAT	Cetyltrimethylammonium tosylate
CuCl	Copper (I) chloride
$CuCl_2$	Copper(II) chloride
CVC	Chemical vapour condensation
DBE	Diethylene glycol monobutyl ether
DEE	Diethylene glycol monoethyl ether
DHE	Diethylene glycol monohexyl ether
DLS	Dynamic Light Scattering
DMAB	Dimethylaminoborane
DME	Diethylene glycol monomethyl ether
DMF	N,N-dimethylformamide
DPE	Diethylene glycol monopropyl ether
EDS	Energy-dispersive X-ray spectroscopy
EG	Ethylene glycol
$FeCl_3$	Iron(III) chloride
FWHM	Full width at half maximum
HCl	Hydrochloric acid
HNO_3	Nitric acid

HRTEM	High-resolution transmission electron microscopy
I-2959	Irgacure 2959
KBr	Potassium bromide
KCl	Potassium chloride
LED	Light Emission Diodes
MTPs	Multiply twinned particles
MW	Molecular weight
$N_2H_4 \cdot H_2O$	Hydrazine hydrate
Na_2S	Sodium sulfide
$NaBH_4$	Sodium borohydride
NaBr	Sodium bromide
NaCl	Sodium chloride
NaOH	Sodium hydroxide
NDs	Nanodecahedron
NMP	N-methyl pyrollidone
NMP	N-methylpyrrolidone
PAA	Poly(acrylic acid, sodium salt)
PAM	Polyacrylamide
PEG	Polyethylene glycol
PEO	Poly(ethylene oxide)
PPO	Poly(propylene oxide)
Pro K	Protease K
PSS	Poly(styrene sulfonate)/poly(sodium styrenesulphonate)/
PSSS	poly (sodium styrenesulphonate) (PSSS)
PVC	Physical vapour condensation
PVDF	Polyvinylidenefluoride
PVE	Poly(methyl vinyl ether)
PVP	Polyvinyl pyrrolidone
SAED	Selected area electron diffraction
$AgClO_4$	Silver perchlorate
SDBS	Sodium dodecyl benzene sulphonate
SDS	Sodium dodecyl sulphate
SDSN	Sodium dodecylsulfonate
SEM	Scanning electron microscopy
SERS	Surface emhanced Raman scattering
SnP	Sn(IV) tetra(N-methyl-4-pyridyl) porphyrin tetratosylate chloride
TA	Tannic acid
TGA	Thioglycolic acid
TSC	Tri-sodium citrate/sodium citrate
TX-100	Poly(oxyethylene isooctyl phenyl ether
UV	Ultraviolet
XRD	X-ray diffraction

References

[1]   Chen X, Schluesener HJ. Nanosilver: A nanoproduct in medical application. Toxicol Lett. 2008;176:1–12.
[2]   Ahamed M, Alsalhi MS, Siddiqui MK. Silver nanoparticle applications and human health. Clin Chim Acta. 2010;411:1841–8.

[3]   Meng XK, Tang SC, Vongehr SA. Review on diverse silver nanostructures. J Mater Sci Technol. 2010;26:487–522.

[4]   Petryayeva E, Krull UJ. Localized surface plasmon resonance: Nanostructures, bioassays and biosensing-a review. Anal Chim Acta. 2011;706:8–24.

[5]   Wang C, Liu B, Dou X. Silver nanotriangles-loaded filter paper for ultrasensitive sers detection application benefited by interspacing of sharp edges. Sensors Actuators, B Chem. 2016;231:357–64.

[6]   Bharti S, Agnihotri S, Mukherji S, Mukherji S. Effectiveness of immobilized silver nanoparticles in inacivation of pathogenic bacteria. J Environ Res Dev. 2015;9:849–56.

[7]   Lu Y, Zhang CY, Zhang DJ, Hao R, Hao YW, Liu YQ. Fabrication of flower-like silver nanoparticles for surface-enhanced raman scattering. Chin Chem Lett. 2016;27:689–92.

[8]   Sharma VK, Yngard RA, Lin Y. Silver nanoparticles: Green synthesis and their antimicrobial activities. Adv Colloid Interface Sci. 2009;145:83–96.

[9]   Iravani S. Bacteria in nanoparticle synthesis: Current status and future prospects. Int Sch Res Not. 2014;2014:1–18. DOI: 10.1155/2014/359316.

[10]   Beyene HD, Werkneh AA, Bezabh HK, Ambaye TG. Synthesis paradigm and applications of silver nanoparticles (AgNPs), a review. Sustain Mater Technol. 2017;13:18–23.

[11]   Rauwel P, Rauwel E, Ferdov S, Singh MP. Silver nanoparticles: Synthesis, properties, and applications. Adv Mater Sci Eng. 2015;2015:2–4.

[12]   Swathy BA. Review on metallic silver nanoparticles. J Pharm. 2014;4:38–44.

[13]   Ruparelia JP, Chatterjee AK, Duttagupta SP, Mukherji S. Strain specificity in antimicrobial activity of silver and copper nanoparticles. Acta Biomater. 2008;4:707–16.

[14]   Khodashenas B, Ghorbani HR. Synthesis of silver nanoparticles with different shapes. Arab J Chem. 2015. https://doi.org/10.1016/j.arabjc.2014.12.014.

[15]   Sajanlal PR, Sreeprasad TS, Samal AK, Pradeep T. Anisotropic nanomaterials: structure, growth, assembly, and functions. Nano Rev. 2011;2:1–62.

[16]   Xia Y, Xiong Y, Lim B, Skrabalak SE. Shape-controlled synthesis of metal nanocrystals: Simple chemistry meets complex physics? Angew Chemie Int Ed. 2009;48:60–103.

[17]   Zeng H, Du XW, Singh SC, Kulinich SA, Yang S, He J, Cai W. Nanomaterials via laser ablation/ irradiation in liquid: A review. Adv Funct Mater. 2012;22:1333–53.

[18]   Saklani V, Suman JV. Microbial synthesis of silver nanoparticles: A review. J Biotechnol Biomater S. 2012;13:2.

[19]   Thakkar KN, Mhatre SS, Parikh RY. Biological synthesis of metallic nanoparticles. Nanomed Nanotechnol, Biol Med. 2010;6:257–62.

[20]   Sharma D, Kanchi S, Bisetty K. Biogenic synthesis of nanoparticles: A review. Arab J Chem. 2015. DOI: doi.org/10.1016/j.arabjc.2015.11.002.

[21]   Firdhouse MJ, Lalitha P. Biosynthesis of silver nanoparticles and its applications. J Nanotechnol. 2015;2015:1–18. https://doi.org/10.1155/2015/829526.

[22]   Singh R, Shedbalkar UU, Wadhwani SA, Chopade BA. Bacteriagenic silver nanoparticles: Synthesis, mechanism, and applications. Appl Microbiol Biotechnol. 2015;99:4579–93.

[23]   Ahmed S, Ahmad M, Swami BL, Ikram S. A review on plants extract mediated synthesis of silver nanoparticles for antimicrobial applications: A green expertise. J Advanced Res. 2016;7:17–28

[24]   Forough M, Farhadi K. Biological and green synthesis of silver nanoparticles *. Turkish J Eng Env Sci. 2010;34:281–7

[25]   Nadagouda MN, Speth TF, Varma RS. Microwave-assisted green synthesis of silver nanostructures. Acc Chem Res. 2011;44:469–78.

[26]   Zhang Y, Zhang R, Liu N, Yuan Z, Yang J, Zhu Y, Cao Y, Zhang A, Yang P. Synthesis of silver particles with various morphologies. J Inorg Organomet Polym Mater. 2012;22: 514–8.

[27] Natsuki JA. Review of silver nanoparticles: Synthesis methods, properties and applications. Int J Mater Sci Appl. 2015;4:325.

[28] Nurani SJ, Saha KC, Rahman Khan MA, Sunny SM. Silver nanoparticles synthesis, properties, applications and future perspectives: A short review. IOSR J Electr Electron Eng Ver I. 2015;10:117–26.

[29] Agnihotri S, Mukherji S, Mukherji S. Size-controlled silver nanoparticles synthesized over the range 5–100 Nm using the same protocol and their antibacterial efficacy. RSC Adv. 2014;4:3974–83.

[30] Desai R, Mankad V, Gupta SK, Jha PK. Size distribution of silver naoparticles: UV-visible spectroscopic assessment. Int J Nanosci. 2012;4:30–4.

[31] Bastys BV, Pastoriza-Santos I, Rodríguez-González B, Vaisnoras R, Liz-Marzán LM. Formation of silver nanoprisms with surface plasmons at communication wavelengths **. Adv Funct Mater. 2006;16:766–73

[32] Yang LC, Lai YS, Tsai CM, Kong YT, Lee CI, Huang CL. One-pot synthesis of monodispersed silver nanodecahedra with optimal SERS activities using seedless photo-assisted citrate reduction method. J Phys Chem C. 2012;116:24292–300.

[33] Xie Z-X, Tzeng W-C, Huang C-L. One-pot synthesis of icosahedral silver nanoparticles by using a photoassisted tartrate reduction method under UV light with a wavelength of 310 nm. Chem Phys Chem. 2016;17:2551–7

[34] Agnihotri S, Mukherji S, Mukherji S. Immobilized silver nanoparticles enhance contact killing and show highest efficacy: Elucidation of the mechanism of bactericidal action of silver. Nanoscale. 2013;5:7328–40.

[35] Eisa WH, Abdel-Moneam YK, Shabaka AA, Hosam AE. In situ approach induced growth of highly monodispersed Ag nanoparticles within free standing PVA/PVP films. Spectrochim Acta Part Mol Biomol Spectrosc. 2012;95:341–6.

[36] Prema P. Chemical mediated synthesis of silver nanoparticles and its potential antibacterial application. Progress in molecular and environmental bioengineering – From analysis and modeling to technology applications. Angelo Carpi (Ed.), ISBN: 978-953-307-268-5, InTech 2011

[37] Jamil Ahmed M, Murtaza G, Mehmood A, Mahmood Bhatti T. Green synthesis of silver nano-particles using leaves extract of *Skimmia laureola*: Characterization and antibacterial activity. Mater Lett. 2015;153:10–3.

[38] Kundu S, Wang K, Liang H. Size-controlled synthesis and self-assembly of silver nanoparticles within a minute using microwave irradiation. J Phys Chem. 2009;113:134–41.

[39] Bhattacharjee S. DLS and zeta potential – what they are and what they are not? J Control Release. 2016;235:337–51.

[40] Chakraborty S, Mukherji S, Mukherji S. Surface hydrophobicity of petroleum hydrocarbon degrading burkholderia strains and their interactions with NAPLs and surfaces. Colloids Surfaces B Biointerfaces. 2010;78:101–8.

[41] Saeb AT, Alshammari AS, Al-Brahim H, Al-Rubeaan KA. Production of silver nanoparticles with strong and stable antimicrobial activity against highly pathogenic and multidrug resistant bacteria. Sci World J. 2014;2014:9. https://doi.org/10.1155/2014/704708.

[42] Ju-Nam Y, Lead JR. Manufactured nanoparticles: An overview of their chemistry, interactions and potential environmental implications. Sci Total Environ. 2008;400:396–414.

[43] Abou El-Nour KM, Eftaiha A, Al-Warthan A, Ammar RA. Synthesis and applications of silver nanoparticles. Arab J Chem. 2010;3:135–40.

[44] Khayati GR, Janghorban K. The nanostructure evolution of Ag powder synthesized by high energy ball milling. Adv Powder Technol. 2012;23:393–7.

[45] Kumar N, Kumar R. Green synthesis of Ag nanoparticles in large quantity by cryomilling XE "cryomilling" †. RSC Adv. 2016;6:111380–8.

[46] Ghorbani HR, Safekordi AA, Attar H, Sorkhabadi SM. Biological and non-biological methods for silver nanoparticles synthesis. Chem Biochem Eng Q J. 2011;25:317–26.

[47] Tien DC, Liao CY, Huang JC, Tseng KH, Lung JK, Tsung TT, Kao WS, et al. Novel technique for preparing a nano-silver water suspension by the arc-discharge method. Rev Adv Mater Sci. 2008;18:752–8.

[48] Rashed HH. Silver nanoparticles prepared by electrical arc discharge method in DIW. J Enginnering Technol. 2016;34:295–301.

[49] Tien DC, Tseng KH, Liao CY, Huang JC, Tsung TT. Discovery of ionic silver in silver nanoparticle suspension fabricated by arc discharge method. J Alloys Compd. 2008;463:408–11.

[50] Tsuji T, Iryo K, Watanabe N, Tsuji M. Preparation of silver nanoparticles by laser ablation in solution: Influence of laser wavelength on particle size. Appl Surf Sci. 2002;202:80–5.

[51] Iravani S, Korbekandi H, Mirmohammadi SV, Zolfaghari B. Synthesis of silver nanoparticles: Chemical, physical and biological methods. Res Pharm Sci. 2014;9:385–406.

[52] Mafuné F, Kohno J, Takeda Y, Kondow T, Sawabe H. Formation and size control of silver nanoparticles by laser ablation in aqueous solution. J Phys Chem B. 2000;104:9111–7.

[53] Bae CH, Nam SH, Park SM. Formation of silver nanoparticles by laser ablation of a silver target in NaCl solution. Appl Surf Sci. 2002;197–198:628–34.

[54] Amendola V, Polizzi S, Meneghetti M. Free silver nanoparticles synthesized by laser ablation in organic solvents and their easy functionalization. Langmuir. 2007;23:6766–70.

[55] Sebastian S, Linslal CL, Vallabhan CP, Nampoori VP, Radhakrishnan P, Kailasnath M. Laser induced augmentation of silver nanospheres to nanowires in ethanol fostered by poly vinyl pyrrolidone. Appl Surf Sci. 2014;320:732–5.

[56] Prabhu S, Poulose EK. Silver nanoparticles: Mechanism of antimicrobial action, synthesis, medical applications, and toxicity effects. Int Nano Lett. 2012;2:32.

[57] Harada M, Katagiri E. Mechanism of silver particle formation during photoreduction using in situ time-resolved saxs analysis. Langmuir. 2010;26:17896–905.

[58] Thanh NT, Maclean N, Mahiddine S. Mechanisms of nucleation and growth of nanoparticles in solution. Chem Rev. 2014;114:7610–30.

[59] De Gusseme B, Sintubin L, Baert L, Thibo E, Hennebel T, Vermeulen G, Uyttendaele M, Verstraete W, Boon N. Biogenic silver for disinfection of water contaminated with viruses. Appl Environ Microbiol. 2010;76:1082–7.

[60] Pantidos N, Horsfall LE. Biological synthesis of metallic nanoparticles by bacteria, fungi and plants. J Nanomed Nanotechnol. 2014;5:10.

[61] Kc S, Sm L, Ts P, Bs L. Preparation of colloidal silver nanoparticles by chemical reduction method. Korean J Chem Eng. 2009;26:153–5.

[62] Van Dong Pham, Ha Chu Hoang, Binh Le Tran, Kasbohm Jörn. Chemical synthesis and anti-bacterial activity of novel-shaped silver nanoparticles. International Nano Letters. 2012;2:9. https://doi.org/10.1186/2228-5326-2-9.

[63] Khaydarov RA, Khaydarov RR, Gapurova O, Estrin Y, Scheper T. Electrochemical method for the synthesis of silver nanoparticles. J Nanoparticle Res. 2008;11:1193–200.

[64] Huang HH, Ni XP, Loy GL, Chew CH, Tan KL, Loh FC, Deng JF, Xu GQ. Photochemical formation of silver nanoparticles in poly(N -Vinylpyrrolidone). Langmuir. 1996;12:909–12.

[65] Long D, Wu G, Chen S. Preparation of oligochitosan stabilized silver nanoparticles by gamma irradiation. Radiat Phys Chem. 2007;76:1126–31.

[66] Shirtcliffe N, Nickel U, Schneider S. Reproducible preparation of silver sols with small particle size using borohydride reduction: For use as nuclei for preparation of larger particles. J Colloid Interface Sci. 1999;211:122–9.

[67] Jana NR, Gearheart L, Murphy CJ. Wet chemical synthesis of silver nanorods and nanowires of controllable aspect ratio. Chem Commun. 2001;7:617–8. DOI: 10.1039/b100521i.

[68] Sun YG, Xia YN. Shape-controlled synthesis of gold and silver nanoparticles. Science (80-). 2002;298:2176–9.

[69] Caswell KK, Bender CM, Murphy CJ. Seedless, surfactantless wet chemical synthesis of silver nanowires. Nano Lett. 2003;3:667–9.

[70] Jin R. Photoinduced conversion of silver nanospheres to nanoprisms. Science (80-). 2001;294:1901–3.

[71] Xiong Y, Siekkinen AR, Wang J, Yin Y, Kim MJ, Xia Y. Synthesis of silver nanoplates at high yields by slowing down the polyol reduction of silver nitrate with polyacrylamide. J Mater Chem. 2007;17:2600.

[72] Aherne D, Ledwith DM, Gara M, Kelly JM. Optical properties and growth aspects of silver nanoprisms produced by a highly reproducible and rapid synthesis at room temperature. Adv Funct Mater. 2008;18:2005–16.

[73] Potara M, Gabudean A-M, Astilean S. Solution-phase, dual LSPR-SERS plasmonic sensors of high sensitivity and stability based on chitosan-coated anisotropic silver nanoparticles. J Mater Chem. 2011;21:3625.

[74] Rao F, Jia F, Lopez-Miranda A, Song S, Lopez-Valdivieso A. Synthesis and characterization of silver nanorods in aqueous sodium dodecylsulfate solutions. J Dispers Sci Technol. 2012;33:799–804.

[75] Sujitha P, Production H. Silver nanoparticles: synthesis, properties, toxicology, applications and perspectives.

[76] Zeng Q, Jiang X, Yu A, Lu GM. Growth mechanisms of silver nanoparticles: A molecular dynamics study. Nanotechnology. 2007;18:35708.

[77] Zhao T, Sun R, Yu S, Zhang Z, Zhou L, Huang H, Du R. Size-controlled preparation of silver nanoparticles by a modified polyol method. Colloids Surfaces Physicochem Eng Asp. 2010;366:197–202.

[78] Leng Z, Wu D, Yang Q, Zeng S, Xia W. Facile and one-step liquid phase synthesis of uniform silver nanoparticles reduction by ethylene glycol. Opt Int J Light Electron Opt. 2018;154:33–40.

[79] Wiley B, Sun YG, Xia YN. Synthesis of silver nanostructures with controlled shapes and properties. Acc Chem Res. 2007;40:1067–76.

[80] Chen D, Qiao X, Qiu X, Chen J, Jiang R. Convenient synthesis of silver nanowires with adjustable diameters via a solvothermal method. J Colloid Interface Sci. 2010;344:286–91.

[81] Siekkinen AR, McLellan JM, Chen J, Xia Y. Rapid synthesis of small silver nanocubes by mediating polyol reduction with a trace amount of sodium sulfide or sodium hydrosulfide. Chem Phys Lett. 2006;432:491–6.

[82] Guo S, Dong S, Wang E. Rectangular silver nanorods: Controlled preparation, liquid-liquid interface assembly, and application in surface-enhanced raman scattering. Cryst Growth Des. 2009;9:372–7.

[83] Korte KE, Skrabalak SE, Xia Y. Rapid synthesis of silver nanowires through a CuCl- or CuCl2-mediated polyol process. J Mater Chem. 2008;18:437–41.

[84] Chang Y-H, Lu Y-C, Chou K-S. Diameter control of silver nanowires by chloride ions and its application as transparent conductive coating. Chem Lett. 2011;40:1352–3.

[85] Zhang P, Wei Y, Ou M, Huang Z, Lin S, Tu Y, Hu J. Behind the role of bromide ions in the synthesis of ultrathin silver nanowires. Mater Lett. 2018;213:23–6.

[86] Dong X, Ji X, Wu H, Zhao L, Li J, Yang W. Shape control of silver nanoparticles by stepwise citrate reduction. J Phys Chem C. 2009;113:6573–6.

[87] Métraux GS, Mirkin CA. Rapid thermal synthesis of silver nanoprisms with chemically tailorable thickness. Adv Mater. 2005;17:412–5.

[88] Si G, Shi W, Li K, Ma Z. Synthesis of PSS-capped triangular silver nanoplates with tunable SPR. Colloids Surfaces Physicochem Eng Asp. 2011;380:257–60.

[89] Quaresma P, Soares L, Contar L, Miranda A, Osório I, Carvalho PA, Franco R, Pereira E. Green photocatalytic synthesis of stable Au and Ag nanoparticles. Green Chem. 2009;11:1889.

[90] Langille MR, Personick ML, Mirkin CA. Plasmon-mediated syntheses of metallic nanostructures. Angew Chemie Int Ed. 2013;52:13910–40.

[91] Maillard M, Huang P, Brus L. Silver nanodisk growth by surface plasmon enhanced photoreduction of adsorbed [Ag +]. Nano Lett. 2003;3:1611–5.

[92] Xue C, Métraux GS, Millstone JE, Mirkin CA. Mechanistic study of photomediated triangular silver nanoprism growth. J Am Chem Soc. 2008;130:8337–44.

[93] Jin R, Cao YC, Hao E, Métraux GS, Schatz GC, Mirkin CA. Controlling anisotropic nanoparticle growth through plasmon excitation. Nature. 2003;425:487–90.

[94] Pietrobon B, Kitaev V. Photochemical synthesis of monodisperse size-controlled silver decahedral nanoparticles and their remarkable optical properties. Chem Mater. 2008;20:5186–90.

[95] Komarneni S, Li D, Newalkar B, Katsuki H, Bhalla AS. Microwave – polyol process for Pt and Ag nanoparticles. Langmuir. 2002;18:5959–62.

[96] Collins AG, Mitra S, Pavlostathis SG. Microwave heating for sludge dewatering and drying. Res J Water Pollut Control Fed. 1991;63:921.

[97] Zhu J, Shi J, Pan Y, Liu X, Zhou L. Synthesis of uniform silver nanoparticles by a microwave method in polyethylene glycol with the assistant of polyvinylpyrrolidone. Wuhan Univ J Nat Sci. 2013;18:530–4.

[98] Navaladian S, Viswanathan B, Varadarajan TK, Viswanath RP. Microwave-assisted rapid synthesis of anisotropic Ag Nanoparticles by solid state transformation. Nanotechnology. 2008;19:045603

[99] Kalimuthu K, Suresh Babu R, Venkataraman D, Bilal M, Gurunathan S. Biosynthesis of silver nanocrystals by *Bacillus licheniformis*. Colloids Surfaces B Biointerfaces. 2008;65:150–3.

[100] Venkataraman D, Kalimuthu, K, Sureshbabu RKP, Sangiliyandi G. An Insight into the Bacterial Biogenesis of Silver Nanoparticles, Industrial Production and Scale-up. In: Metal Nanoparticles in Microbiology, Ed. Rai M. and Duran N (Eds.) Springer, 2011;17–36, ISBN 978-3-642-18312-6.

[101] Li G, He D, Qian Y, Guan B, Gao S, Cui Y, Yokoyama K, Wang L. Fungus-mediated green synthesis of silver nanoparticles using *Aspergillus terreus*. Int J Mol Sci. 2012;13:466–76.

[102] Kumar SA, Abyaneh MK, Gosavi SW, Kulkarni SK, Pasricha R, Ahmad A, Khan MI. Nitrate reductase-mediated synthesis of silver nanoparticles from AgNO3. Biotechnol Lett. 2007;29:439–45

[103] Lee PC, Meisel D. Adsorption and surface-enhanced raman of dyes on silver and gold sols. J Phys Chem. 1982;86:3391–5.

[104] Van Hyning DL, Zukoski CF. Formation mechanisms and aggregation behavior of borohydride reduced silver particles. Langmuir. 1998;14:7034–46.

[105] He B, Tan JJ, Liew KY, Liu H. Synthesis of size controlled Ag nanoparticles. J Mol Catal Chem. 2004;221:121–6.

[106] Guzmán MG, Dille J, Godet S. Synthesis of silver nanoparticles by chemical reduction method and their antibacterial activity. Int Sholarly Sci Reserarch Innov. 2008;2:91–8.

[107] Liang H, Wang W, Huang Y, Zhang S, Wei H. Controlled synthesis of uniform silver nanospheres Vol. 114. J. Phys. Chem. C, 2010:7427–31.

[108] Won HI, Nersisyan H, Won CW, Lee J-M, Hwang J-S. Preparation of porous silver particles using ammonium formate and its formation mechanism. Chem Eng J. 2010;156:459–64.

[109] Pyatenko A, Yamaguchi M, Suzuki M. Synthesis of spherical silver nanoparticles with controllable sizes in aqueous solutions. J Phys Chem C. 2007;111:7910–7.

[110] Ranoszek-Soliwoda K, Tomaszewska E, Socha E, Krzyczmonik P, Ignaczak A, Orlowski P, Krzyzowska M, Celichowski G, Grobelny J. The role of tannic acid and sodium citrate in the synthesis of silver nanoparticles. J Nanoparticle Res. 2017;19:273. DOI 10.1007/s11051-017-3973-9.

[111] Qin Y, Ji X, Jing J, Liu H, Wu H, Yang W. Colloids and surfaces a: Physicochemical and engineering aspects size control over spherical silver nanoparticles by ascorbic acid reduction. Colloids Surfaces Physicochem Eng Asp. 2010;372:172–6.

[112] Li K, Jia X, Tang A, Zhu X, Meng H, Wang Y. Preparation of spherical and triangular silver nanoparticles by a convenient method. Integr Ferroelectr. 2012;136:9–14.

[113] Shervani Z, Ikushima Y, Sato M. Morphology and size-controlled synthesis of silver nanoparticles in aqueous surfactant polymer solutions. Colloid and Polymer Science. 2008;286: 403–10

[114] Rivero PJ, Goicoechea J, Urrutia A, Arregui FJ. Effect of both protective and reducing agents in the synthesis of multicolor silver nanoparticles. Nanoscale Res Lett. 2013;8:101.

[115] Zhang S, Liu Q, Yang K. Formation of silver nanoparticles induced by ultrasonically treated sodium dodecyl benzene sulphonate. Micro Nano Lett. 2015;10:422–6.

[116] Dong X, Ji X, Jing J, Li M, Li J, Yang W. Synthesis of triangular silver nanoprisms by stepwise reduction of sodium borohydride and trisodium citrate. J Phys Chem C. 2010;114:2070–4.

[117] Haber J, Sokolov K. Synthesis of stable citrate-capped silver nanoprisms. Langmuir. 2017;33:10525–30.

[118] Washio I, Xiong Y, Yin Y, Xia Y. Reduction by the end groups of poly(vinyl pyrrolidone): A new and versatile route to the kinetically controlled synthesis of Ag triangular nanoplates. Adv Mater. 2006;18:1745–9.

[119] Kim MH, Lee -J-J, Lee J-B, Choi K-Y. Synthesis of silver nanoplates with controlled shapes by reducing silver nitrate with poly(vinyl pyrrolidone) in N-methylpyrrolidone. Cryst Eng Comm. 2013;15:4660.

[120] Mansouri SS, Ghader S. Experimental study on effect of different parameters on size and shape of triangular silver nanoparticles prepared by a simple and rapid method in aqueous solution. Arab J Chem. 2009;2:47–53.

[121] Liu M, Leng M, Yu C, Wang X, Wang C. Selective synthesis of hexagonal Ag nanoplates in a solution-phase chemical reduction process. Nano Res. 2010;3:843–51.

[122] Li K, Wu Q, Shan Y, Qiu S, Cui F, Lin Y, Chen Z, Guo C, Zheng T. Shape transformation of Ag nanospheres to triangular Ag nanoplates: Hydrogen peroxide is a magic reagent. Integr Ferroelectr. 2016;169:22–8.

[123] Le Guével X, Wang FY, Stranik O, Nooney R, Gubala V, McDonagh C, MacCraith BD. Synthesis, stabilization, and functionalization of silver nanoplates for biosensor applications. J Phys Chem C. 2009;113:16380–6.

[124] Lee CL, Syu CM, Chiou HP, Chen CH, Yang HL. High-yield, size-controlled synthesis of silver nanoplates and their applications as methanol-tolerant electrocatalysts in oxygen reduction reaction. Int J Hydrogen Energy. 2011;36:10502–12.

[125] Ni C, Hassan PA, Kaler EW. Structural characteristics and growth of pentagonal silver nanorods prepared by a surfactant method. Langmuir. 2005;21:3334–7.

[126] Wiley BJ, Chen Y, McLellan JM, Xiong Y, Li ZY, Ginger D, Xia Y. Synthesis and optical properties of silver nanobars and nanorice. Nano Lett. 2007;7:1032–6.

[127] Hu J-Q, Chen Q, Xie Z-X, Han G-B, Wang R-H, Ren B, Zhang Y, Yang Z-L, Tian Z-Q. A simple and effective route for the synthesis of crystalline silver nanorods and nanowires. Adv Funct Mater. 2004;14:183–9.

[128] Liu S, Yue J, Gedanken A. Synthesis of long silver nanowires from AgBr nanocrystals. Adv Mater. 2001;13:656–8.

[129] Sun Y, Xia Y. Large-scale synthesis of uniform silver nanowires through a soft, self-seeding, polyol process. Adv Mater. 2002;14:833–7.

[130] Zhang W, Chen P, Gao Q, Zhang Y, Tang Y. High-concentration preparation of silver nanowires: Restraining in situ nitric acidic etching by steel-assisted polyol method. Chem Mater. 2008;20:1699–704.

[131] Nekahi A, Marashi SP, Fatmesari DH. High yield polyol synthesis of round- and sharp-end silver nanowires with high aspect ratio. Mater Chem Phys. 2016;184:130–7.

[132] Mdluli PS, Revaprasadu N. An improved N, N-dimethylformamide and polyvinyl pyrrolidone approach for the synthesis of long silver nanowires. J Alloys Compd. 2009;469:519–22.

[133] Cong F-Z, Wei H, Tian X-R, Xu H-X. A facile synthesis of branched silver nanowire structures and its applications in surface-enhanced raman scattering. Front Phys. 2012;7:521–6.

[134] Sang HI, Yun TL, Wiley B, Xia Y. Large-scale synthesis of silver nanocubes: The role of HCl in promoting cube perfection and monodispersity. Angew Chemie Int Ed. 2005;44:2154–7.

[135] Zhang WC, Wu XL, Chen HT, Gao YJ, Zhu J, Huang GS, Chu PK. Self-organized formation of silver nanowires, nanocubes and bipyramids via a solvothermal method. Acta Mater. 2008;56:2508–13.

[136] Zhou S, Li J, Gilroy KD, Tao J, Zhu C, Yang X, Sun X, Xia Y. Facile synthesis of silver nanocubes with sharp corners and edges in an aqueous solution. ACS Nano. 2016;10:9861–70.

[137] Zaheer Z. Rafiuddin. Multi-branched flower-like silver nanoparticles: Preparation and characterization. Colloids Surfaces Physicochem Eng Asp. 2011;384:427–31.

[138] Yang J, Dennis RC, Sardar DK. Room-temperature synthesis of flowerlike Ag nanostructures consisting of single crystalline Ag nanoplates. Mater Res Bull. 2011;46:1080–4.

[139] Sakamoto M, Fujistuka M, Majima T. Light as a construction tool of metal nanoparticles: Synthesis and mechanism. J Photochem Photobiol C Photochem Rev. 2009;10:33–56.

[140] Stamplecoskie K, Scaiano J. Light emitting diode irradiation can control the morphology and optical properties of silver nanoparticles – supporting info. J Am Chem Soc. 2010;132:1825–7. DOI: 10.1021/ja910010b.

[141] Huang L, Zhai ML, Long DW, Peng J, Xu L, Wu GZ, Li JQ, Wei GS. UV-induced synthesis, characterization and formation mechanism of silver nanoparticles in alkalic carboxymethylated chitosan solution. J Nanoparticle Res. 2008;10:1193–202.

[142] Annadhasan M, SankarBabu VR, Naresh R, Umamaheswari K, Rajendiran N. A sunlight-induced rapid synthesis of silver nanoparticles using sodium salt of N-cholyl amino acids and its antimicrobial applications. Colloids Surfaces B Biointerfaces. 2012;96:14–21.

[143] Pu F, Ran X, Guan M, Huang Y, Ren J, Qu X. Biomolecule-templated photochemical synthesis of silver nanoparticles: Multiple readouts of localized surface plasmon resonance for pattern recognition. Nano Research. 2018;11:3213–21.

[144] Sudeep PK, Kamat PV. Photosensitized growth of silver nanoparticles under visible light irradiation: A mechanistic investigation. Chem Mater. 2005;17:5404–10.

[145] Zhang J, Langille MR, Mirkin CA. Photomediated synthesis of silver triangular bipyramids and prisms: The effect of pH and BSPP. J Am Chem Soc. 2010;132:12502–10.

[146] Saade J, De Araújo CB. Synthesis of silver nanoprisms: A photochemical approach using light emission diodes. Mater Chem Phys. 2014;148:1184–93.

[147] Wang H, Cui X, Guan W, Zheng X, Zhao H, Xue T, Zheng W. Synthesis of silver nanoprisms and nanodecahedra for plasmonic modulating surface-enhanced raman scattering. J Nanosci Nanotechnol. 2016;16:6829–36.

[148] Tang B, Sun L, Li J, Zhang M, Wang X. Sunlight-driven synthesis of anisotropic silver nanoparticles. Chem Eng J. 2015;260:99–106.

[149] Tanimoto H, Ohmura S, Maeda Y. Size-selective formation of hexagonal silver nanoprisms in silver citrate solution by monochromatic-visible-light irradiation. J Phys Chem C. 2012;116:15819–25.

[150] Zheng X, Peng Y, Lombardi JR, Cui X, Zheng W. Photochemical growth of silver nanoparticles with mixed-light irradiation. Colloid Polym Sci. 2016;294:911–6.

[151] Sun Y, Xia Y. Triangular nanoplates of silver: Synthesis, characterization, and use as sacrificial templates for generating triangular nanorings of gold. Adv Mater. 2003;15:695–9.

[152] Kim BH, Lee JS. One-pot photochemical synthesis of silver nanodisks using a conventional metal-halide lamp. Mater Chem Phys. 2015;149:678–85.

[153] Bera RK, Raj CR. A facile photochemical route for the synthesis of triangular Ag nanoplates and colorimetric sensing of h2o2. J Photochem Photobiol Chem. 2013;270:1–6.

[154] Zhang J, Langille MR, Mirkin CA. Synthesis of silver nanorods by low energy excitation of spherical plasmonic seeds. Nano Lett. 2011;11:2495–8.

[155] Murshid N, Keogh D, Kitaev V. Optimized synthetic protocols for preparation of versatile plasmonic platform based on silver nanoparticles with pentagonal symmetries. Part Part Syst Charact. 2014;31:178–89.

[156] Pietrobon B, McEachran M, Kitaev V. Synthesis of Size-controlled faceted pentagonal silver nanorods with tunable plasmonic properties and self-assembly of these nanorods. ACS Nano. 2009;3:21–6.

[157] Jakab A, Rosman C, Khalavka Y, Becker J, Trügler A, Hohenester U, Sönnichsen C. Highly sensitive plasmonic silver nanorods. ACS Nano. 2011;5:6880–5.

[158] Lu H, Zhang H, Yu X, Zeng S, Yong KT, Ho HP. Seed-mediated plasmon-driven regrowth of silver nanodecahedrons (NDs). Plasmonics. 2012;7:167–73.

[159] Cardoso-Ávila PE, Pichardo-Molina JL, Kumar KU, Arenas-Alatorre JA. Temperature and amino acid-assisted size- and morphology-controlled photochemical synthesis of silver decahedral nanoparticles. J Exp Nanosci. 2014;9:639–51.

[160] Cardoso-Avila PE, Pichardo-Molina JL, Krishna CM, Castro-Beltran R. Photochemical transformation of silver nanoparticles by combining blue and green irradiation. J Nanoparticle Res. 2015;17:160.

[161] Lee SW, Chang SH, Lai YS, Lin CC, Tsai CM, Lee YC, Chen JC, Huang CL. Effect of temperature on the growth of silver nanoparticles using plasmon-mediated method under the irradiation of green LEDs. Materials (Basel). 2014;7:7781–98.

[162] Ye S, Song J, Tian Y, Chen L, Wang D, Niu H, Qu J. Photochemically grown silver nanodecahedra with precise tuning of plasmonic resonance. Nanoscale. 2015;7:12706–12.

[163] Keunen R, Cathcart N, Kitaev V. Plasmon mediated shape and size selective synthesis of icosahedral silver nanoparticles via oxidative etching and their 1-D transformation to pentagonal pins. Nanoscale. 2014;6:8045–51.

[164] Chen P, Song L, Liu Y, Fang Y. Synthesis of silver nanoparticles by γ-ray irradiation in acetic water solution containing chitosan. Radiat Phys Chem. 2007;76:1165–8.

[165] Zheng X, Xu W, Corredor C, Xu S, An J, Zhao B, Lombardi JR. Laser-induced growth of monodisperse silver nanoparticles with tunable surface plasmon resonance properties and a wavelength self-limiting effect. J Phys Chem C. 2007;111:14962–7.

[166] Yoksan R, Chirachanchai S. Silver nanoparticles dispersing in chitosan solution: Preparation by γ-ray irradiation and their antimicrobial activities. Mater Chem Phys. 2009;115:296–302.

[167] Liu Y, Chen S, Zhong L, Wu G. Preparation of high-stable silver nanoparticle dispersion by using sodium alginate as a stabilizer under gamma radiation. Radiat Phys Chem. 2009;78:251–5.

[168] Salkar RA, Jeevanandam P, Aruna ST, Koltypin Y, Gedanken A. The sonochemical preparation of amorphous silver nanoparticles. J Mater Chem. 1999;9:1333–5.

[169] Zhu YP, Wang XK, Guo WL, Wang JG, Wang C. Sonochemical synthesis of silver nanorods by reduction of sliver nitrate in aqueous solution. Ultrason Sonochem. 2010;17:675–9.

[170] Kumar B, Smita K, Cumbal L, Debut A, Pathak RN. Sonochemical synthesis of silver nanoparticles using starch: A comparison. Bioinorganic Chemistry and Applications. 2014;2014:8.

[171] Zhu J, Liu S, Palchik O, Koltypin Y, Gedanken A. Shape-controlled synthesis of silver nanoparticles by pulse sonoelectrochemical methods. Langmuir. 2000;16:6396–9.

[172] Tang S, Meng X, Lu H, Zhu S. PVP-assisted sonoelectrochemical growth of silver nanostructures with various shapes. Mater Chem Phys. 2009;116:464–8.

[173] Rodríguez-Sánchez L, Blanco MC, López-Quintela MA. Electrochemical synthesis of silver nanoparticles. J Phys Chem B. 2000;104:9683–8.

[174] Johans C, Clohessy J, Fantini S, Kontturi K, Cunnane VJ. Electrosynthesis of polyphenylpyrrole coated silver particles at a liquid-liquid interface. Electrochem Commun. 2002;4:227–30.

[175] Reicha FM, Sarhan A, Abdel-Hamid MI, El-Sherbiny IM. Preparation of silver nanoparticles in the presence of chitosan by electrochemical method. Carbohydr Polym. 2012;89:236–44.

[176] Rabinal MK, Kalasad MN, Praveenkumar K, Bharadi VR, Bhikshavartimath AM. Electrochemical synthesis and optical properties of organically capped silver nanoparticles. J Alloys Compd. 2013;562:43–7.

[177] Yin B, Ma H, Wang S, Chen S. Electrochemical synthesis of silver nanoparticles under protection of poly(N -vinylpyrrolidone). J Phys Chem B. 2003;107:8898–904.

[178] Ma H, Yin B, Wang S, Jiao Y, Pan W, Huang S, Chen S, Meng F. Synthesis of silver and gold nanoparticles by a novel electrochemical method. Chem Phys Chem. 2004;5:68–75.

[179] Lim PY, Liu RS, She PL, Hung CF, Shih HC. Synthesis of Ag nanospheres particles in ethylene glycol by electrochemical-assisted polyol process. Chem Phys Lett. 2006;420:304–8.

[180] Yin H, Yamamoto T, Wada Y, Yanagida S. Large-scale and size-controlled synthesis of silver nanoparticles under microwave irradiation. Mater Chem Phys. 2004;83:66–70.

[181] Yamamoto T, Yin H, Wada Y, Kitamura T, Sakata T, Mori H, Yanagida S. Morphology-control in microwave-assisted synthesis of silver particles in aqueous solutions. Bull Chem Soc Jpn. 2004;77:757–61.

[182] Kundu S, Maheshwari V, Niu S, Saraf RF. Polyelectrolyte mediated scalable synthesis of highly stable silver nanocubes in less than a minute using microwave irradiation. Nanotechnology. 2008;19:5.

[183] Darmanin T, Nativo P, Gilliland D, Ceccone G, Pascual C, De Berardis B, Guittard F, Rossi F. Microwave-assisted synthesis of silver nanoprisms/nanoplates XE "nanoplates" using a "modified polyol process. Colloids Surfaces Physicochem Eng Asp. 2012;395:145–51.

[184] Dzido G, Markowski P, Małachowska-Jutsz A, Prusik K, Jarzębski AB. Rapid continuous micro-wave-assisted synthesis of silver nanoparticles to achieve very high productivity and full yield: From mechanistic study to optimal fabrication strategy. J Nanoparticle Res. 2015;17:27

[185] Klaus T, Joerger R, Olsson E, Granqvist C-G. Silver-based crystalline nanoparticles, microbially fabricated. Proc Natl Acad Sci. 1999;96:13611–4.

[186] Park TJ, Lee KG, Lee SY. Advances in microbial biosynthesis of metal nanoparticles. Appl Microbiol Biotechnol. 2016;100:521–34.

[187] Gurunathan S, Kalishwaralal K, Vaidyanathan R, Venkataraman D, Pandian SR, Muniyandi J, Hariharan N, Eom SH. Biosynthesis, purification and characterization of silver nanoparticles using *Escherichia coli*. Colloids Surfaces B Biointerfaces. 2009;74:328–35.

[188] Sintubin L, De Gusseme B, Van Der Meeren P, Pycke BF, Verstraete W, Boon N. The antibacterial activity of biogenic silver and its mode of action. Appl Microbiol Biotechnol. 2011;91:153–62.

[189] Ganesh Babu MM, Gunasekaran P. Production and structural characterization of crystalline silver nanoparticles from *Bacillus cereus* isolate. Colloids Surfaces B Biointerfaces. 2009;74:191–5.

[190] Saifuddin N, Wong CW, Yasumira AA. Rapid biosynthesis of silver nanoparticles using culture supernatant of bacteria with microwave irradiation. E-Journal Chem. 2009;6:61–70.

[191] Sharma S, Ahmad N, Prakash A, Singh VN, Ghosh AK, Mehta BR. Synthesis of crystalline Ag nanoparticles (AgNPs) from microorganisms. Mater Sci Appl. 2010;1:1–7.

[192] Ahmad A, Senapati S, Khan MI, Kumar R, Ramani R, Srinivas V, Sastry M. Intracellular synthesis of gold nanoparticles by a novel alkalotolerant actinomycete, *Rhodococcus* species. Nanotechnology. 2003;14:824–8.

[193] Samundeeswari A, Dhas SP, Nirmala J, John SP, Mukherjee A, Chandrasekaran N. Biosynthesis of silver nanoparticles using actinobacterium *Streptomyces albogriseolus* and its antibacterial activity. Biotechnol Appl Biochem. 2012;59:503–7.

[194] Thamilselvi V, Radha KV. Synthesis of silver nanoparticles from *Pseudomonas piutida* NCIM 2650 in silver nitrate supplemented growth medium and optimization using response surface methodology. Dig J Nanomater Biostructures. 2013;8:1101–11.

[195] Manivasagan P, Kang K-H, Kim DG, Kim S-K. Production of polysaccharide-based bioflocculant for the synthesis of silver nanoparticles by *Streptomyces* Sp. Int J Biol Macromol. 2015;77:159–67.

[196] El-Naggar NE, Abdelwahed NA. Application of statistical experimental design for optimization of silver nanoparticles biosynthesis by a nanofactory *Streptomyces viridochromogenes*. J Microbiol. 2014;52:53–63.

[197] Tsibakhashvili N, Kalabegishvili T, Gabunia V, Gintury E, Kuchava N, Bagdavadze N, Pataraya D, et al. Synthesis of silver nanoparticles using bacteria. Nano. 2010;4:179–82.

[198] Shanthi S, David Jayaseelan B, Velusamy P, Vijayakumar S, Chih CT, Vaseeharan B. Biosynthesis of silver nanoparticles using a probiotic *Bacillus licheniformis* dahb1 and their antibiofilm activity and toxicity effects in ceriodaphnia cornuta. Microb Pathog. 2016;93:70–7.

[199] Shahverdi AR, Minaeian S, Shahverdi HR, Jamalifar H, Nohi AA. Rapid synthesis of silver nanoparticles using culture supernatants of enterobacteria: A novel biological approach. Process Biochem. 2007;42:919–23.

[200] Faghri Zonooz N, Salouti M. Extracellular biosynthesis of silver nanoparticles using cell filtrate of *Streptomyces* Sp. ERI-3. Sci Iran. 2011;18:1631–5.

[201] Mohanta YK, Behera SK. Biosynthesis, characterization and antimicrobial activity of silver nanoparticles by *Streptomyces* Sp. SS2. Bioprocess Biosyst Eng. 2014;37:2263–9.

[202] Wei X, Luo M, Li W, Yang L, Liang X, Xu L, Kong P, Liu H. Synthesis of silver nanoparticles by solar irradiation of cell-free *Bacillus amyloliquefaciens* extracts and AgNO3. Bioresour Technol. 2012;103:273–8.

[203] Kalishwaralal K, Deepak V, Pandian RK, Kottaisamy M, BarathManiKanth S, Kartikeyan B, Gurunathan S. Biosynthesis of silver and gold nanoparticles using *Brevibacterium casei*. Colloids Surfaces B Biointerfaces. 2010;77:257–62.

[204] Priyadarshini S, Gopinath V, Meera Priyadarshini N, MubarakAli D, Velusamy P. Synthesis of anisotropic silver nanoparticles using novel strain, *Bacillus flexus* and its biomedical application. Colloids Surfaces B-Biointerfaces. 2013;102:232–7.

[205] Xie J, Lee JY, Wang DI, Ting YP. Identification of Active biomolecules in the high-yield synthesis of single-crystalline gold nanoplates in algal solutions. Small. 2007;3:672–82.

[206] Mukherjee S, Chowdhury D, Kotcherlakota R, Patra S, Vinothkumar B, Bhadra MP, Sreedhar B, Patra CR. Potential theranostics application of bio-synthesized silver nanoparticles (4-in-1 system). Theranostics. 2014;4:316–35.

[207] Ahmad A, Mukherjee P, Senapati S, Mandal D, Khan MI, Kumar R, Sastry M. Extracellular biosynthesis of silver nanoparticles using the fungus *Fusarium oxysporum*. Colloids Surfaces B-Biointerfaces. 2003;28:313–8.

[208] Bhainsa KC, D'Souza SF. Extracellular biosynthesis of silver nanoparticles using the fungus *Aspergillus fumigatus*. Colloids Surfaces B Biointerfaces. 2006;47:160–4.

[209] Vahabi K, Mansoori GA, Karimi S. Biosynthesis of silver nanoparticles by fungus *Trichoderma reesei* (A route for large-scale production of AgNPs). Insciences J. 2011;1:65–79.

[210] Abd El-Aziz AR, Al-Othman MR, Alsohaibani SA, Mahmoud MA, Sayed SR. Extracellular biosynthesis and characterization of silver nanoparticles using *Aspergillus niger* isolated from Saudi Arabia (Strain KSU-12). Dig J Nanomater Biostructures. 2012;7:1491–9.

[211] Fayaz AM, Balaji K, Girilal M, Yadav R, Kalaichelvan PT, Venketesan R. Biogenic synthesis of silver nanoparticles and their synergistic effect with antibiotics: A study against gram-positive and gram-negative bacteria. Nanomed Nanotechnol, Biol Med. 2010;6:103–9.

[212] Basavaraja S, Balaji SD, Lagashetty A, Rajasab AH, Venkataraman A. Extracellular biosynthesis of silver nanoparticles using the fungus *Fusarium semitectum*. Mater Res Bull. 2008;43:1164–70.

[213] Chan YS, Don MM. Optimization of process variables for the synthesis of silver nanoparticles by *Pycnoporus sanguineus* using statistical experimental design. J Korean Soc Appl Biol Chem. 2013;56:11–20.

[214] Goswami AM, Sarkar TS, Ghosh S. An ecofriendly synthesis of silver nano-bioconjugates by *Penicillium citrinum* (MTCC9999) and its antimicrobial effect. AMB Express. 2013;3:16.

[215] Abdel-Hafez SI, Nafady NA, Abdel-Rahim IR, Shaltout AM, Mohamed MA. Biogenesis and optimisation of silver nanoparticles by the endophytic fungus *Cladosporium sphaerospermum*. Int J Nanomater Chem. 2016;2:11–9.

[216] Kowshik M, Ashtaputre S, Kharrazi S, Vogel W, Urban J, Kulkarni SK, Paknikar KM. Extracellular synthesis of silver nanoparticles by a silver-tolerant yeast strain MKY3. Nanotechnology. 2002;14:95–100.

[217] Vigneshwaran N, Kathe AA, Varadarajan PV, Nachane RP, Balasubramanya RH. Biomimetics of silver nanoparticles by white rot fungus, *Phaenerochaete chrysosporium*. Colloids Surfaces B Biointerfaces. 2006;53:55–9.

[218] Mukherjee P, Roy M, Mandal BP, Dey GK, Mukherjee PK, Ghatak J, Tyagi AK, Kale SP. Green synthesis of highly stabilized nanocrystalline silver particles by a non-pathogenic and agri-culturally important fungus *T. asperellum*. Nanotechnology. 2008;19:75103.

[219] Mukherjee P, Ahmad A, Mandal D, Senapati S, Sainkar SR, Khan MI, Parishcha R, et al. Fungus mediated synthesis of silver nanoparticles and their immobilization in the mycelial matrix: A novel biological approach to nanoparticle synthesis. Nano Lett. 2001;1:515–9.

[220] Chen JC, Lin ZH, Ma XX. Evidence of the production of silver nanoparticles via pretreatment of *Phoma* sp.3.2883 with silver nitrate. Lett Appl Microbiol. 2003;37:105–8.

[221] Sharma A, Sharma S, Sharma K, Chetri SP, Vashishtha A, Singh P, Kumar R, Rathi B, Agrawal V. Algae as crucial organisms in advancing nanotechnology: A systematic review. J Appl Phycol. 2016;28:1759–74.

[222] Jena J, Pradhan N, Dash B P, Sukla L B, Panda P K, et al. Biosynthesis and Characterization of Silver Nanoparticles using Microalga *Chlorococcum humicola* and its Antibacterial Activity. International Journal of Nanomaterials and Biostructures. 2013;3:1–8.

[223] Vincy W, Mahathalana TJ, Sukumaran S. Algae as a sources for synthesis of nanoparticles-a review. Int J Latest Trends Eng Technol. 2017;5–9.

[224] Aziz N, Faraz M, Pandey R, Shakir M, Fatma T, Varma A, Barman I, Prasad R. Facile algae-derived route to biogenic silver nanoparticles: Synthesis, antibacterial, and photocatalytic properties. Langmuir. 2015;31:11605–12.

[225] Barwal I, Ranjan P, Kateriya S, Yadav SC. Cellular oxido-reductive proteins of *Chlamydomonas reinhardtii* control the biosynthesis of silver nanoparticles. J Nanobiotechnol. 2011;9:1–12.

[226] Karemore A, Pal R, Sen R. Strategic enhancement of algal biomass and lipid in *Chlorococcum infusionum* as bioenergy feedstock. Algal Res. 2013;2:113–21.

[227] Dhanalakshmi PK, Azeez R, Rekha R, Poonkodi S, Nallamuthu T. Synthesis of silver nanopar-ticles using green and brown seaweeds. Phykos. 2012;42:39–45.

[228] Mohandass C, Vijayaraj AS, Rajasabapathy R, Satheeshbabu S, Rao SV, Shiva C. Biosynthesis of silver nanoparticles from marine seaweed *Sargassum cinereum* and their antibacterial activity. Indian J Pharm Sci. 2013;75:606.

[229] Azizi S, Namvar F, Mahdavi M, Ahmad MB, Mohamad R. Biosynthesis of silver nanoparticles using brown marine macroalga, *Sargassum muticum* aqueous extract., 2013;6:5942–50

[230] Vijayan SR, Santhiyagu P, Singamuthu M, Ahila NK, Jayaraman R, Ethiraj K. Synthesis and characterization of silver and gold nanoparticles using aqueous extract of seaweed, *Turbinaria conoides*, and their antimicrofouling activity. Sci World J. 2014;2014:1–10.

[231] Vivek M, Kumar PS, Steffi S, Sudha S. Biogenic silver nanoparticles by *Gelidiella acerosa* extract and their antifungal effects. Med Biotech. 2011;3:143–8.

[232] Shukla MK, Singh RP, Reddy CR, Jha B. Synthesis and characterization of agar-based silver nanoparticles and nanocomposite film with antibacterial applications. Bioresour Technol. 2012;107:295–300.

[233] Lengke MF, Fleet ME, Southam G. Biosynthesis of silver nanoparticles by filamentous cyanobacteria from a silver (I) nitrate complex. Langmuir. 2007;23:2694–9.

[234] Ali DM, Sasikala M, Gunasekaran M, Thajuddin N. Biosynthesis and characterization of silver nanoparticles using marine cyanobacterium, *Oscilatoria willei* NTDM01. Dig J Nanomater Biostructures. 2011;6:385–90.

[235] Mahdieh M, Zolanvari A, Azimee AS, Mahdieh M. Green biosynthesis of silver nanoparticles by *Spirulina platensis*. Sci Iran. 2012;19:926–9.

[236] Sudha SS, Rajamanickam K, Rengaramanujam J. Microalgae mediated synthesis of silver nanoparticles and their antibacterial activity against pathogenic bacteria. Indian J Exp Biol. 2013;51:393–9.

[237] Bar H, Bhui DK, Sahoo GP, Sarkar P, De SP, Misra A. Green synthesis of silver nanoparticles using latex of *Jatropha curcas*. Colloids Surfaces Physicochem Eng Asp. 2009;339:134–9.

[238] Krishnaraj C, Jagan EG, Rajasekar S, Selvakumar P, Kalaichelvan PT, Mohan N. Synthesis of silver nanoparticles using *Acalypha indica* leaf extracts and its antibacterial activity against water borne pathogens. Colloids Surfaces B Biointerfaces. 2010;76:50–6.

[239] Logeswari P, Silambarasan S, Abraham J. Synthesis of silver nanoparticles using plants extract and analysis of their antimicrobial property. J Saudi Chem Soc. 2012;19:311–7.

[240] Ahmad N, Sharma S. Green synthesis of silver nanoparticles using extracts of *Ananas comosus*. Green Sustain Chem. 2012;2:141–7.

[241] Panigrahi T. Synthesis and characterization of silver nanoparticles using leaf extract of Azadirachta indica 2013.

[242] Mani U, Dhanasingh S, Arunachalam R, Paul E, Shanmugam P, Rose C, Mandal AB, Simple A. Green Method for the synthesis of silver nanoparticles using *Ricinus communis* leaf extract. Prog Nanotechnol Nanomater. 2013;2:21–5.

[243] Selvam K, Sudhakar C. Eco-friendly biosynthesis and characterization of silver nanoparticles using *Tinospora cordifolia* (Thunb.) miers and evaluate its antibacterial, antioxidant potential. J Radiat Res Appl Sci. 2016;10:6–12

[244] Logaranjan K, Raiza AJ, Gopinath SC, Chen Y, Pandian K. Shape- and size-controlled synthesis of silver nanoparticles using *Aloe vera* plant extract and their antimicrobial activity. Nanoscale Res Lett. 2016;11:520.

[245] Talib A, Hui-Fen-Wu. Biomimetic synthesis of lotus leaf extract-assisted silver nanoparticles and shape-directing role of cetyltrimethylammonium bromide. J Mol Liq. 2016;220:795–801.

[246] Song JY, Kim BS. Rapid biological synthesis of silver nanoparticles using plant leaf extracts. Bioprocess Biosyst Eng. 2009;32:79–84.

[247] Ashraf JM, Ansari MA, Khan HM, Alzohairy MA, Choi I. Green synthesis of silver nanoparticles and characterization of their inhibitory effects on AGEs formation using biophysical techniques. Sci Rep. 2016;6:20414.

[248] Singh P, Kim YJ, Zhang D, Yang DC. Biological synthesis of nanoparticles from plants and microorganisms. Trends Biotechnol. 2016;34:588–99

[249] Banerjee P, Satapathy M, Mukhopahayay A, Das P. Leaf extract mediated green synthesis of silver nanoparticles from widely available indian plants: Synthesis, characterization, antimicrobial property and toxicity analysis. Bioresour Bioprocess. 2014;1:1–10.

[250] Khalil MM, Ismail EH, El-Baghdady KZ, Mohamed D. Green synthesis of silver nanoparticles using olive leaf extract and its antibacterial activity. Arab J Chem. 2013;7:1131–9.

[251] Shaik S, Kummara MR, Poluru S, Allu C, Gooty JM, Kashayi CR, Chinna M, Subha S. A Green approach to synthesize silver nanoparticles in starch-co-poly (acrylamide) hydrogels by *Tridax procumbens* leaf extract and their antibacterial activity., 2013;2013.

[252] Swamy VS, Prasad RA. Green synthesis of silver nanoparticles from the leaf extract of *Santalum album* and its antimicrobial activity. J Pharm Sci Res. 2015;7:690–5.

[253] Manikprabhu D, Lingappa K. Microwave assisted rapid and green synthesis of silver nanoparticles using a pigment produced by *Streptomyces coelicolor* klmp33. Bioinorg Chem Appl. 2013;2013:1–5.

[254] Roy E, Patra S, Saha S, Madhuri R, Sharma PK. Shape-specific silver nanoparticles prepared by microwave-assisted green synthesis using pomegranate juice for bacterial inactivation and removal. RSC Adv. 2015;5:95433–42.

[255] Smalley PJ. Laser safety: Risks, hazards, and control measures. Laser Ther. 2011;20: 95–106.

[256] Hartemann Philippe, Hoet Peter, Proykova Ana, Farnandes Teresa, Baun Anders, Jong Wim De, Filser Juliane, Hensten Arne, Kneuer Carsten, Miallard Jean-Yves, Norppa Hannu, Scheringer Martin, Wijnhoven Susan. Scenihr. Nanosilver: Safety, health and environmental effects and role in antimicrobial resistance. Materials Today. 2015 4;18:122–3

[257] Schäfer B, Vom Brocke J, Epp A, Götz M, Herzberg F, Kneuer C, Sommer Y, et al. State of the art in human risk assessment of silver compounds in consumer products: A conference report on silver and nanosilver held at the BfR in 2012. Arch Toxicol. 2013;87: 2249–62.

[258] Batley GE, Kirby JK, McLaughlin MJ. Fate and risks of nanomaterials in aquatic and terrestrial environments. Acc Chem Res. 2013;46:854–62.

Bionotes

Dr Suparna Mukherji completed her B.Tech. in Energy Engineering from Indian Institute of Technology Kharagpur (Kharagpur, India); M.S. in Civil and Environmental Engineering from Clarkson University (Potsdam, USA); and Ph.D. in Environmental Engineering from University of Michigan, Ann Arbor, USA. She joined IIT Bombay in 1998 and is currently Professor in the Centre for Environmental Science and Engineering (CESE) at IIT Bombay. She received the DBT Women Bioscientist Award in 2009 and AICTE Career Award for Young Teachers in 2000. Her research interests include: Environmental application of nanomaterials; biodegradation & bioremediation and water and wastewater treatment.

Ms. Sharda Bharti completed her M. Tech. degree from the Centre for Environmental Science and Engineering (CESE), IIT Bombay in 2014. She is currently pursuing PhD at CESE, IIT Bombay under the supervision of Prof Suparna Mukherji and Prof Soumyo Mukherji. Her current research focuses on the synthesis of shape-controlled silver nanoparticles using various synthesis methods including chemical, photochemical and biological approaches, characterization and their application as an alternative antimicrobial agent for point-of-use water disinfection.

Ms. Gauri M. Shukla received her M. Tech degree from department of Bioscience and Bioengineering, IIT Bombay in 2010. Presently, she is pursuing PhD at Bioscience and Bioengineering, IIT Bombay under the supervision of Prof Soumyo Mukherji and Prof Tapanendu Kundu, Department of Physics, IIT Bombay. Her current research focuses on the synthesis of silver nanoparticles and their applications as an optical sensor.

Dr Soumyo Mukherji did his B.Tech. in Instrumentation Engineering, Indian Institute of Technology (Kharagpur), MS in Mechanical Engineering, Colorado State University (Fort Collins, USA) and Ph.D. in Biomedical Engineering, University of North Carolina (Chapel Hill, USA). After his PhD, he joined IIT Bombay in 1997, where he is now a Professor in the Department of Biosciences and Bioengineering. He was the Head of the Centre for Research in Nanotechnology and Sciences at IIT Bombay from 2010 to 2013. His research interests are in sensors and instruments for wide-scale deployment in resource-constrained locales for medical and environmental applications, mobile health, security, etc.

Pei Zhang, Shudong Lin and Jiwen Hu

2 Synthesis and characterization of size-controlled silver nanowires

Abstract: Silver nanowires (AgNWs) have attracted attentions form both academia and industry due to their outstanding electronic and optical properties. The AgNW-based devices for various uses were invented in recent years. It is well known that the sizes of AgNWs have a crucial effect on the performance of AgNW-based devices. However, how to synthesize AgNWs with controlled sizes is still unsolved. Researchers reported many methods to synthesize AgNWs with different sizes in the past decade. However, a review that focuses on the synthetic methods of AgNWs is very rare. The aim of this review is to summarize the recent developments that have been achieved with AgNWs, and many procedure details and results and discuss ions will be provided for practical use.

Graphical Abstract:

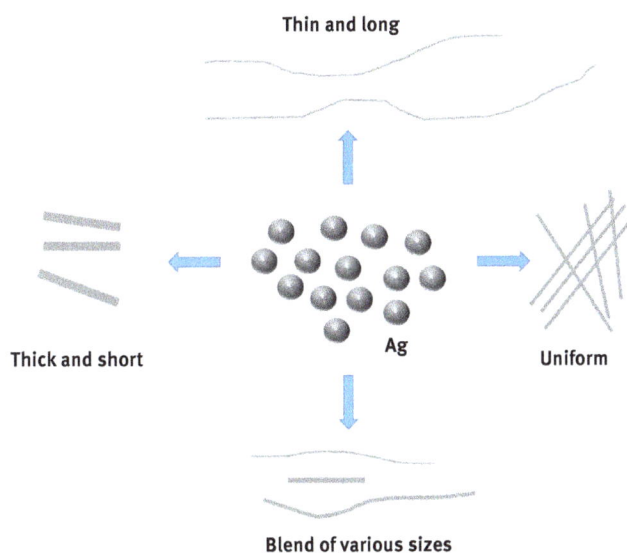

Keywords: silver nanowires, synthesis method, nanomaterial

This article has previously been published in the journal *Physical Sciences Reviews*. Please cite as: Zhang, P., Lin, S, Hu, J. Synthesis and characterization of size-controlled silver nanowires. *Physical Sciences Reviews* [Online] **2018**, 3. DOI: 10.1515/psr-2017-0084

https://doi.org/10.1515/9783110636666-002

2.1 Introduction

Silver nanowires (AgNWs) are one-dimensional silver nanostructures with diameters that are typically in a range of 10–200 nm, and lengths in a range of 5–100 μm. In the past decade, AgNWs have attracted many attentions due to their unique electronic and optical properties. The potential application in smart electronic devices makes AgNWs one of most promising nanomaterials in the twenty-first century. For example, the traditional electronic material in transparent films and glasses is tin-doped indium oxide; however, the brittleness and the manufacturing methods limit the use of indium tin oxide (ITO) in smart electronic devices. Finding the replacement of ITO is one of the hottest material topics in the past decade; conductive polymers, carbon nanotubes, metallic nanowires and graphene have been used to replace ITO, and among these candidates, AgNWs step ahead of the competition.

The large and extensive research of AgNWs started from 2002, when Murphy et al. reported a wet-chemistry synthesis method [1], and by using this method, uniform AgNWs were synthesized in high yield. Xia et al. proposed polyol method [2], which is more efficient than Murphy's method, and this method has been widely used by researchers. The morphologies of AgNWs are usually investigated by transmission electron microscopy (TEM) and scanning electron microscopy (SEM). From these microscopy images, we can measure the diameter and length of AgNWs, and the aspect ratio which is defined as the ratio of length to diameter. These three parameters are the most important physical parameters of AgNWs.

In the past decade, the application of AgNWs in various fields has made a rapid development, such as flexible transparent conductors [3, 4], flexible touch panel [5, 6], stretchable electronics [7] and stretchable energy devices [8]. Many researchers have proposed that the dimensions of the AgNWs, such as the diameter, length and aspect ratio, have a crucial influence on the performance of AgNW-based applications. For example, the nanowires with smaller diameter show better performance in the AgNW-based transparent films, but the nanowires with large diameter have a better mechanical property. The length of nanowires could be controlled by a successive multistep method [9]. How the size of AgNWs influences the electronic and optical properties was studied by many researchers. With the help of these research results, smart devices with high performance have been designed and manufactured. How to synthesis AgNWs with different sizes in high yield is a very important topic both in academic research fields and in industrial fields. Thanks to the contributions of many researchers, various sizes of AgNWs could be synthesized by different methods. These synthesis conditions reported by researchers provide a valuable reference to synthesize size-controlled AgNWs. Although some reviews of AgNWs have been published in the past decade, the review which focuses on the synthesis method and the size of AgNWs is rare.

In this paper, we will provide a brief description of different preparation methods of AgNWs in Chapter 2; for practical use, many synthesis details and results will be provided in this part. The characterization methods of AgNWs will be provided in Chapter 3, and the methods of sample preparation and the use of different instruments will be introduced in this part. To help readers to further understand the synthetic procedure, we will discuss more examples in Chapter 4, and we will discuss the typical examples from formula to results. In the last chapter of this paper, conclusions of this paper and the outlook of AgNWs will be provided.

2.2 Preparation methods

In the early stage, AgNWs were usually synthesized by electrochemical methods, but the AgNWs synthesized by this method were not uniform, and the low yield of this method is another drawback. Therefore, the study of AgNWs was in stagnation before a more efficient method was proposed. The wet-chemistry method avoided the disadvantages of electrochemical methods, and AgNWs synthesized by this method were uniform and in a relative high yield. In the past decade, this wet-chemistry method was modified by many researchers, and more and more simple and efficient methods were proposed. The wet-chemistry method can be divided into soft-template method and hard-template method based on the type of template the method exploit, and we will discuss these two methods in the following sections.

2.2.1 Hard-template method

The hard-template method was proposed before soft-template method. In this method, water and alcoholic solvents were used as solvent, and sodium borohydride, citrate, ascorbic acid and so on were used as reductant. Nanoporous membranes and carbon nanotubes are usually exploited as hard template, and the AgNWs can be grown within the pores of the membrane or the tubes of carbon nanotubes, and therefore, the diameter can be varied by tuning the sizes of the pores and the tubes. Martin's group proposed many different types of nanoporous membranes to use as hard templates, and AgNWs synthesized by this method have well-defined morphologies and relatively thin diameter. However, the purification process of this method is too complex, and this method was gradually abandoned by researchers; thus, the details of this method will not be discussed in this chapter.

2.2.2 Wet-chemistry method

Due to the complex purification process, hard-template method was eliminated by researchers; however, the soft templates overcome these shortcomings. Various kinds of surfactants, micelles, and many other polymers were used for soft templates. These soft templates can easily dissolve in the solvent, since most wet-chemistry methods were conducted in solvents.

The rod-like micelles and surfactants were first chosen as soft templates, and the researchers proposed that, the rod-like shape helps the AgNW anisotropic growth, and the soft templates were also called capping agents due to its capping function. Murphy's group reported a seed-mediated method which use cetyltrimethyl ammonium bromide (CTAB) as micellar soft template [1]. In this method, the silver seeds were first produced, the same amounts of silver nitrate ($AgNO_3$) and trisodium citrate were added into water solution, and the silver nitrate was reduced by sodium borohydride ($NaBH_4$) and the silver seeds with diameter 4 nm were finally generated. In the second step, the $AgNO_3$ and CTAB were added into the water solution, the silver ions were reduced by ascorbic acid in this step, the NaOH was also added into the solution to adjust the pH value. The reaction was conducted at room temperature for 15 min when a yellow colour appeared. The nanowires were purified by centrifugation process. The AgNWs synthesized by this method were short, only 1–4 μm long with an aspect ratio of 50–350. Although the length and aspect ratio of nanowires were low, the method first use micelles as templates, which is relatively easier to perform than hard-template method.

Other micelles and polymers were also used as soft templates by researchers, but the length of AgNWs synthesized by this method was less than 10 μm and was too short to be applied in many electronic devices. Based on the method, Xia's group proposed a new soft-template method which uses polyvinyl pyrrolidone (PVP) as capping agent; in this method, ethylene glycol (EG) acted both as solvent and as reductant, which is the reason why this method is also called polyol method. The polyol method has been the most successful route to produce AgNWs both at large scales and of high yield. Most of the researchers use a modified polyol method to synthesize AgNWs, and by tuning the parameters, different sizes of nanowires were generated.

Xia's group first reported the polyol method in 2002, which was previously used to synthesize other metallic nanostructures. Xia extended this method to the synthesis of AgNWs. PVP is a common dispersant, but in the synthesis of AgNWs, Xia found that PVP could direct the anisotropic growth of the wires. In the early stage, $PtCl_2$ was added in the first step [10], and the Pt nanoparticles were generated at 160 °C and acted as seeds. In the second step, the $AgNO_3$ and PVP were added at the same time, and the reaction mixture was heated at 160 °C for 1 h. Based on this method, Xia developed a self-seeding process [11], which did not require Pt seeds. In this self-seeding process, the $PtCl_2$ was eliminated, but a two-syringe pump was introduced to control the adding rate of $AgNO_3$ and PVP. The adding rate was crucial for the formation of AgNWs. The silver nanoparticles reduced by EG in the early stage act as seeds for the subsequent formation of AgNWs. The AgNWs synthesized by this method contains less level of impurities, but the use of syringe pump limits its application in a large scale. The AgNWs synthesized by these two methods have diameters in the range of 30–40 nm and lengths up to ~50 μm. Hu et al. reported a similar method which use AgCl as seeds [12]. In this method, AgCl is ground finely and added to the flask for initial nucleation, and the $AgNO_3$ was titrated in 10 min, the temperature was kept at 170 °C for 30 min, but the diameter of $AgNO_3$ was in a range of 40–100 nm.

Schuette et al. used AgCl nanocubes as seeds for the growth of AgNWs [13] and found the larger nanocubes promoted the growth of AgNWs with larger diameters. Based on that findings, the diameter of AgNWs was controlled by tuning the size of nanocubes. Xia et al. introduced Fe^{3+} into the method [14] and found that the rate of the polyol reduction was accelerated upon the addition of Fe^{3+}. The Fe^{3+} was reduced by EG, and then the Fe^{2+} reacted with the oxygen on the surface of silver nanostructures. The reaction mechanism Xia proposed is illustrated in Figure 2.1. A typical Fe-mediated polyol synthesis was conducted as follows. EG was first added in the flask and heated at 160 °C for 10 min under a light nitrogen flow. The NaCl and $FeCl_3$ were added into the flask, and then the $AgNO_3$ solution and PVP solution were titrated simultaneously by a pump; the reaction temperature was set at 148 °C and the reaction time was set at 1 h.

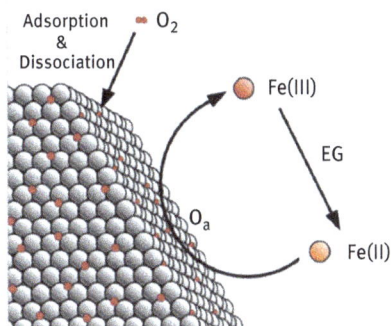

Figure 2.1: Illustration of the proposed mechanism by which Fe(II) removes atomic oxygen from the surface of silver nanostructures. Reduction by ethylene glycol (EG) competes with oxidation by atomic oxygen to form an equilibrium between Fe(III) and Fe(II) (The figure 2.1 was reprinted with permission from Ref.[14], Copyright 2005 American Chemistry Society).

Xia's group introduced copper ions into the method [15] and found the role of copper ions to be similar to iron ions; the existence of copper ions can remove the oxygen on the silver surface, thus facilitating the growth of nanowires. They also found that the addition of chloride ions could reduce the amount of free Ag^+ during the formation of initial seeds. The formation of AgCl avoids the quick supersaturation, and the release of Ag^+ from AgCl was very slow. By well controlling the amount of chloride ions, the pump became unnecessary in this method. Moreover, the addition of copper ions could reduce the diameter of nanowires. The Cu-mediated reaction mechanism is similar to Fe-mediated reaction. A typical Fe-mediated polyol synthesis was conducted as follows. EG was first added to a disposable glass vial, heated for 1 h at 150 °C. The $CuCl_2$ solution was added to the vial and heated for 15 min, and the PVP and $AgNO_3$ were added to the vial and heated for 1.5 h. The advantage of this method is that the $AgNO_3$ could be added without pump due to the role of chloride ions. Xia proposed the role of chloride ions, stabilized the silver seed particles via electrostatic interactions and reduced the concentrations of free Ag^+ ions.

Xia's method was further modified by Wiley et al. In Wiley's method [16], EG was first added to a flask, heated for 1 h at 140 °C, and all the reagents were added to the flask in sequence at approximately the same time. A range of different sizes of AgNWs were synthesized by tuning the parameters of this method.

Although this modified method was widely used by many researchers, the two-step process limited its large-scale production. Therefore, the one-pot process became a hot topic in AgNWs in recent years. In the one-pot process, all the reagents were added at one time and required no pump or dropping funnel. Since Xia et al. proposed that, by controlling the amount of chloride ions, many researchers made an effort to work out one-pot process. Zhan et al. reported a one-pot stirring-free method [17], which uses Fe^{3+} and Cl^- as co-nucleation agents. In this method, EG, PVP, $AgNO_3$ and $FeCl_3$ were added to the flask subsequently and stirred thoroughly at room temperature. Afterwards, the flask was transferred to an oven preheated at 140 °C and heated for 4.5 h. By tuning the amount of Fe^{3+} and Cl^-, the size of AgNWs in a broad range (D = 45–220 nm, L = 10–230 μm) could be produced.

Ding et al. reported a large-scale one-pot method [18], and more than 10 g nanowires could be produced at one time. A typical experiment was conducted as follows: EG, PVP, $AgNO_3$ and $FeCl_3$ were added to a 5 L reaction kettle, the mixture was electrically heated at 100 °C and held there for 30 min. Then, the solution was heated at 140 °C, and the temperature was kept at 140 °C for a period of time. By tuning six parameters the author provided, the size of AgNWs in a broad range (D = 40–110 nm, L = 5–35 μm) could be produced. The author also proposed the crucial role of chloride ions, and the releasing rate of Ag^+ from the AgCl nanocubes determined the yield and morphology of AgNWs.

Wang et al. reported a rapid one-step polyol method [19], and long AgNWs were synthesized by this method within only 30 min. In a typical synthesis process, EG, $CuCl_2$, PVP and $AgNO_3$ were subsequently added to a glass vial and heated the mixture at 160 °C for 30 min to complete the reaction. The author found that both the length and the diameter of AgNWs were increased as the concentration of $CuCl_2$ increases. Therefore, by simply tuning the concentration of $CuCl_2$, AgNWs with different size could be produced. Ran et al. reported a one-pot method to synthesize AgNWs with an aspect ratio above 1,000 [20], and the process was similar to Wang et al. and Ding et al.'s method; all the reagents were added to the flask, then the mixture was heated at a certain temperature for different time durations. In this method, the weight of PVP was a new parameter, and the size of AgNWs in a range (D = 25–160 nm, L = 5–50 μm) could be produced. Jiu et al. reported a one-pot method to synthesize very long AgNWs by only adjusting the stirring speeds [21]. For a typical synthesis of this method, EG, FeCl$_3$, PVP and AgNO$_3$ were added to the flask, and the mixture was stirred for completely dissolved at room temperature, then the contents of the flask were transferred into a reactor preheated at 130 °C for 5 h. AgNWs synthesized by this method have a diameter of 60 nm and a length of over 60 μm. Wiley et al. reported a bromide-mediated one-pot method. AgNWs with aspect ratios more than 2,000, and diameters of 20 nm were produced by this method. For a typical synthesis, EG, NaCl, NaBr, PVP and AgNO$_3$ were added to the flask, and the mixture was stirred vigorously for 30 min, and then heated at 170 °C with stirring for 15 min and reacted without stirring for 1 h. Zhang et al. also report a bromide-mediated one-pot method using KBr as co-nucleant [22]. By adjusting the exact concentration of KBr, AgNWs with an average length of 21 lm and average diameter of 26 nm could be synthesized conveniently. Hu et al. reported a similar method using NaCl and KBr as nucleation events [23]. In this method, all the reagents were added in the flask, and the mixture was heated at 75 °C with stirring for 30 min and subsequently heated at 120 °C with stirring for 8 h. By tuning the ratio of Cl$^-$ to Br$^-$, the size of AgNWs in a range (D = 25–45 nm, L = 5–30 μm) could be produced.

All these one-pot methods introduced above proposed the crucial role of chloride ions. The formation of the AgCl replaces the pump or dropping funnel, and by adjusting the amount of chloride ions, different sizes of nanowires were produced. However, the other parameters such as the amount of silver precursor and the ratio of PVP to Ag$^+$ were also used to adjust the size of AgNWs.

2.2.3 Factors influencing the polyol synthesis of AgNWs

2.2.3.1 Reaction temperature and time

From the introduction of the polyol methods, we can learn that some parameters were used to adjust the size of AgNWs. The reducing power increases with the temperature. In a typical polyol process, the reaction time decreases with the increase of temperature. In Xia's method [2], the temperature was set at 160 °C, and the method demonstrated that when the reaction temperature was higher or lower than 160 °C, the lengths of AgNWs decreased from 50 μm to 2 μm. However, in the Cu-mediated method, the temperature was tuned to 150 °C. In the modified method modified by Wiley [16], the temperature dropped to 140 °C. Although the temperature was changed in different methods, the length and diameter were not changed significantly. Wiley et al. studied how the temperature influenced the size of AgNWs. A series of control experiments were conducted, and the author found that, nanowires grow faster at higher temperature. AgNWs with length of 5 μm were obtained within 30 min at 160 °C, but it took 2 h at 160 °C. Although the nanowires grow faster at higher temperature, the longest nanowires were obtained at lower temperature. From the results, the author also found that longer and thicker nanowires were obtained at lower temperature (e.g. lengths and diameters of 20 μm and 65 nm obtained at 130 °C and 7.5 μm and 45 nm obtained at 160 °C). The AgNW length and diameter versus reaction time for different reaction temperatures are illustrated in the Figure 2.2.

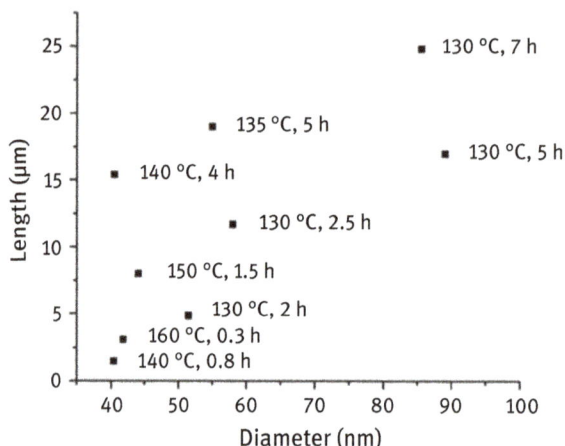

Figure 2.2: Reaction conditions for synthesizing nanowires with distinct lengths and diameters (reproduced from Ref. [16] with permission from The Royal Society of Chemistry).

Jiu et al. conducted a similar research [21]; AgNWs were synthesized at 110 °C and 180 °C, respectively. Similar to Wiley's conclusion, it took over 12 h to complete the reaction at 110 °C and the reaction was completed within 45 min at 110 °C. The length of AgNWs synthesized at 110 °C was greater than 80 μm; however, the length of AgNWs synthesized at 180 °C was only 20–30 μm. Based on these results, the author proposed that a low temperature and long time correspond to the formation of longer and thicker AgNWs. The AgNWs synthesized by this method had a longer length than Wiley's method, and the author attributed this to low-speed stirring.

Unalan et al. did a similar research [24], but concluded the opposite conclusion. AgNWs were synthesized at 110, 130, 150, 170, 190 and 200 °C, respectively, and only micrometre-sized silver structures are formed at 110 °C. The author attributed this to the lack of conversion of EG to glycolaldehyde, which reduces Ag^+ ions to Ag atoms. Above 130 °C, the length of the nanowires increased with temperature, and the diameter of the nanowires decreased with temperature, which was opposite to Wiley and Jiu's conclusion. However, the longest length was obtained at 190 °C, which was less than 10 μm. The aspect ratio of AgNWs synthesized by this method was too low. The contrasting results may have been due to the different synthetic conditions employed by each research group.

We also conducted a set of experiments to investigate the influence of reaction temperature. EG, NaCl, PVP and $AgNO_3$ were added to the flask, and then the mixture was heated at various temperatures for a period of time. We measured the average diameter and length of AgNWs. The average diameter increased with the temperature and the average length decreased with the temperature. Our conclusion indicated that a low temperature and long time correspond to the formation of longer and thinner AgNWs.

2.2.3.2 The influence of PVP

The amount and weight of PVP determine the final morphology of silver nanostructures. In the synthesis of silver nanoparticles, PVP acted as dispersant. Tuning the amount of PVP, various morphologies of silver nanostructures were produced. In Xia's method [2], when the molar ratio of PVP to Ag^+ was more than 15:1, only silver nanoparticles were obtained. A mixture of short nanowires and nanoparticles were produced when the ratio was 6:1. AgNWs with aspect ratio more than 1,000 were obtained when the ratio was 1.5:1, and when the ratio was further decreased to less than 1:1, the AgNWs were thick and non-uniform. Unalan et al. did a similar study [24] in which AgNWs were synthesized with different $PVP/AgNO_3$ molar ratios of 3:1, 4.5:1, 6:1, 7.5:1, 9:1 and 11:1. When the ratio was less than 6:1, large amounts of micrometer-sized Ag particles were found in the products, and the diameter was very large. The diameter decreased gradually when the ratio increased, and micrometer-sized Ag particles were found when the ratio was more than 9:1. Therefore, the optimum ratio was 7.5:1, and this ratio was much greater than Xia's 1.5:1. Zhan et al. conducted a series of experiments to investigate the role of PVP [17]. When the ratio increased from 1 to 2, both the diameter and the length decreased, and the silver nanoparticles were obtained. Further increased the ratio to 3, the silver nanoparticles were drastically increased. The author concluded that a low ratio leads to thick and long nanowires and a high ratio generates more nanoparticles; the optimum ratio was 1.5, which was the same as Xia's method. Ding et al. also studied the effect of $PVP/AgNO_3$ molar ratio on the morphology of AgNWs [18], and when the ratio was less than 6, large amounts of by-products were obtained. When the ratio was larger than 9, no nanowires were obtained. The optimum ratio the author proposed was 6.

Xia et al. studied how the polymerization of PVP influences the morphology of nanowires [2], and when the molar ratio was kept at 6, a series of PVP polymers of various degrees of polymerization were used to synthesize AgNWs. When the degree was 1, only irregular nanoparticles were obtained, and when the degree was increased to 90, rod-shaped nanostructures were obtained. The uniform nanowires without nanoparticles were formed when the degree was 11,700. The results indicated that the polymerization is as important as the molar ratio of PVP to Ag^+. As an extension of these results, Ran et al. studied the effect of a mixture of PVP molecules of different molecular weights to the nanowires [20]. PVP with molecular weight of 55,000, 360,000 and 1,300,000 were used to synthesize nanowires, respectively. When the molecular weight was 55,000, the diameter of the resultant AgNWs was

100 nm and the length was 25 μm. The diameter was 60 nm and the length was to 46 μm, when the molecular weight was increased to 3,600,000. Last, when the molecular weight was increased to 1,300,000, the diameter increased to 80 nm and the length grew to 50 μm. From these results, we can conclude that the longer nanowires were obtained when the PVP with larger molecular weight was used. Furthermore, AgNWs with high aspect ratio were obtained by using a mixture of PVP molecules. When the weight ratio between PVP chains with molecular weights of 360,000 and 55,000 was tuned to 2:1, more than 80 % of the AgNWs had diameters that were smaller than 30 nm, and more than 50 % had lengths exceeding 40 μm.

We conducted similar control experiments to investigate the role of PVP in the synthesis of nanowires; we found that, when the molecular ratio of PVP to Ag^+ was in the range of 2:1 to 6:1, the size of AgNWs changed little. When the ratio was less than 2:1, only irregular nanoparticles were obtained; when the ratio was more than 6, the diameter increased and the length dropped dramatically, and when the ratio was more than 10, no nanowires were obtained in the products. Due to the O–Ag and N–Ag coordination bond, the PVP could bind to Ag facet. Xia et al. demonstrated that the PVP macromolecules interacted more strongly with the {100} planes than with the {111} planes of silver. Therefore, by tuning the amount of PVP, the {100} facets were covered with PVP while leaving the {111} facets largely uncovered, and the anisotropic growth was facilitated.

2.2.3.3 The influence of AgNO3

The concentration of $AgNO_3$ is another parameter determining the morphology of AgNWs. In Xia's method [2, 10, 11], the concentration of $AgNO_3$ was controlled within a range of 0.018 to 0.023 M, but the reason why the concentration was set in this range was not provided. Although in Wiley's modified method, the scale was increased to 1 g $AgNO_3$ per reaction, the concentration was 0.025 M. In the one-step method proposed by Ye et al. [20]., the concentration of $AgNO_3$ was increased to 0.036 M. The concentration of $AgNO_3$ was increased to 0.058 M in Jiu's method [21]. AgNWs with diameter of 60 nm, with length of 70 μm, were synthesized under this concentration. Zhan et al. conducted a series experiments to investigate the effects of $AgNO_3$ concentration on the morphology of products [17]. AgNWs were synthesized under different $AgNO_3$ concentrations while maintaining the other parameters. The author found that the diameter of AgNWs increased from 142 nm to 290 nm, while the length gradually decreased from 14.3 μm to 7.8 μm, as the $AgNO_3$ concentration increased from 0.1 M to 0.6 M. Although the concentration of $AgNO_3$ was 10 times greater than typical method, the diameter was too large and the length was too short. The SEM images of AgNWs synthesized under different concentration are shown in Figure 2.3.

Huang et al. studied how the ratio of PVP to $AgNO_3$ concentration of $AgNO_3$ influenced the final morphology of AgNWs, when the concentration of $AgNO_3$ increased [25] and the ratio of PVP to $AgNO_3$ decreased. From the results, the author concluded

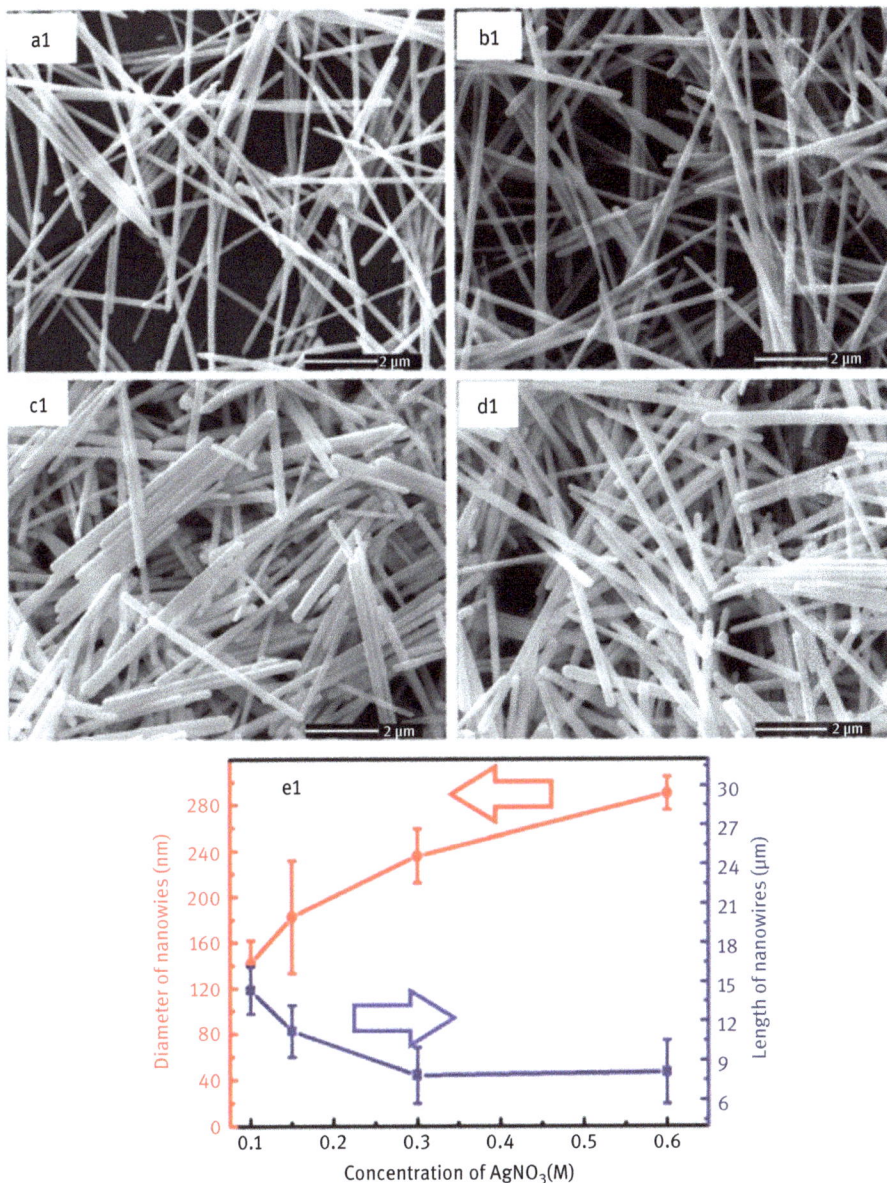

Figure 2.3: SEM images of silver nanowires synthesized with different concentrations of AgNO₃: (a1) 0.1 M, (b1) 0.15 M, (c1) 0.3 M, (d1) 0.6 M. (e1) Changes in nanowire diameter and length as a function of the AgNO₃ concentration (reproduced from Ref. [17] with permission from The Royal Society of Chemistry).

that when the ratio was increased from 2:1 to 16:1, the diameters of the synthesized nanowires decreased from 400 to 100 nm, respectively, and their aspect ratios increased. When the ratio was further increased, a lot silver nanoparticles were obtained in the products. Zhang et al., like a similar researcher, increased the concentration while keeping the amount of PVP constant [18]. When the ratio increased, shorter and thicker nanowires were obtained, which was similar to Huang's conclusion. There is a common mistake the researchers usually make when designing a control experiment. When the concentration of $AgNO_3$ was increased, the ratio of PVP to $AgNO_3$ was decreased at the same time. Both the concentration $AgNO_3$ of and the ratio of PVP to $AgNO_3$ have a crucial effect on the final morphology of products. If the researchers want to study how the concentration of $AgNO_3$ solely affects the final morphology of products, the amount of $AgNO_3$ and the amount of PVP must increase at the same time to keep the ratio of PVP to $AgNO_3$ constant.

Besides the concentration of $AgNO_3$, the pretreatment of $AgNO_3$ is another factor which influences the final morphology of AgNWs. Lee et al. studied the effects of $AgNO_3$ sonication before injection [26].

2.2.3.4 The influence of chloride ions

The amount of chloride ions was usually used to tune the morphology of AgNWs. However, in the early stage of Xia's method [2], the Cl^- was absent in the process, and a syringe pump was used to control the injection rate of $AgNO_3$. In Xia's Cu-mediated method [15], the addition of Cl^- replaces the syringe pump. The author proposed that a little amount of Cl^- is necessary in the polyol synthesis to provide electrostatic stabilization for the initially formed Ag seeds. Another role of Cl^- is to reduce the concentration of free Ag^+ ions in the solution in the way of the formation of AgCl nanocrystallites. The slow release of Ag^+ from AgCl mimics the use of a syringe pump.

Unalan et al. investigated how the amount of Cl^- influences the morphology of AgNWs [24]. Various amounts of NaCl were added to the reaction, and only nanoparticles were obtained when the Cl^- was absent. When the concentration of NaCl increased to 8.5 μm, nanowires with nanoparticles were obtained, and with a further increase to 12 μm, only nanowires were obtained in the product. When the concentration of NaCl increased to 17.1 and 25.6 μm, the nanoparticles were formed with nanowires, AgCl particles were formed in the absence of any nanowires when the concentration of NaCl was increased to 85.5 μm. The SEM images of the products synthesized under different conditions were shown in Figure 2.4.

Ding et al. used $NiCl_2$ as the control agent to synthesis AgNWs with different sizes [18]. When the $NiCl_2/AgNO_3$ molar ratio was 1: 2,000, the irregular particles were main product. When the ratio increased from 1: 1,000 to 1: 100, the diameter decreased from 90 nm to 47 nm and the length increased from 5 μm to 35 μm. When the ratio was further increased to 1: 25, no nanowires were obtained. Figure 2.5a–f show the SEM images of the AgNW products prepared at different $NiCl_2/AgNO_3$ molar ratio.

Zhan et al. used $FeCl_3$ as the control agent to synthesize AgNWs with different sizes [27]. The concentration of $FeCl_3$ was varied from 0.05 mM to 0.25 mM, while all the other parameters were constant. When the concentration of $FeCl_3$ was too low, only

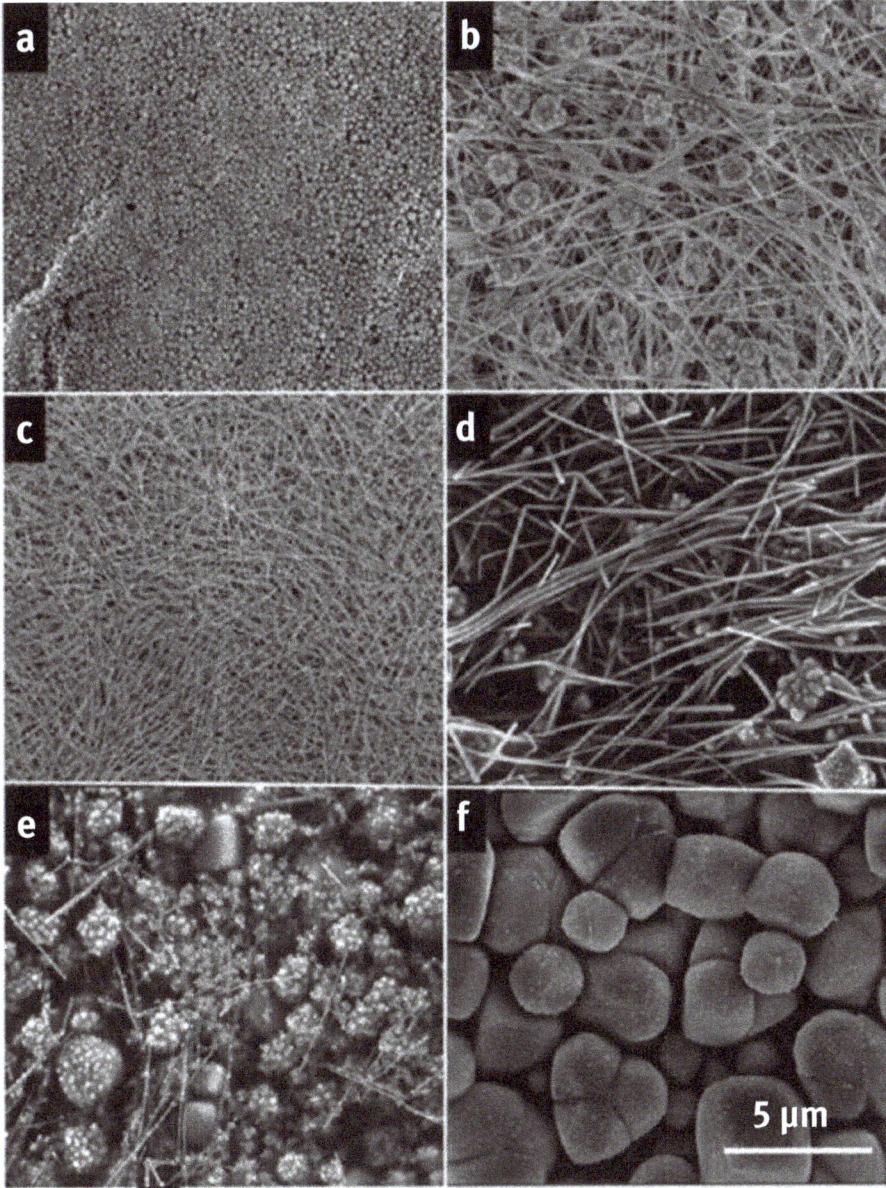

Figure 2.4: SEM images of the polyol process products produced with different NaCl amounts of (a) 0, (b) 8.5, (c) 12, (d) 17.1, (e) 25.6 and (f) 85.5 μm. All scales are the same (The Figure 2.4 was reprinted with permission from Ref. [24], copyright 2011 American Chemistry Society).

nanorods were obtained. When the concentration of $FeCl_3$ was increased to 0.1 mM and 0.15 mM, nanowires without any by-products were obtained. When the concentration further increased to 0.2 mM, nanorods with large amounts of irregular particles were

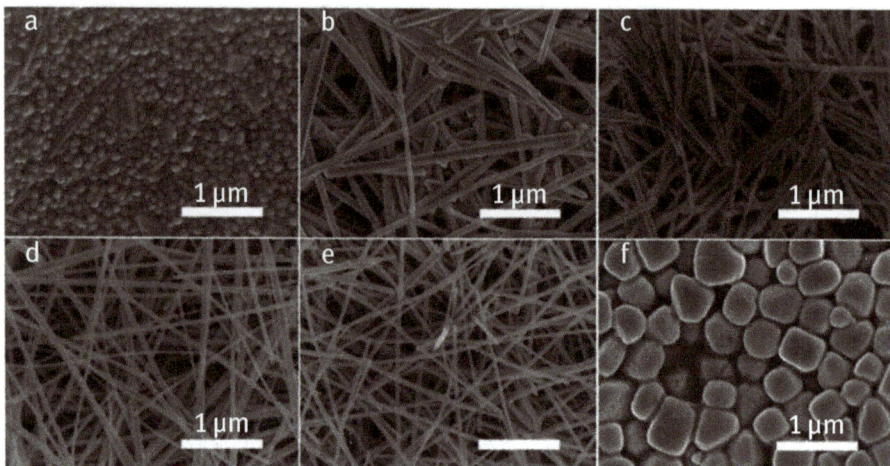

Figure 2.5: SEM images of silver nanowires synthesized at different NiCl$_2$/AgNO$_3$ molar ratio: (a) 1:2,000, (b) 1: 1,000, (c) 1: 500, (d) 1: 250, (e) 1: 100 and (f) 1: 25 (reproduced from Ref. [18] with permission from The Royal Society of Chemistry).

obtained. Only irregular particles were obtained when the concentration of FeCl$_3$ was increased to 0.1 mM. The author conducted another series of experiments using different amounts of NaCl as the control agent. When the concentration of NaCl increased from 0.15 mM to 0.45 mM, the diameter increased from 80 nm to 139 nm, while the length decreased from 9.4 μm to 5.5 μm.

From these reports, we can conclude that the diameter increases and the length decreases as the concentration of Cl$^-$ increases in a range. When the concentration of NaCl was too low, there was not enough Cl$^-$ to provide the electrostatic stabilization for the initially formed Ag seeds. When the concentration of NaCl was too high, there was not enough free Ag$^+$ for the growth of nanowires.

2.2.3.5 The influence of bromide ions

The addition of bromide ions has been found to promote the formation of AgNWs with narrow diameters in recent years. The bromide ions was used in wet-chemistry method proposed by Murphy et al. [1]. In this method, the cetyltrimethylammonium bromide was used as the soft template; AgNWs with diameter of less than 40 nm were synthesized by this method. However, in Murphy's conclusion, the important role of the rod-like shape of cetyltrimethylammonium bromide was demonstrated, and the role of bromide ions was not studied. Gedanken et al. used crystalline AgBr along with a photographic developer solution incorporating AgNO$_3$ as precursors to synthesize AgNWs [28]. The diameter of nanowires synthesized by this method was below 30 nm, but the author also did not research the role of bromide ions. Kim reported that the addition of bromide ions could reduce the diameter of nanowires effectively [29].

By tuning the ratio of tetrapropylammonium chloride to tetrapropylammonium bromide, the diameters of the AgNWs could be conveniently tuned between 20 and 50 nm.

Zhang et al. used KBr as co-nucleant to synthesis ultrathin nanowires [22]. AgNWs with an average diameter of 26 nm and an aspect ratio exceeding 800 were synthesized by this method. When the KBr was used alone, only nanoparticles were obtained, when the NaCl or $FeCl_3$ was used alone, the diameter of nanowires was larger than 35 nm. AgNWs with diameter below 30 nm could be synthesized by using the mixture of Br^- and Cl^-. Wiley proposed a similar conclusion that the addition of Br^- could reduce the diameter of nanowires. AgNWs with diameters of 20 nm and aspect ratios of up to 2,000 were obtained by using NaBr as co-nucleant. When the experiment was conducted without NaBr, the diameter increased to 72 nm. When the concentration of NaBr was too high, large amount of nanoparticles were present in the final product.

Xia et al. proposed a bromide-mediated method to synthesize sub-20 nm AgNWs [30]. Although many related papers demonstrated that when the Br^- was used alone, no nanowires were obtained in the products. However, in Xia's bromide-mediated method, the syringe pump was used again to control the injection rate of $AgNO_3$. By tuning the ratio of $AgNO_3$ to Br^-, different morphology of silver nanostructures were obtained. When the ratio of PVP to $AgNO_3$ was 1.5 and the ratio of $AgNO_3$ to Br^- was 60, AgNWs with diameters below 20 nm and aspect ratio greater than 1,000 in high morphology purity were obtained.

Kim et al. proposed a NaCl- and KBr-mediated polyol method to synthesize AgNWs with aspect ratio of over 1,300 [31]. AgNWs with 40–80 nm diameters and 10–30 μm lengths were obtained when 0.4 mM NaCl were added alone, and the diameter decreased to 20–50 nm and the length increased to 30–60 μm when 0.4 mM KBr was added. The results also proved that the addition of Br^- could reduce the diameter effectively. Hu et al. investigated the role of bromide ions in the synthesis of ultrathin AgNWs [23]. The author conducted a series of experiments in which the amount of halide ions is kept constant while the ratio of Cl^- to Br^- ions is varied. The plots of the average AgNW diameter and length versus KBr concentration are shown in Figure 2.6. The diameter decreased when the concentration of KBr increased, and it reached the lowest value when the KBr concentration was 0.084 mM, the diameter increased when the KBr concentration further increased. The length decreased readily when the concentration of KBr increased.

2.2.3.6 The influence of other factors

Many other factors that influence the morphology of AgNWs were proposed by different researchers. However, these factors were not widely studied by researchers, and there is not enough results for us to reference and discuss. Unalan et al. proposed that the method of the addition of $AgNO_3$ and PVP determines the final morphology of products [24]. In Xia's method, the syringe pump was used when the chloride ions were absent, but Unalan proposed that the rate of the addition of $AgNO_3$ determines

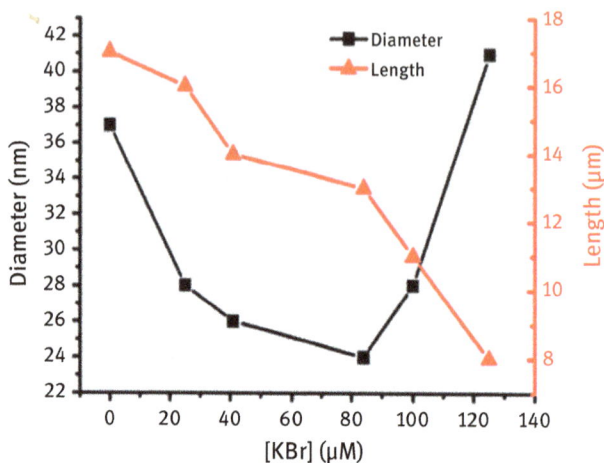

Figure 2.6: Plots of AgNW size (average length and diameter) versus concentration of KBr (reprinted from Ref. [23] with permission from Elsevier).

the size of nanowires, and by tuning the rate of addition, various sizes of nanowires were obtained. In our own experience, the using of syringe usually generate nanowires with more uniform diameter, but the using of syringe pump limits its large-scale production. The stirring rate and the length of stir bar were also used to tune the size of nanowires, and the related experiments were conducted by Zhan et al. to study how the stir bar influences the morphology of the final products [17].

2.3 Characterization methodologies and instrumentation techniques

2.3.1 Morphological investigations

The AgNWs were usually investigated by TEM. The TEM images were widely used to measure the size of nanowires, especially for the diameter. The samples were normally prepared by placing a drop of diluted solution of samples on the copper grids coated with amorphous carbon film. The samples were allowed to dry at room temperature in a desiccator connected to vacuum pump. The solution samples were usually purified by centrifugation process, and most of the small nanoparticles were removed with the supernatant. However, in the research of the yield of nanowires, we need to count all the products, and the solutions were diluted without centrifugation process. The products contain a large amount of PVP and EG, and these impurities could be removed by centrifugation with acetone and ethanol. When the samples were diluted without centrifugation process, the samples have to dry at relatively high temperature for the efficient evaporation of the EG solution. Although the TEM images were also used to measure the length of nanowires, the diameter of hole of copper grid was usually less than 100 μm, it was not easy to measure the length of long nanowires. It is noteworthy that the surfaces of AgNWs are usually bound to PVP, and the diameters of AgNWs measured from SEM images are usually larger than those measured by TEM images. Moreover, by using the selected area electron diffraction pattern, we can determine the type of crystal facets present on the surfaces of the AgNWs. Meanwhile, by studying the TEM images taken from the microtome samples, we can observe the cross-section of the nanowires.

SEM is another widely used characterization method of AgNWs. The SEM images were usually used to study the morphology of silver nanostructures. The SEM images could provide more details on the surfaces of silver nanostructures. Moreover, due to the preparation method of samples, some micro-sized particles are usually lost in the preparation process of TEM samples. However, the SEM samples were normally prepared by placing a drop of diluted solution of samples on the silicon wafer, and almost all the products were stayed on the samples. The SEM images were also used to measure the length of ultralong nanowires. The typical TEM image of AgNWs was shown in Figure 2.7 (a), the typical SEM image of AgNWs was shown in Figure 2.7 (b).

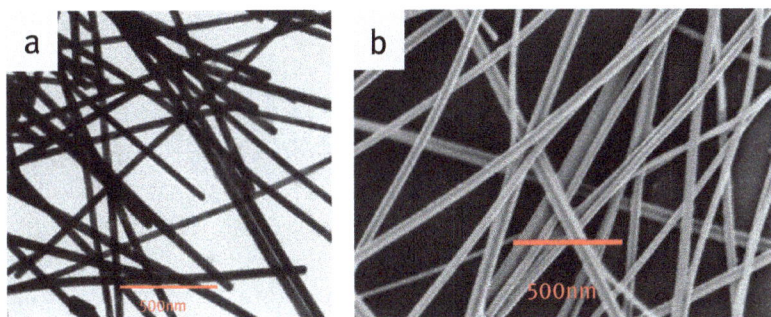

Figure 2.7: The typical (a) TEM images and (b) SEM images of AgNWs.

Optical microscope is another instrument used to study the morphology of AgNWs; the samples were prepared by dropping the solution directly on the slides. It is a facile way to roughly assess the impurity content of products. For instance, the optical microscope images were used by Wiley et al. to investigate the nanoparticle content of products, and the dark-field optical microscope images are shown in the Figure 2.8. Zhang et al. also used the optical microscope images to measure the length of nanowires.

It is noteworthy that, although the aspect ratio is defined as the ratio of length to diameter, the aspect ratio was usually mean to an average value. The distribution of

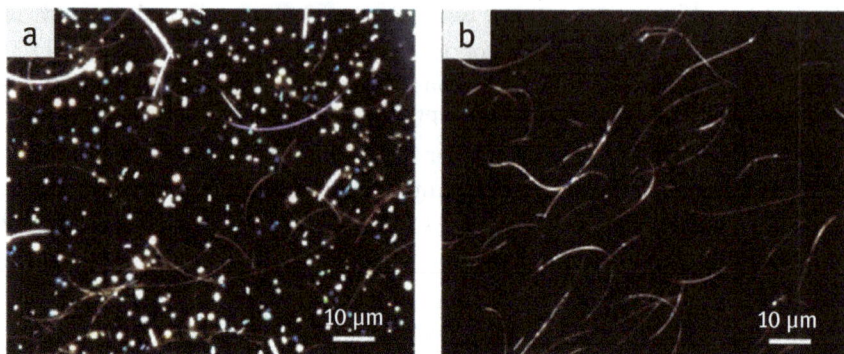

Figure 2.8: Dark-field optical microscope images of Ag nanowires before (a) and after (b) purification (reprinted with permission from Ref. [34] Copyright (2015) American Chemical Society).

diameter and length of AgNWs was studied by many researchers. However, it is difficult to measure both the diameter and length of a single nanowire precisely. Therefore, the researchers usually measured the average diameter of nanowires using TEM images and measured the length of nanowires using optical microscope images or SEM images. It is very difficult to count the distribution of aspect ratio, which is the reason why there is no related report so far.

2.3.2 Optical and spectroscopic properties

It is well known that the frequencies of particular localized surface plasmon resonance bands as well as the surface-enhanced Raman scattering behaviour of a metal nanoparticle are influenced by its morphology. Therefore, the morphologies of silver nanostructures could be determined by the optical prosperities. UV-visible spectroscopy is widely used to determine the morphologies of silver nanostructures, especially for the products in the early stage of reaction. Xia et al. used the UV-visible spectroscopy to trace the reaction. The UV-vis absorption spectra of the reaction mixture in the different stage of the reaction are shown in the Figure 2.9. The X-ray diffraction (XRD) is used to study the facet of AgNWs, which is similar to the selected area electron diffraction pattern in the TEM. Figure 2.10 is a typical XRD pattern of AgNWs, and all the peaks in this figure could be indexed to the face-centred cubic phase of silver.

The four-probe method was used by Xia et al. to test the electrical continuity of AgNWs [10], but the electronic properties were usually measured by fabricating a AgNWs-based device, such as transparent conductive films (TCFs). The optical properties of AgNWs are usually measured by the transmittance of TCFs, and the electronic properties of AgNWs are usually measured by the sheet resistance of TCFs.

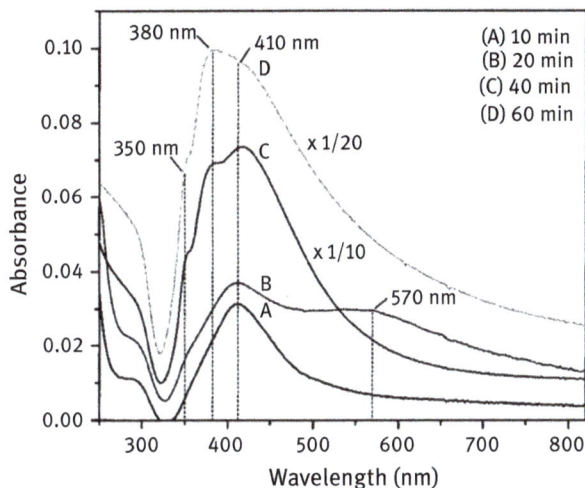

Figure 2.9: UV-vis absorption spectra of the reaction mixture after $AgNO_3$ and PVP were added for (A) 10, (B) 20, (C) 40 and (D) 60 min (reprinted with permission from Ref. [10] Copyright (2002) American Chemical Society).

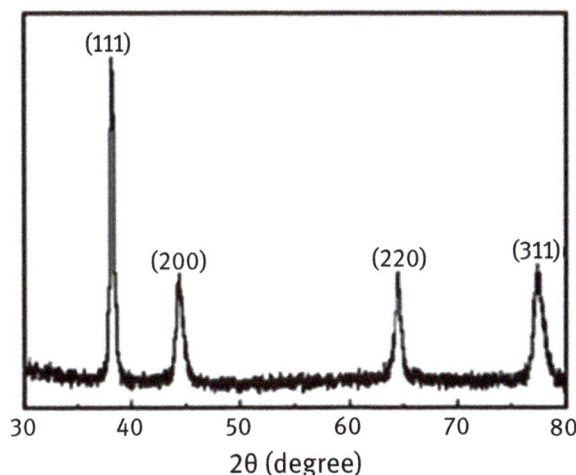

Figure 2.10: A typical XRD pattern of AgNWs (reprinted with permission from Ref. [2] Copyright (2002) American Chemical Society).

2.4 A broad category of examples and discussion

In this chapter, we will introduce and discuss some typical polyol methods from the design of recipe to the synthetic procedure. Most of the examples were chosen from the published papers.

2.4.1 Two-step method with a syringe pump

Xia et al. first proposed the method in 2002 [11]. The nucleation step was well controlled by the syringe pump, and the exotic seeds that act as seeds were excluded in this method. The method was widely used to synthesize AgNWs with low level of impurities.

AgNO$_3$ was used as silver source; PVP with a molecular weight of 55,000 was used as capping agent. A two-channel syringe pump (KDS-200, Stoelting Co., Wood Dale, IL, USA) was used to inject the PVP and AgNO$_3$.

In a typical synthesis, a 25 mL flask with 5 mL EG solution was heated at 160 °C, and 3 mL EG solution of AgNO$_3$ (0.1 M) and 3 mL EG solution of PVP (0.6 M) were injected to the flask by the pump within 10 min. The reaction mixture was further refluxed at 160 °C for 1 h, and the stirring was continuously applied in the entire process. The equipment employed in this synthesis is shown in Figure 2.11.

Figure 2.11: The equipment employed by Xia's group for the synthesis of AgNWs (reprinted from Ref. [32] with permission from John Wiley and Sons).

The final concentration of AgNO$_3$ in this synthesis was 27 mM, and the final concentration of PVP was 41 mM. The molar ratio of PVP to AgNO$_3$ was 1.5.

The products of AgNWs synthesized by this method might contain some nanoparticles, and the AgNWs could be purified by centrifugation process. The reaction mixture was diluted 10 times by volume with acetone, and then centrifuged at 2,000 rpm for 20 min. The supernatant was removed and the process was repeated three times.

The AgNWs synthesized by this method had a diameter of 60 nm and an aspect ratio of 1,000.

The typical synthesis introduced above was the optimal condition in this method, and the key of this method is the choice of molecular weight of PVP and the molar ratio of PVP to $AgNO_3$. The best condition of this method was introduced above. When the ratio of PVP to $AgNO_3$ was reduced to 1:1, the amount of nanoparticles was too large, and when the ratio was increased to 15:1, only nanoparticles were obtained. When the degree of polymerization of PVP was 1, only irregular particles were obtained, and when the degree of polymerization increased to 90, rod-like nanoparticles were formed. The molecular weight of PVP the author proposed was 55,000; correspondingly the degree of polymerization was about 495.

In addition, the author also proposed that AgNWs with diameter of 55 nm and length of 4.8 μm could be synthesized when the degree of polymerization of PVP was 11,700 and the molar ration of PVP to $AgNO_3$ was 6:1. Although the nanowires were too short, the products synthesized under this condition were free of nanoparticles.

2.4.2 Two-step method without syringe pump

The method was first reported by Xia et al. in 2002 [2, 10]. Many other two-step methods were modified based on this method.

$AgNO_3$ was used as silver source; PVP with a molecular weight of 55,000 was used as capping agent. Pt nanoparticles could serve as seeds for the heterogeneous nucleation.

In a typical synthesis, a 25 mL flask with 5.5 mL EG solution and 2×10^{-5} g of $PtCl_2$ was heated at 160 °C, and after 4 min, the 2.5 mL EG solution of $AgNO_3$ (0.05 g) and 5 mL EG solution of PVP (0.2 g) were added to the reaction mixture. The reaction mixture was further refluxed at 160 °C for 1 h, and the stirring was continuously applied in the entire process. The products of AgNWs synthesized by this method might contain some nanoparticles, and the AgNWs could be purified by centrifugation process. The reaction mixture was diluted 10 times by volume with acetone and then centrifuged at 2,000 rpm for 20 min. The supernatant was removed and the process was repeated three times.

The final concentration of $AgNO_3$ in this synthesis was 22.3 mM, and the final concentration of PVP was 138 mM. The final concentration of Pt^+ in this synthesis was 0.06 mM. The molar ratio of PVP to $AgNO_3$ was 6.2. The molar ratio of $AgNO_3$ to Cl^- was 1922.

The AgNWs synthesized by this method had uniform diameters in the range of 30–40 nm and length up to 50 μm.

The author demonstrated the importance of the reaction temperature; when the reaction temperature was higher or lower than 160 °C, the lengths of AgNWs dropped dramatically from 50 μm to 2 μm. When the temperature was lower than 100 °C, few nanowires were obtained. The SEM images of nanowires synthesized under different temperatures are shown in Figure 2.12. When the amount of $PtCl_2$ was increased 10 times, the diameter was decreased from 40 nm to 30 nm.

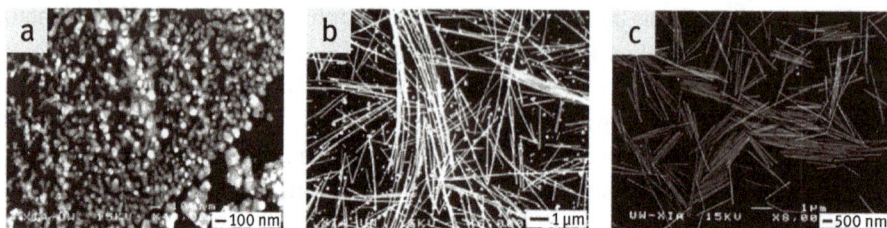

Figure 2.12: (a) SEM image of the product obtained after the solution was heated at 100 °C for 20 h. (b) SEM image of the product obtained when the solution was heated at 160 °C for 1 h. (c) SEM image of the product obtained when the solution was heated at 185 °C for 1 h (reprinted with permission from Ref. [2] Copyright (2002) American Chemical Society).

When the molar ratio of PVP to $AgNO_3$ increased from 6:1 to 18:1, only nanoparticles with an average size of 20 nm were formed. When the molar ratio of PVP to $AgNO_3$ decreased from 6:1 to 1.5:1, AgNWs with diameter of 100 nm were obtained. The SEM images of nanowires synthesized under different ratios are shown in Figure 2.13.

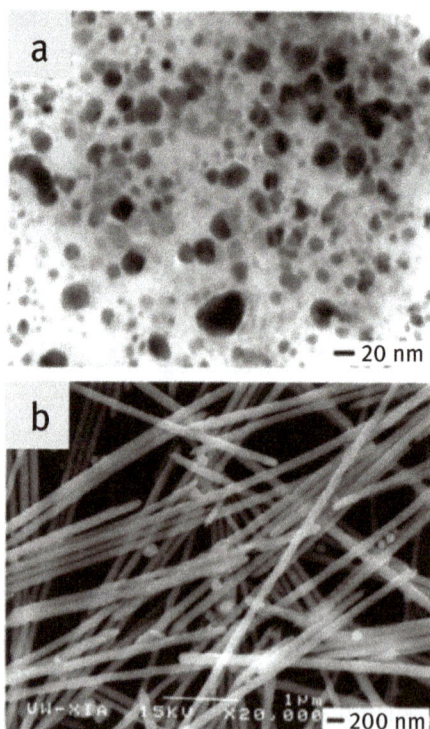

Figure 2.13: SEM images of two as-synthesized products, showing the variation of morphology when the molar ratio between PVP and $AgNO_3$ was changed. (a) nPVP/nAgNO$_3$ = 18; (b) nPVP/nAgNO3 = 1.5 (reprinted with permission from Ref. [2] Copyright (2002) American Chemical Society).

2.4.3 Cu-mediated method

It is well known that the addition of trace amounts of salts has a crucial effect on the final morphology of products. The copper ions were widely used by researchers to tune the size of nanowires, but the Cu-mediated method was first reported by Xia et al. in 2008 [15]. The copper ions were found to facilitate the growth of nanowires. Moreover, the experiment could be conducted in a disposable glass vial.

$AgNO_3$ was used as silver source; PVP with molecular weight of 55,000 was used as capping agent. $CuCl_2$ or CuCl was used to tune the size of nanowires, as Cl^- could reduce the level of free Ag^+ in the reaction mixture and Cu(I) could scavenge the adsorbed atomic oxygen from the surface of seeds.

In a typical synthesis, 5 mL of EG was added to a disposable glass vial and heated at 151.5 °C for 1 h under stirring. 40 μL of a 4 mM $CuCl_2$ solution in EG was added to the reaction, and the mixture was heated for an additional 15 min. Then, 1.5 mL of a 0.147 M PVP solution in EG and 1.5 mL of a 0.094 M $AgNO_3$ solution in EG were added to the glass vial, and the mixture solution was heated for 1 h.

The final concentration of $AgNO_3$ in this synthesis was 17.6 mM, and the final concentration of PVP was 27.6 mM. The final concentration of Cu^+ in this synthesis was 0.02 mM, and the final concentration of Cl^- in this synthesis was 0.04 mM. The molar ratio of PVP to $AgNO_3$ was 1.5. The molar ratio of $AgNO_3$ to Cl^- was 442.

The AgNWs synthesized by this method had uniform diameters of 100 nm and length of 10 to 50 μm.

The author demonstrated that when the concentration of increased to 0.04 mM or decreased to 0.01 mM, only irregular particles were formed in the final products. When the temperature increased to 160 °C or decreased to 140 °C, the amount of nanoparticles increased.

When the $CuCl_2$ was replaced by $Cu(NO_3)_2$, only nanoparticles were observed in the final products, and the SEM image is shown in Figure 2.14(a). However, when a trace amount of NaCl was added to the reaction, the products were also nanoparticles, and when the amount of NaCl increased to 0.02 mM, few nanowires were obtained.

When 0.02 mM NaCl was added without $Cu(NO_3)_2$, only nanoparticles were obtained and the SEM image is shown in Figure 2.14(b). However, when the reaction was conducted under argon atmosphere, nanowires were produced.

When the typical experiment was conducted under argon atmosphere, the diameter of nanowires decreased. The SEM images of AgNWs synthesized under these two conditions are shown in Figure 2.14(c–d). Moreover, when the typical experiment was performed under argon, the reaction was reduced to less than 1 h.

Figure 2.14: (a) SEM image of the product obtained in the presence of Cu(NO3)$_2$ but no Cl$_2$; (b) SEM image of the product obtained in the presence of NaCl but no Cu(II); (c) SEM image of the product obtained in the presence of both CuCl and NaCl; and (d) SEM image of the product obtained under an argon atmosphere in the presence of both CuCl and NaCl (reproduced from Ref. [15] with permission from The Royal Society of Chemistry).

2.4.4 Fe-mediated method

The Fe^{2+} was first used to tune the size of AgNWs by Xia et al. in 2005 [14]. However, the syringe pump was used in Xia's Fe-mediated method. Wiley et al. further modified the method [16, 33] and the modified method was applied to produce nanowires by many researchers.

(a) 20 mL EG solution of NaCl (0.257 g), (b) 10 mL EG solution of Fe(NO$_3$)$_3$ (0.081 g), (c) 25 mL EG solution of PVP (1.05 g) and (d) 25 mL EG solution of AgNO$_3$ (1.05 g) were prepared.

In a typical synthesis, a 500 mL flask with 158.4 mL EG solution was heated at 140 °C for 1 h. 0.2 mL of solution (1), 0.1 mL of solution (2), 20.76 mL of solution (3) and 20.76 mL of solution (4) were subsequently added to the reaction mixture. The flask was stoppered and allowed to react for 2 h at 140 °C for 1 h, and the stirring was continuously applied in the entire process.

The final concentration of $AgNO_3$ in this synthesis was 25.8 mM, and the final concentration of PVP was 39.3 mM. The final concentration of Fe^{3+} in this synthesis was 0.0165 mM, and the final concentration of Cl^- in this synthesis was 0.22 mM. The molar ratio of PVP to $AgNO_3$ was 1.5. The molar ratio of $AgNO_3$ to Cl^- was 117.

The AgNWs synthesized by this method had uniform diameters of 60 nm and length of 10 to 20 μm.

When the temperature increased to 160 °C, the diameter decreased to 42 nm and the length decreased to 3 μm. When the temperature decreased to 130 °C, the diameter decreased to 85 nm and the length decreased to 25 μm. The SEM images of nanowires synthesized under these two temperatures are shown in Figure 2.15. The author demonstrated that longer nanowires with larger diameters could be produced at lower reaction temperatures. Based on these conclusions, different sizes of AgNWs could be produced by tuning the reaction temperature and time.

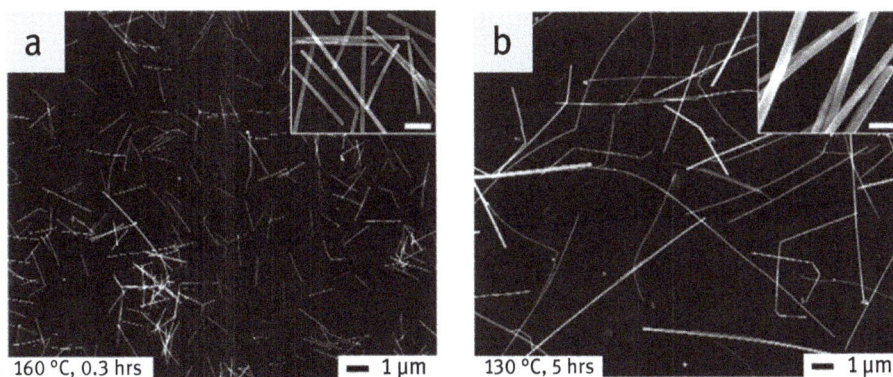

Figure 2.15: (a) SEM image of silver nanowire product after growth for 0.3 h at 160 °C. The nanowires are 42 ± 5 nm in diameter and 3 ± 0.5 μm in length. (b) SEM image of silver nanowire product after growth for 5 h at 130 °C. The nanowires are 85 ± 25 nm in diameter and 25 ± 5 μm in length. The scale bar in the inset is 200 nm (reproduced from Ref. [6] with permission from The Royal Society of Chemistry).

2.4.5 Stirring-control method

Stirring speed is a crucial factor that influences the morphology of AgNWs. Jiu et al. conducted a systematic research on how the stirring speed influences the morphology of AgNWs [21]. By tuning the stirring speed and other parameters, ultralong nanowires with length more than 100 μm could be synthesized.

In a typical synthesis, 25 mL EG solution, 0.2 g PVP (Mw = 360, 000), 0.25 g $AgNO_3$ and 3.5 g $FeCl_3$ were added to a 50 mL flask. When the reagents were dissolved well by stirring under room temperature, the mixture was immediately transferred into a reactor preheated at 130 °C, heated for 5 h.

The final concentration of $AgNO_3$ in this synthesis was 54 mM, and the final concentration of PVP was 64 mM. The final concentration of Fe^{3+} in this synthesis was 0.067 mM, and the final concentration of Cl^- in this synthesis was 0.201 mM. The molar ratio of PVP to $AgNO_3$ was 1.2. The molar ratio of $AgNO_3$ to Cl^- was 270.

When the stirring speed was 700 rpm, the length of most of the nanowires was in a range of 15–20 μm, and a large amount of irregular particles were formed. When the stirring speed decreased to 300 rpm, the length of most of the nanowires was 25–30 μm, and the amount of nanoparticles decreased. When no stirring or any other agitation was applied to the reaction, the length was in a range of 65–75 μm.

AgNWs with lengths in a range of 15–75 μm could be produced by only tuning the stirring speed.

When the experiment was conducted at 110 °C without any stirring, the reaction time increased to 12 h, but the length increased to 80 μm. When the experiment was conducted at 180 °C without any stirring, the reaction time decreased to 45 min, and the length decreased to 20–30 μm. The SEM images of AgNWs synthesized under these two temperatures are shown in Figure 2.16.

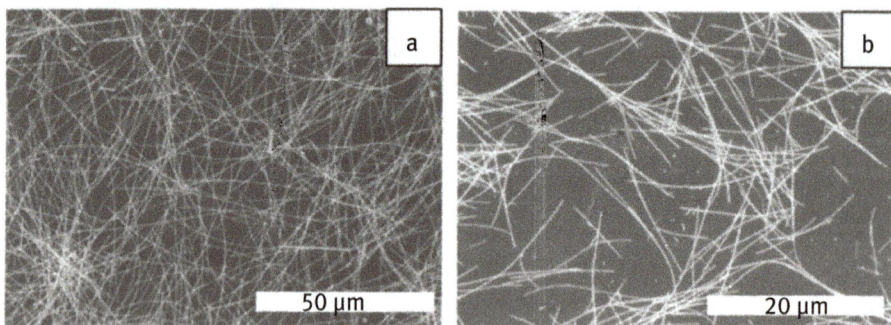

Figure 2.16: The SEM images of AgNWs prepared at 110 °C (a) and 180 °C (b) without agitation (reproduced from Ref. [21] with permission from The Royal Society of Chemistry).

AgNWs synthesized under different stirring speed were count. The dominant diameter of nanowires synthesized under different stirring speeds was 60 nm. The results are shown in Figure 2.17(d). From the results, the author proposed that the agitation favours a uniform AgNWs diameter.

2.4.6 Stirring-free FeCl₃-mediated method

In Section 2.4.5, the author concluded that longer nanowires with larger diameters could be produced at lower stirring speed. Zhan et al. reported a similar method to synthesize AgNWs with tunable sizes [17].

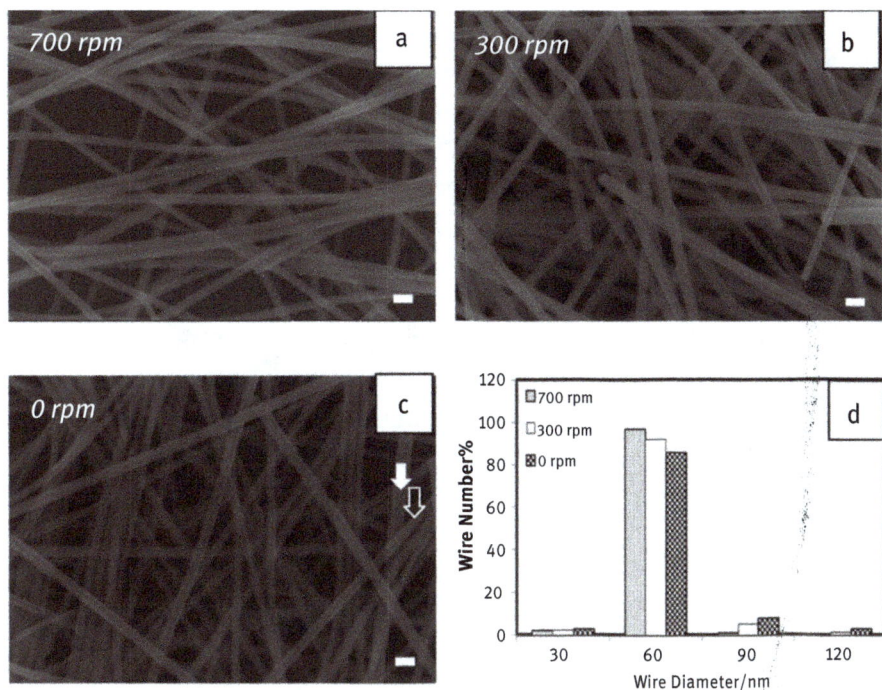

Figure 2.17: The SEM images of AgNWs prepared at 130 °C with stirring speeds (a, b and c with the scale bar of 100 nm) and related diameter distribution (d) (reproduced from Ref. [21] with permission from The Royal Society of Chemistry).

In a typical synthesis, 40 mL EG solution, 0.485 g PVP, 0.725 g $AgNO_3$ and various amounts of $FeCl_3$ were added to a 50 mL flask. When the reagents were dissolved well by stirring under room temperature, the mixture was immediately transferred into a reactor preheated at 140 °C, heated for 4.5 h without stirring.

When the $FeCl_3$ was absent, only nanoparticles were formed, and when Fe^{3+} was added without Cl^-, nanoparticles and sparse Ag wires or rods were formed. When Fe^{3+} and Cl^- were added, high-quality nanowires were obtained. The SEM images of products synthesized under different conditions are shown in Figure 2.18.

When the KCl was solely used, to tune the size of nanowires. When the amount of KCl increased steadily, both the length and diameter of nanowires decreased fast at first and then slowed down gradually and was kept steady at a small size finally. When $FeCl_3$ was used to tune the size of nanowires, both the length and diameter of nanowires increased at first and then decreased when the amount of $FeCl_3$ increased steadily. AgNWs with size in a broad range ($D = 80$–220 nm, $L = 13$–230 μm) could be produced by tuning the amount of $FeCl_3$. The synthetic conditions and statistical results of different reactions are listed in Table 2.1.

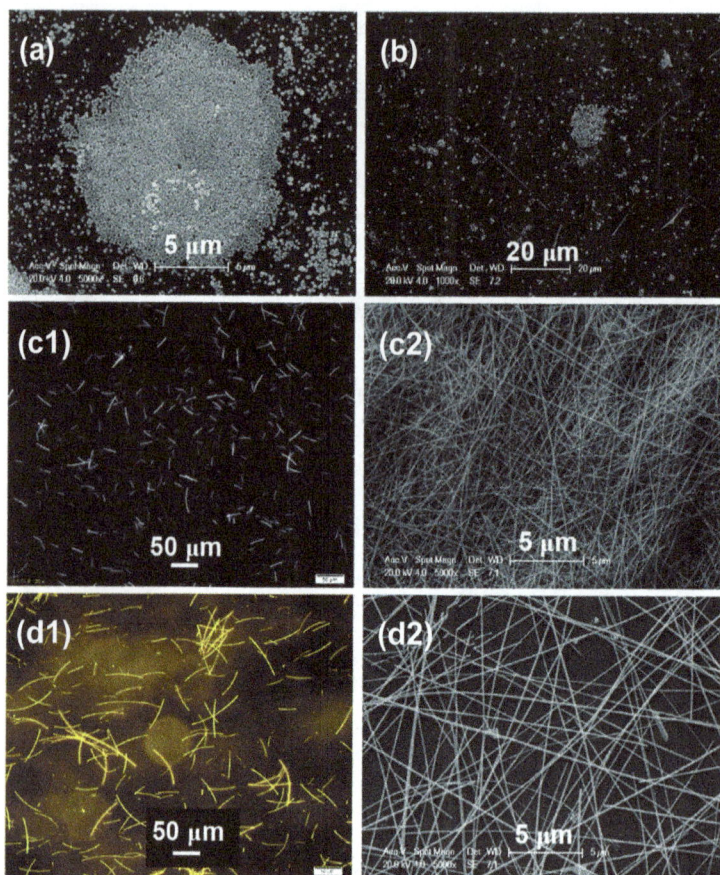

Figure 2.18: Product comparison of four different recipes, including reaction (a) without any additive, (b) with only $Fe(NO_3)_3$, (c1, c2) with only KCl and (d1, d2) with the combined addition of $Fe(NO_3)_3$ and KCl. The results verify that combined addition of $Fe(NO_3)_3$ and KCl is equal to single addition of $FeCl_3$ with the same amount of Fe^{3+} and Cl^- (reproduced from Ref. [17] with permission from The Royal Society of Chemistry).

2.4.7 PVP-control method

It is well known that the amount of PVP has a crucial effect on the morphology of AgNWs, and the molar ratio of PVP to Ag^+ and the molecular weight of PVP are discussed in Chapter 3. Ran et al. proposed a method to produce AgNWs with controllable sizes by utilizing a mixture of poly(vinylpyrrolidone) with different molecular weights [20]. AgNWs with an average diameter of 25 nm and an aspect ratio larger than 1,000 were synthesized by this method.

Table 2.1: A list of synthetic conditions and statistical results of different reactions (reproduced from Ref. [17] with permission from The Royal Society of Chemistry).

Trail names	KCl (mL)	Fe(NO$_3$)$_3$ (mL)	FeCl$_3$ (mL)	Sizes D (nm), L (µm)	Aspect ratio (AR = L/D)	NW yield (%)	Remarks
FeCl$_3$-STD	0	0	0.4	94.5, 38.0	402	97.3	–
NONE	0	0	0	–	–	0	Only NPs
Cl$^-$ only	0.4	0	0	61.3, 15.6	255	> 95	–
Fe^{3+} only	0	0.4	0	–	–	< 20	Non-uniform NWs
Fe^{3+} and Cl$^-$	0.4	0.4	0	93.7, 38.5	411	> 95	Equal to FeCl$_3$-STD
KCl control	0.1–10	0	0	45–77, 11–18	235–270	> 90	–
FeCl$_3$ control	0	0	0.1–1.0	77–221, 13–230	174–1,423	> 90	–
FeCl$_3$ control	0	0	1.5–3.0	125–158, 92–147	730–927	< 80	NWs and irregular NPs
Fe4Cln	0.1–1.0	0.4	0	72–130, 26–38	280–402	> 95	–
FenCl4	0.4	0.1–1.0	0	61–200, 13–226	210–1,618	> 90	–

In a typical synthesis, 27.5 mL EG solution, 0.163 g PVP, 0.18 g AgNO$_3$ and 0.0015 m mole FeCl$_3$ were added to a 50 mL flask. The reaction was performed at a certain temperature for different time durations.

When the PVP with molecular weight of 55,000 was used as capping agent, AgNWs with a diameter of 100 nm and a length of 25 µm were formed. When the PVP with molecular weight of 360,000 was used as the capping agent, AgNWs with a diameter of 60 nm and a length of 46 µm were formed. When the PVP with a molecular weight of 1,300,000 was used as capping agent, AgNWs with a diameter of 80 nm and a length of 50 µm were formed. From the results, the author proposed that the length of nanowires corresponds to the molecular weight of PVP.

When the PVP with molecular weight of 1,300,000 was used as the capping agent, the reaction was conducted under different temperatures. The author found that relatively thin and long AgNWs were obtained at 140 °C. Many researchers reported that the longer nanowires were obtained at lower temperature. However, the length of nanowires increased when the temperature increased. The details are listed in Table 2.2.

The diameter of nanowires decreased from 60 nm to 30 nm, when PVP-360,000 mixed with PVP-55,000 at a weight ratio of 1: 1. However, the length of nanowires decreased from 46 µm to 29 µm. Based on these results, the author

Table 2.2: Preparation conditions of AgNWs under different temperatures (reproduced from Ref. [20] with permission from The Royal Society of Chemistry).

Sample	$AgNO_3$ (g)	PVP (g)	EG (mL)	Temperature (°C)	Reaction time (min)	Final product (yields) and size of nanowires
1	0.180	0.163 (Mw = 1,300,000)	25	110	720	NPs (~85 %) and NWs (~15 %); D: ~160 nm, L: ~5 µm
2	0.180	0.163 (Mw = 1,300,000)	25	120	300	NPs (~35 %) and NWs (~65 %); D: ~140 nm, L: ~20 µm
3	0.180	0.163 (Mw = 1,300,000)	25	130	120	NPs (~20 %) and NWs (~80 %); D: ~90 nm, L: ~60 µm
4	0.180	0.163 (Mw = 1,300,000)	25	140	50	NPs (~15 %) and NWs (~85 %); D: ~80 nm, L: ~50 µm
5	0.180	0.163 (Mw = 1,300,000)	25	150	30	NPs (~60 %) and NWs (~40 %); D: ~70 nm, L: ~15 µm

demonstrated that the diameter could be reduced by mixing with PVP with different molecular weights.

Different sizes of could be produced by adjusting the weight ratio between PVP-360,000 or PVP-1,300,000 and PVP-55 000. AgNWs with an average diameter of 25 nm and an aspect ratio larger than 1,000 were synthesized when the weight ratio of PVP-360,000 to PVP-1,300,000 is close to 2 : 1.

2.4.8 Rapid Cu-mediated method

We have discussed the Cu-mediated method in Section 2.4.3, and the method was modified by Wang in 2016 [19]. AgNWs with mean length of 49.4 µm were synthesized by this method within only 30 min. The diameter of nanowires could be tuned by adjusting the concentration of Cu^{2+}.

In a typical synthesis, 20 mL EG solution in a 50 mL flask was heated at 160 °C for 5 min. 2 mL of 4 mM $CuCl_2$, 0.4 g PVP and 0.2 g $AgNO_3$ were added to the mixture and then the mixture was heated for 30 min to complete the reaction.

When the $CuCl_2$ was replaced by NaCl, a mixture of particles and wires with length of 8 µm was produced. When the concentration of $CuCl_2$ increased to 0.19 mM, the length of nanowires increased to 12.6 µm. When the concentration of $CuCl_2$ increased to 0.36 mM, the length of nanowires reached 49.4 µm. The SEM images are shown in Figure 2.19. When the concentration of Cu^{2+} was increased, the diameter of AgNWs increased from 48 nm to 72 nm.

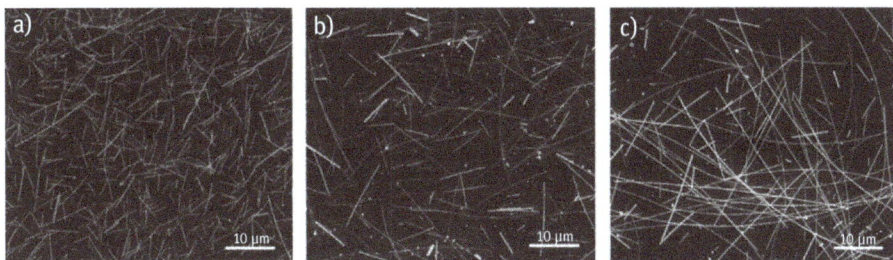

Figure 2.19: SEM images of AgNWs synthesized with (a) 0.38 mMNaCl, (b) 0.19 mM and (c) 0.36 mM CuCl$_2$, respectively (reprinted from Ref. [19] with permission from Elsevier).

2.4.9 Bromide-mediated method

Xia et al. reported a method to synthesis AgNWs with diameters below 20 nm and aspect ratios over 1,000 [30]. The Br$^-$ acted as capping agent to inhibit the lateral growth of nanowires and the Cl$^-$ was absent in this method.

In a typical synthesis, 100 mM AgNO$_3$ and 50 mM NaBr solutions in EG were prepared. 25 mg PVP (molecular weight = 1,300,000) and 4 mL EG were added to a 20 mL vial and heated at 160 °C for 1 h under stirring. Then, 50 µL of the NaBr solution was added to the vial, 1.5 mL of the AgNO$_3$ solution was injected to the vial by a syringe pump within 10 min and the mixture was heated at 160 °C for 35 min.

The final concentration of AgNO$_3$ in this synthesis was 27 mM, and the final concentration of PVP was 40 mM. The final concentration of Br$^-$ in this synthesis was 0.45 mM. The molar ratio of PVP to AgNO$_3$ was 1.5. The molar ratio of AgNO$_3$ to Br$^-$ was 60.

AgNWs synthesized by this method had an average diameter of 20 ± 4 nm and an average aspect ratio greater than 1,000. The molar ratio of PVP to AgNO$_3$ was set to 3 and 6, and the diameters of AgNWs were 22 ± 5 and 24 ± 6 nm, respectively. Further increasing the ratio, the aspect ratio decreased and large amount of nanoparticles were formed.

When the Br$^-$ was absent in the synthesis, the diameter of nanowires increased to 100 nm. When the molar ratio of Ag$^+$ to Br$^-$ decreased from 60 to 6, only irregular particles were formed. The products obtained at different Ag/Br and PVP/Ag molar ratios are illustrated in Figure 2.20.

2.4.10 Bromide-control method

Wiley et al. reported a method to synthesis AgNWs with diameters of 20 nm and aspect ratios up to 2,000 [34]. The key of this method is using Br$^-$ as co-nucleant, and the diameter could be reduced effectively, but the products were contaminated by nanoparticles.

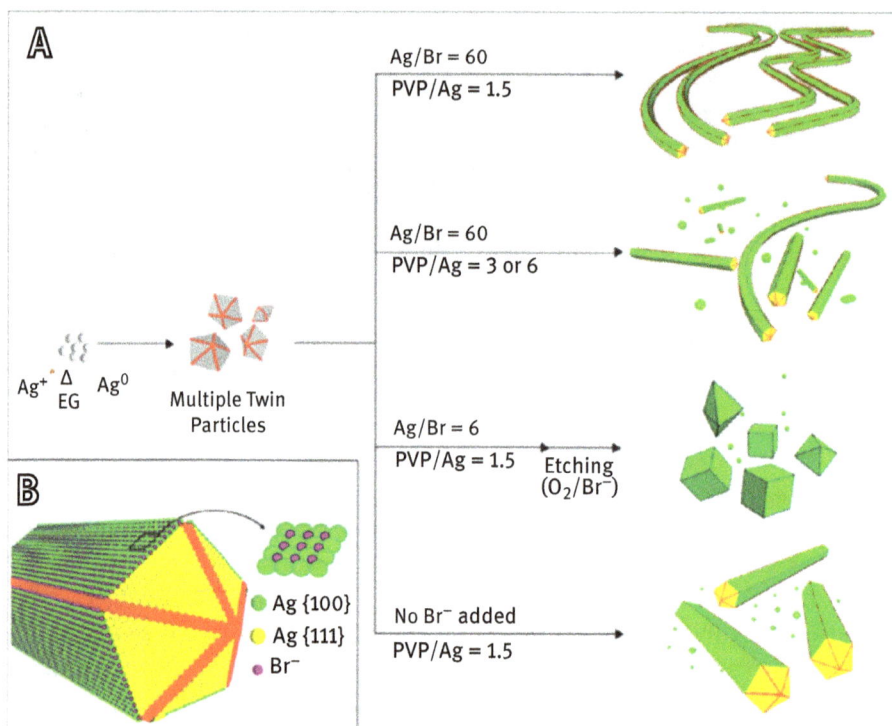

Figure 2.20: (A) Schematic illustration showing the distinct products obtained at different Ag/Br and PVP/Ag molar ratios. A high concentration of PVP or Br⁻ induces the formation of more isotropic particles. (B) Due to their small size relative to PVP, the Br⁻ ions can limit the lateral growth of a Ag nanowire by effectively capping the {100} facets on the side surface. Green and yellow colours denote the {100} and {111} facets on a Ag nanowire. Twin planes are outlined with red lines (reprinted from Ref. [30] with permission Copyright (2016) American Chemical Society).

In a typical synthesis, 0.1 mL NaBr (220 mM), 0.2 mL NaCl (210 mM), 1 mL PVP (505 mM) and 7.7 mL EG were added to a 50 mL flask. The mixture was under stirring in room temperature, and then heated at 170 °C, and nitrogen gas was bubbled through the reaction during heating. The flask was stoppered and allowed to react for 1 h without stirring.

The final concentration of $AgNO_3$ in this synthesis was 26.5 mM, and the final concentration of PVP was 50.5 mM. The final concentration of Cl^- in this synthesis was 4.2 mM. The final concentration of Br^- in this synthesis was 2.2 mM. The molar ratio of PVP to $AgNO_3$ was 1.9. The molar ratio of $AgNO_3$ to Br^- was 12. The molar ratio of $AgNO_3$ to Cl^- was 6.3. The molar ratio of Cl^- to Br^- was 1.9.

When the Br^- was absent, AgNWs with diameters of 72 nm and length of 63 μm are formed. When the concentration of Br^- increased to 1.1 mM, AgNWs with diameters of 36 nm and length of 48 μm were formed. When the concentration was further increased to 2.2 mM, the diameter decreased to 20 nm, and the length decreased to 40 μm. When the concentration of Br⁻ was set to 4.4 mM, most of the products were nanoparticles, and when the concentration further increased to 8.8 mM, only nanoparticles were obtained. The SEM images of AgNWs synthesized under different conditions are shown in Figure 2.21.

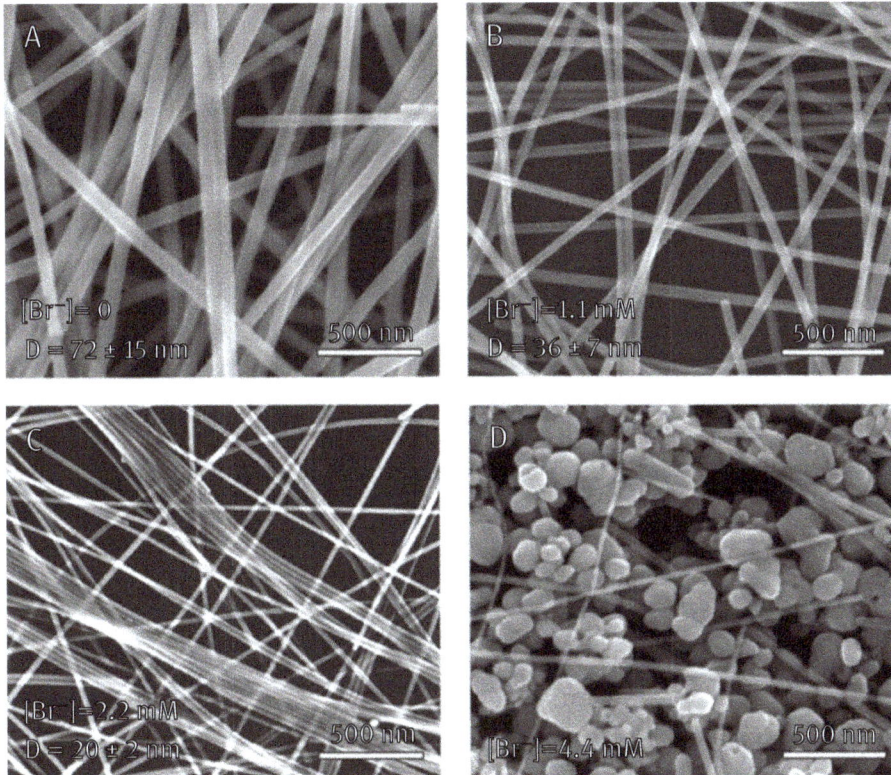

Figure 2.21: Typical high-resolution SEM images of purified products obtained from different concentrations of NaBr: (a) 0, (b) 1.1, (c) 2.2 and (d) 4.4 mM. The concentrations of AgNO₃, PVP and NaCl in each of these reactions were 26.5, 50.5 and 4.2 mM, respectively, and the reaction temperature was 170 °C (reprinted from Ref. [34] with permission Copyright (2015) American Chemical Society).

2.5 Critical Safety Considerations

In this paper, all of the chemicals purchased from reagent corporations for the synthesis of AgNWs are nontoxic in the condition of one can strictly compliance with experimental safety regulations.

Some experiments should pay attention to the following:

1. All the experiments must be conducted in fume hood, especially for the cases that need sampling in the process.
2. When a syringe pump is used in experiments, the reagents bottles should be tightly sealed.
3. When the nitrogen or argon gas bubbling is applied in the reaction, the rate of flow should be controlled well.
4. During the centrifugation process, the acetone or ethanol is usually used as solvent. Since the organic solvent would hurt the skins, the proper protection measure should be taken.
5. Some experiments require transferring the flask to an oil bath at elevated temperature, so an oven mitt is must be worn during this operation.

2.6 Conclusions and Future Perspective

The polyol method has become widely accepted by researchers to synthesize AgNWs, and various parameters were used to tune the final morphology of nanowires. Most of the common parameters in the synthesis method were introduced and discussed in Chapter 2. However, two researchers conducted experiment with the same recipe and procedure, the size of final products were different. For example, in Section 2.4.4, Wiley reported that longer and thicker nanowires could be produced at lower temperature. However, in Section 2.4.7, the author concluded that thicker and shorter nanowires were obtained at lower temperature. The differences are probably due to the different capping agent used in the methods.

Therefore, the common parameters influence the final morphology of products concluded by researchers were only for reference. Most of the reported recipe and method cannot be directly used without any modifications. When a researcher starts to design a method to synthesize nanowires, a series of exploratory experiments must be conducted. Then, the modifications could be made based on the experiments data.

Herein, we provide a common designing procedure for reference. The reaction temperature and reaction time are the most important parameters in a method. We usually first study how the temperature influences the morphology of nanowires. In the designing of recipe, only the $AgNO_3$, PVP and NaCl were used to minimize the number of variables in the experiment. When we have figured out how the temperature influences the morphology of AgNWs, the other parameters were tuned one by one to find the optimum choice.

Although the polyol method was proved the most successful strategy to synthesize AgNWs, many disadvantages are yet to be solved. Moreover, the exploration of new methods to synthesize AgNWs has seen only modest growth in recent years.

AgNWs synthesized by polyol method are usually coated with PVP, and the coating cannot be removed by purification process. When the nanowires were applied to conductive films, a physical press or annealing process is needed to ensure the contact between different nanowires. In addition, the PVP coatings are hydrophilic and limit the nanowires disperse in oil solvents. When the nanowires were applied to composite materials, various polymers are usually used to blend with nanowires. The hydrophilic surface limits the homogenous mixing of AgNWs with polymers. Some researchers try to modify the surface of nanowires, but the complicate process limits its large production. Therefore, how to synthesize AgNWs with oleophilic coating is an important issue that is to be solved.

The purification process is another crucial technic that needs improvement, and centrifugation process is widely used to purify the nanowires, and however, only small-size nanoparticles are removed. The nanowires settled down in the bottom, and most of the nanowires are aggregated. It is very difficult to disperse the aggregated nanowires when high centrifugation speeds are applied. Moreover, the purification process usually means to separate the nanowires from the particles. However, nanowires usually have a wide size distribution, and how to separate the long or thin nanowires from nanowires with other sizes is still unsolved.

The AgNWs synthesized by wet-chemistry method usually have a wide size distribution. As we know, the size of nanowires have a crucial effect on the performance of AgNW-based devices. The wide distribution of nanowires will limit the nanowires applied in many areas. The parameters that influence the morphology of nanowires we discussed in Chapter 2 can only use to tune the size of nanowires, and how to synthesis AgNWs with a very narrow distribution of size is still yet to be solved.

References

[1] Jana NR, Gearheart L, Murphy CJ. Wet chemical synthesis of silver nanorods and nanowires of controllable aspect ratio. Chem Commun. 2001;7:617–18.

[2] Sun Y, Yin Y, Mayers BT, Herricks T, Xia Y. Uniform silver nanowires synthesis by reducing agno$_3$ with ethylene glycol in the presence of seeds and poly(Vinyl Pyrrolidone). Chem Mater. 2002;14:4736–45.

[3] Phillip Lee JH, Lee J, Hong S, Han S, Suh YD, Lee SE, et al. Highly stretchable or transparent conductor fabrication by a hierarchical multiscale hybrid nanocomposite. Adv Funct Mater. 2014;24:10.

[4] Cho JH, Kang DJ, Jang N-S, Kim K-H, Won P, Ko SH, et al. Metal nanowire-coated metal woven mesh for high-performance stretchable transparent electrodes. ACS Appl Mater Interfaces. 2017;9:40905–13.

[5] Lee J, Lee P, Lee HB, Hong S, Lee I, Yeo J, et al. Room-temperature nanosoldering of a very long metal nanowire network by conducting-polymer-assisted joining for a flexible touch-panel application. Adv Funct Mater. 2013;23:4171–76.

[6] Lee J, An K, Won P, Ka Y, Hwang H, Moon H, et al. A dual-scale metal nanowire network transparent conductor for highly efficient and flexible organic light emitting diodes. Nanoscale. 2017;9:1978–85.

[7]   Lee P, Lee J, Lee H, Yeo J, Hong S, Nam KH, et al. Highly stretchable and highly conductive metal electrode by very long metal nanowire percolation network. Adv Mater. 2012;24:3326–32.

[8]   Chang I, Lee J, Lee Y, Lee YH, Ko SH, Cha SW. Thermally stable Ag@ZrO2 core-shell via atomic layer deposition. Mater Lett. 2017;188:372–74.

[9]   Lee J, Lee P, Lee H, Lee D, Lee SS, Ko SH. Very long Ag nanowire synthesis and its application in a highly transparent, conductive and flexible metal electrode touch panel. Nanoscale. 2012;4:6408–14.

[10] Sun Y, Gates B, Mayers B, Xia Y. Crystalline silver nanowires by soft solution processing. Nano Lett. 2002;2:165–68.

[11] Sun Y, Xia Y. Large-scale synthesis of uniform silver nanowires through a soft, self-seeding, polyol process. Adv Mater. 2002;14:833–37.

[12] Hu L, Kim HS, Lee J-Y, Peumans P, Cui Y. Scalable coating and properties of transparent, flexible, silver nanowire electrodes. ACS Nano. 2010;4:2955–63.

[13] Schuette WM, Buhro WE. Silver chloride as a heterogeneous nucleant for the growth of silver nanowires. ACS Nano. 2013;7:3844–53.

[14] Wiley B, Sun Y, Xia Y. Polyol synthesis of silver nanostructures: control of product morphology with Fe(II) or Fe(III) species. Langmuir. 2005;21:8077–80.

[15] Korte KE, Skrabalak SE, Xia Y. Rapid synthesis of silver nanowires through a CuCl- or CuCl$_2$-mediated polyol process. J Mater Chem. 2008;18:437–41.

[16] Bergin SM, Chen Y-H, Rathmell AR, Charbonneau P, Li Z-Y, Wiley BJ. The effect of nanowire length and diameter on the properties of transparent, conducting nanowire films. Nanoscale. 2012;4:1996–2004.

[17] Zhan K, Su R, Bai S, Yu Z, Cheng N, Wang C, et al. One-pot stirring-free synthesis of silver nanowires with tunable lengths and diameters via a Fe3+ & Cl- co-mediated polyol method and their application as transparent conductive films. Nanoscale. 2016;8:18121–33.

[18] Ding H, Zhang Y, Yang G, Zhang S, Yu L, Zhang P. Large scale preparation of silver nanowires with different diameters by a one-pot method and their application in transparent conducting films. RSC Adv. 2016;6:8096–102.

[19] Wang S, Tian Y, Ding S, Huang Y. Rapid synthesis of long silver nanowires by controlling concentration of Cu^{2+} ions. Mater Lett. 2016;172:175–78.

[20] Ran Y, He W, Wang K, Ji S, Ye C. A one-step route to Ag nanowires with a diameter below 40 nm and an aspect ratio above 1000. Chem Commun. 2014;50:14877–80.

[21] Jiu J, Araki T, Wang J, Nogi M, Sugahara T, Nagao S, et al. Facile synthesis of very-long silver nanowires for transparent electrodes. J Mater Chem. 2014;2:6326–30.

[22] Zhang K, Du Y, Chen S. Sub 30 nm silver nanowire synthesized using KBr as co-nucleant through one-pot polyol method for optoelectronic applications. Org Electron. 2015;26:380–85.

[23] Zhang P, Wei Y, Ou M, Huang Z, Lin S, Tu Y, et al. Behind the role of bromide ions in the synthesis of ultrathin silver nanowires. Mater Lett. 2018;213:23–26.

[24] Coskun S, Aksoy B, Unalan HE. Polyol synthesis of silver nanowires: an extensive parametric study. Cryst Growth Des. 2011;11:4963–69.

[25] Lin J-Y, Hsueh Y-L, Huang -J-J, Wu J-R. Effect of silver nitrate concentration of silver nanowires synthesized using a polyol method and their application as transparent conductive films. Thin Solid Films. 2015;584:243–47.

[26] Lee JH, Lee P, Lee D, Lee SS, Ko SH. Large-scale synthesis and characterization of very long silver nanowires via successive multistep growth. Cryst Growth Des. 2012;12:5598–605.

[27] Ma J, Wang K, Zhan M. A comparative study of structure and electromagnetic interference shielding performance for silver nanostructure hybrid polyimide foams. RSC Adv. 2015;5:65283–96.

[28] Liu S, Yue J, Gedanken A. Synthesis of long silver nanowires from AgBr nanocrystals. Adv Mater. 2001;13:656–58.
[29] Chang M-H, Cho H-A, Kim Y-S, Lee E-J, Kim J-Y. Thin and long silver nanowires self-assembled in ionic liquids as a soft template: electrical and optical properties. Nanoscale Res Lett. 2014;9:330.
[30] Da Silva RR, Yang M, Choi S-I, Chi M, Luo M, Zhang C, et al. Facile synthesis of sub-20 nm silver nanowires through a bromide-mediated polyol method. ACS Nano. 2016;10:7892–900.
[31] Trung TN, Arepalli VK, Gudala R, Kim E-T. Polyol synthesis of ultrathin and high-aspect-ratio Ag nanowires for transparent conductive films. Mater Lett. 2017;194:66–69.
[32] Wiley B, Sun Y, Mayers B, Xia Y. Shape-controlled synthesis of metal nanostructures: the case of silver. Chem – Eur J. 2005;11:454–63.
[33] Yang L, Zhang T, Zhou H, Price SC, Wiley BJ, You W. Solution-processed flexible polymer solar cells with silver nanowire electrodes. ACS Appl Mater Interfaces. 2011;3:4075–84.
[34] Li B, Ye S, Stewart IE, Alvarez S, Wiley BJ. Synthesis and purification of silver nanowires to make conducting films with a transmittance of 99 %. Nano Lett. 2015;15:6722–26.

Bionotes

Pei Zhang received his B.S. degree from South China University of Technology in 2011 and M.A. degree from South China Normal University in 2014. Then, he continued working on his Ph.D. degree at Guangzhou Institute of Chemistry, Chinese Academy of Science under the supervision of Prof. Jiwen Hu. His research focuses on the synthesis of silver nanowires and the fabrication of silver nanowire-based conductive films.

Shudong Lin received his M.S. degree from Shantou University in 2008, majoring in applied chemistry. After obtaining his Ph.D. degree in 2012 at Sun Yat-sen University, majoring in polymer chemistry & physics, he then worked in Guangzhou Institute of Chemistry, Chinese Academy of Sciences. Since 2015, he holds the position of associate researcher. His current research focuses on the synthesis and application of functional polymer composites.

Jiwen Hu received his B.S. degree from Xiangtan University in 1989, M.A. degree from Guangzhou Institute of Chemistry, Chinese Academy of Science in 1992 and Ph.D. degree from South China University of Technology in 2001. Since 2001, he holds the position of Professor of polymer chemistry at Guangzhou Institute of Chemistry, Chinese Academy of Science. He has published more than 150 peer-reviewed journal publications and obtained more than 200 invention patents. His current research interests are focused on the design and synthesis nanomaterials and functional membrane materials.

Jiawei Zhang, Huiqi Li, Zhiyuan Jiang and Zhaoxiong Xie

3 Size and Shape Controlled Synthesis of Pd Nanocrystals

Abstract: Palladium (Pd) has attracted substantial academic interest due to its remarkable properties and extensive applications in many industrial processes and commercial devices. The development of Pd nanocrystals (NCs) would contribute to reduce overall precious metal loadings, and allow the efficient utilization of energy at lower economic costs. Furthermore, some of the important properties of Pd NCs can be substantially enhanced by rational designing and tight controlling of both size and shape. In this review, we have summarized the state-of-the-art research progress in the shape and size-controlled synthesis of noble-metal Pd NCs, which is based on the wet-chemical synthesis. Pd NCs have been categorized into five types: (1) single-crystalline Pd nano-polyhedra with well-defined low-index facets (e.g. {100}, {111} and {110}); (2) single-crystalline Pd nano polyhedra with well-defined high-index facets, such as Pd tetrahexahedra with {hk0} facets; (3) Pd NCs with cyclic penta-twinned structure, including icosahedra and decahedra; (4) monodisperse spherical Pd nanoparticles; (5) typical anisotropic Pd NCs, such as nanoframes, nanoplate, nanorods/wires. The synthetic approach and growth mechanisms of these types of Pd NCs are highlighted. The key factors that control the structures, including shapes (surface structures), twin structures, single-crystal nanostructures, and sizes are carefully elucidated. We also introduce the detailed characterization tools for analysis of Pd NCs with a specific type. The challenges faced and perspectives on this promising field are also briefly discussed. We believe that the detailed studies on the growth mechanisms of NCs provide a powerful guideline to the rational design and synthesis of noble-metal NCs with enhanced properties.

This article has previously been published in the *journal Physical Sciences Reviews*. Please cite as: Zhang, J., Li, H, Jiang, Z., Xie, Z. Size and Shape Controlled Synthesis of Pd Nanocrystals. *Physical Sciences Reviews* [Online] **2018**, 3. DOI: 10.1515/psr-2017-0101

https://doi.org/10.1515/9783110636666-003

Graphical Abstract:

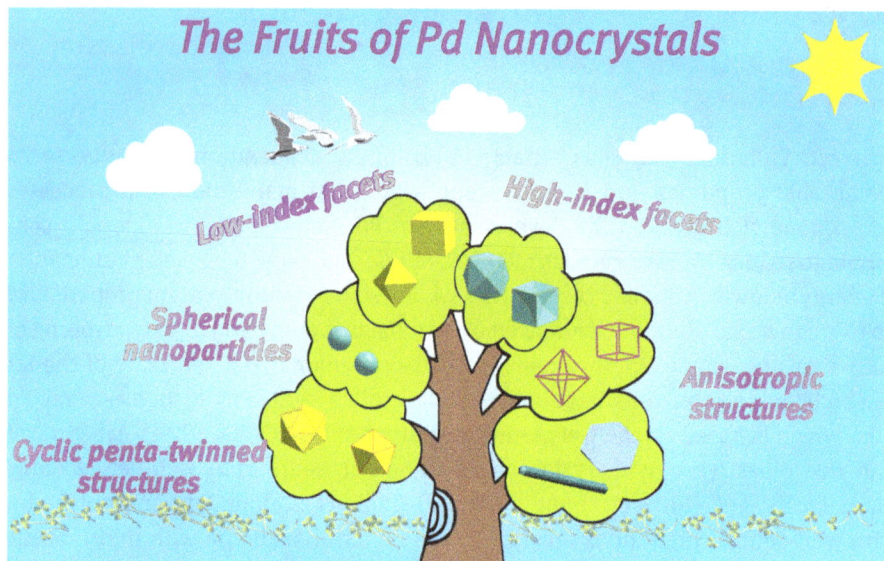

Keywords: Pd, shape, size, nanocrystal, crystal growth

3.1 Introduction

Palladium (Pd) has attracted substantial academic interest due to its remarkable properties and extensive applications in many industrial processes and commercial devices [1–7]. For example, Pd exhibits great activity toward a myriad of carbon–carbon coupling reactions, including Suzuki, Heck, and Negishi reactions, which provide a powerful methodology for the formation of carbon–carbon bonds [8, 9]. Thus, the 2010 Nobel Prize in Chemistry was awarded to Akira Suzuki, Richard Heck and Ei-ichi Negishi "for palladium catalyzed cross couplings in organic synthesis". Pd has a great absorptive affinity for hydrogen (H_2) and thus plays an important role in hydrogen storage, sensor, and purification [10, 11]. In addition, owing to the chemical similarity to Pt, Pd shows great performance in CO oxidation and fuel cells [12–15]. For example, Pd is well known for its superior performance than Pt in a direct formic acid oxidation fuel cell for its great CO-tolerance ability [13–15]. Besides, similar to Ag and Au, Pd can display unique localized surface plasmon resonance property by precisely controlling their shapes [16–18]. For example, the ultrathin Pd nanosheets exhibit strong SPR absorption in the NIR region [16]. The concave Pd nanocubes with sharp sub-10 nm edge and corner display unique tunable plasmonic properties in visible spectrum [18]. Over the past decades, numerous approaches have been reported to improve the performance of the Pd-based nanocrystals (NCs) [19–21]. The development of Pd-based NCs would contribute to reduce overall precious metal loadings, and allow the efficient utilization of energy at lower economic costs. It is well known that the size and shape have profound effects on tailoring the properties of noble metal NCs [22–29]. For example, the specific surface area, the ratio of surface atoms to bulk atoms, and the catalytic activity can be adjusted via controlling the size of noble metal NCs. More important, different facets of noble metal NCs may have different atom arrangements and electronic structures, and therefore exhibit distinct physical and chemical properties. It is possible to further tailor the corresponding properties via controlling the shape and the surface structure of noble metal NCs. The past decade has witnessed the successful synthesis of Pd NCs in various shapes with controllable sizes, including cube, octahedron, rhombic dodecahedron (RD) with low-index {100}, {111}, {110} facets, respectively [30–42], as well as tetrahexahedron (THH), concave nanocube with high-index {hk0} facets [43–52]. These Pd NCs exhibited outstanding performance in lots of applications, such as carbon–carbon coupling, hydrogenation, and fuel cells [43, 45, 53–59]. Usually, the performance over Pd NCs is highly size- and/or shape (surface structure)-dependent. Some of the important properties of Pd NCs can be substantially enhanced by rational designing and tight controlling of both size and shape. These works clearly demonstrate the importance of controlling the size and shape of Pd NCs.

Among various synthetic routes, solution-phase-based wet-chemical method is a powerful tool widely used for the size and shaped-controlled synthesis of noble metal NCs [19, 60–72]. During the growth of NCs, both size and shape can be well controlled by manipulating the thermodynamic and kinetic parameters. The growth thermodynamics determine the growth trends, while the different growth pathways that related to the energetic barriers of growth are governed by growth kinetics [27]. Therefore, a comprehensive understanding of the fundamental growth mechanisms of Pd NCs with different surface structures is highly desired. In this article, we begin with the summary of state-of-the-art progress in the synthesis of Pd NCs with different sizes and shapes via wet-chemical methods. Then, we introduce the detailed characterization tools for analysis of Pd NCs of a specific type. Following, we highlight the growth mechanisms of Pd NCs with various structural parameters. Finally, we discuss the challenges faced and perspectives on this promising field.

3.2 Preparation methods

In this section, we described wet-chemical synthetic approaches for different kinds of Pd NCs with well-defined surface structures and controllable sizes in detail. They are easy to be followed and repeated in laboratory. In addition, we precisely classify the preparation methods for Pd NCs into five types: (1) Synthetic approach for single-crystalline Pd nano polyhedra with well-defined low-index facets, including cube with {100} facets, octahedron with {111} facets, RD with {110} facets; (2) Synthetic approach for single-crystalline Pd nano polyhedra with well-defined high-index facets, such as Pd THH with {hk0} facets; (3) Synthetic approach for Pd NCs with cyclic penta-twinned (CPT) structure, including icosahedra and decahedra; (4) Synthetic approach for monodisperse spherical Pd nanoparticles; (5) Synthetic approach for typical anisotropic Pd NCs.

3.2.1 Synthetic approach for single-crystalline Pd nano-polyhedra with well-defined low-index facets

The NCs with specific surface structures are crucial for harvesting high catalytic activity and selectivity. Previously, the surface science on elucidating the relationship between surface structures and their corresponding properties mainly focused on single crystals, which are very expensive and have small surface areas, greatly impede their use in industrial applications. To fill the gap between practical applications and traditional surface science studies, the controlled synthesis of NCs with well-defined surface structures is highly needed. During the crystal's growth, in order to minimize the total surface energy, it requires the crystal facets should be bounded by low surface energies. Thus, Pd NCs with low-index facets are usually thermodynamically stable. The precise control of these shapes can provide fundamental knowledge on how to tune the surface structures of NCs thermodynamically. In addition, by controlling their sizes, the specific surface areas and some important properties, such as the activity/selectivity of the catalytic performance can be finely adjusted.

3.2.1.1 Pd nanocubes with {100} facets

3.2.1.1.1 Sizes in the range of ~6 nm to ~18 nm based on one-pot synthesis [33]
In a typical synthesis, 105 mg of poly(vinyl pyrrolidone) (PVP, MW ~55000), 60 mg of L-ascorbic acid (AA) and different amounts of KBr and KCl powders were dissolved in 8 mL of ultra-pure water, the homogeneous solution was pre-heated in air under magnetic stirring at 80 °C for 10 min. Then, 3 mL of an aqueous solution containing 57 mg of Na_2PdCl_4 was added using a pipette. The reaction solution was kept at 80 °C for 3 h in a 20 mL vial. The sizes of as-prepared Pd nanocubes are dependent on the amount of KBr and KCl. 5 mg of KBr and 185 mg of KCl result ~6 nm Pd nanocubes (Figure 3.1a), 75 mg of KBr and 141 mg of KCl result ~7 nm Pd nanocubes (Figure 3.1b), 300 mg of KBr yields ~10 nm Pd nanocubes (Figure 3.1c), 600 mg KBr yields ~18 nm Pd nanocubes (Figure 3.1d). The product was collected by centrifugation (16000 rpm, 10 min) and washed several times with ultra-pure water to remove excess PVP, and finally dispersed in 11 mL ultra-pure water.

This one pot-approach is usually used for the synthesis of Pd nanocubes with size smaller than 20 nm. In this protocol, the PVP mainly acts as dispersant, while the KBr acts as capping agent. These two agents can be replaced by cetyltrimethylammonium bromide (CTAB), a cationic surfactant. The detailed synthetic approach is described

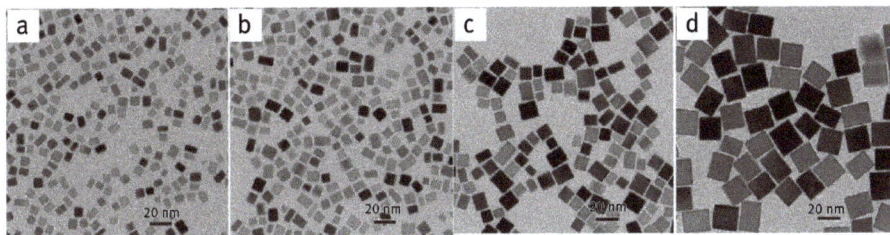

Figure 3.1: TEM images of Pd nanocubes with sizes about (a) ~6 nm (b) ~7 nm, (c) ~10 nm, and (d) ~18 nm (Adapted with permission from ref. 33. Copyright 2011, Tsinghua University Press and Springer-Verlag Berlin Heidelberg).

in Section 3.2.1.1.2. The method is versatile and can be modified for synthesis of Pd nanocubes with other sizes.

3.2.1.1.2 Pd nanocubes prepared in the presence of CTAB via one-pot synthesis

~22 nm Pd nanocubes [31]: In a typical synthesis, H_2PdCl_4 aqueous solution (10 mmol/L, 0.5 mL) was added to CTAB aqueous solution (12.5 mmol/L, 10 mL) under stirring at 95 °C, after about 5 min, freshly prepared AA aqueous solution (0.1 mol/L, 80 µL) was quickly injected using pipette. After proceeding for about 30 min, the reaction solution was cooled naturally at room temperature and washed with ultra-pure water.

~11 nm Pd nanocubes [41]: CTAB aqueous solution (12.5 mmol/L, 10 mL) was added into a 20 mL glass vial under stirring at 95 °C. Then, H_2PdCl_4 aqueous solution (20 mmol/L, 0.25 mL) was added. After stirred for 5 min, AA solution (0.1 mol/L, 0.2 mL) was quickly injected, and the solution was stirred for additional 10 min. After that, the reaction solution was cooled down to 30 °C, and allowed to age for 1 h.

~12 nm Pd nanocubes [51]: Typically, the Na_2PdCl_4 aqueous solution (30 mmol/L, 0.6 mL and NaOH solution (0.5 mol/L, 0.2 mL) were added to CTAB solution (12.5 mmol/L, 10 mL) under stirring at 95 °C. After about 10 min, freshly prepared AA solution (0.5 mol/L, 0.2 mL) was added. After continuing reaction about 30 min, the products were collected using centrifugation and washed with ultra-pure water.

These as-prepared nanocubes can also be used as seeds for further growth of Pd nanocubes into large sizes or various shapes, such as octahedra, RD, concave Pd nanocubes.

3.2.1.1.3 Sizes in the range of ~22 to ~109 nm by seeded growth [31]

At 40 °C, CTAB aqueous solution (50 mmol/L, 5 mL) was mixed with H_2PdCl_4 aqueous solution (10 mmol/L, 125 µL), and different volumes of as-prepared ~22 nm Pd nanocubes seed solution (see Section 3.2.1.1.2, Figure 3.2a) were added. Then, freshly prepared ascorbic acid solution (0.1 mol/L, 25 µL) was added and mixed thoroughly. The resulting solution was placed in a water bath at 40 °C for about 14 h. The final sizes are highly dependent on the added volume of seed solution, 0.4 mL for 37 nm (Figure 3.2b), 0.2 mL for 44 nm (Figure 3.2c), 80 µL for 56 nm (Figure 3.2d), 40 µL for 76 nm

Figure 3.2: (a) SEM image of ~22 nm Pd nanocube seeds. (b–f) SEM images of ~37 nm, ~44 nm, ~56 nm, ~76 nm, and ~109 nm Pd nanocubes (Adapted with permission from ref. 31. Copyright 2008, American Chemical Society).

(Figure 3.2e), 10 μL for 109 nm nanocubes (Figure 3.2f). The final products can be collected by centrifugation (12000 rpm, 10 min), and washed twice with ultra-pure water.

3.2.1.1.4 Large-scale synthesis of ~37 nm Pd nanocube [41]

With modified methods by Section 3.2.1.1.2, large-scale (about 60-fold high than Section 3.2.1.1.2) synthesis of Pd nanocube about 37 nm can be achieved.

In a typical synthesis, 1 g of CTAB, 3.2 mL of 0.1 mol/L H_2PdCl_4 and 6.8 mL of deionized water was mixed together under stirring in a water bath at 45 °C. After the formation of a turbid orange suspension, 10 mL of aged ~11 nm Pd nanocube seed solution (see Section 3.2.1.1.2) was added to the growth solution under vigorous stirring, followed by the addition of 5 mL of 1 mol/L AA. After vigorously stirring for 6 h, the products were collected by centrifugation at 12000 rpm for 10 min to remove excess CTAB and was subsequently redispersed in deionized water.

This protocol can be further modified to yield of up to 70 mg of Pd nanocubes (about 500-fold high than Section 3.2.1.1.2) with size about 72 nm for reaction volumes of 1 L [73].

3.2.1.2 Size controlled Pd nanooctahedra with {111} facets

3.2.1.2.1 Pd nanooctahedra with an edge length of ~6 nm based on one-pot synthesis [74]

In a typical synthesis, 105 mg PVP (MW~55 000), 180 mg of citric acid, and 57 mg of Na_2PdCl_4 were dissolved in a mixture solution containing 3 mL of ethanol and 8 mL of ultra-pure water. The resulting homogeneous solution was heated at 80 °C in air under magnetic stirring for 3 h before cooled down to room temperature. The product was collected by centrifugation (16,000 rpm, 10 min) and washed several times with ultra-pure water and finally dispersed in 10 mL ultra-pure water.

3.2.1.2.2 Pd nanooctahedra with an edge length of about 37 nm based on one-pot synthesis [75]

In a typical synthesis, 60 mg of AA, 60 mg of citric acid were dissolved in an aqueous solution of 8 mL 0.1 mol/L CTAC. The solution was heated to 100 °C. Then 3 mL of an aqueous solution containing 60 mg of Na_2PdCl_4 was added under vigorously stirring. After continually stirring about 3 h, the solution was cooled down to room temperature and the products were collected by centrifugation (12,000 rpm, 10 min), washed several times with ultra-pure water.

3.2.1.2.3 Pd nanooctahedra in the range of 14–37 nm by seeded growth [76]

This method is based on the seeded growth of well-defined Pd nanocubes, which are pre-prepared according to the above Section 3.2.1.1.1. For the synthesis of nanoocta-hedra, in a 20 mL glass bottle, 0.3 mL of Pd nanocubes aqueous solution (Pd mass concentration was about 2.0 mg/mL), 105 mg of PVP-55K, 0.1 mL of formaldehyde solution and 7.6 mL of ultra-pure water were mixed evenly. The solution was pre-heated at 60 °C for about 10 min under magnetic stirring. After that, 3 mL aqueous solution containing 29 mg of Na_2PdCl_4 was rapidly added to the preheated reaction solution with a pipette. The reaction was allowed to proceed at 60 °C for additional 3 h. The sizes of as-prepared Pd octahedra can be tuned by the sizes of Pd nanocube seeds. For example, the Pd octahedra of ~37, ~21 and ~14 nm in edge length (see corresponding TEM images in Figure 3.3a–c, respectively) were obtained when Pd cubes of ~18, ~10 and ~6 nm in edge length were used, respectively. The products were collected by centrifugation at 15,000 rpm for 15 min, and washed with deionized water, finally dispersed in ultra-pure water.

3.2.1.2.4 Pd nanooctahedra in the range of ~9 to ~23 nm via controlling the rates of etching and regrowth of Pd nanocubes [77]

This method describes the oxidative etching and regrowth of Pd nanocubes into nanooctahedra without involving additional Pd precursors. In a standard procedure, 33.3 mg of PVP, and 0.1 mL of Pd nanocubes (see Section 3.2.1.1.1. The final Pd nanocubes were dissolved in 11 mL of triethylene glycol for further use) were mixed with 8.0 mL of triethylene glycol. The suspension was placed in an oil bath under magnetic stirring, and pre-heated to 130 °C for about 10 min. Then, a certain amount of aqueous HCl (the etchant) was added and kept at 130 °C for additional 2 h to produce Pd nanooctahedra with different sizes. By using ~18 nm Pd nanocubes as starting materials, ~13, ~18 and ~23 nm Pd nanooctahedra can be obtained when 15 μL (180 μmol), 10 μL (120 μmol) and 5 μL (60 μmol) of HCl were added into the reaction, respectively (see corresponding TEM images in Figure 3.3d–f, respectively).

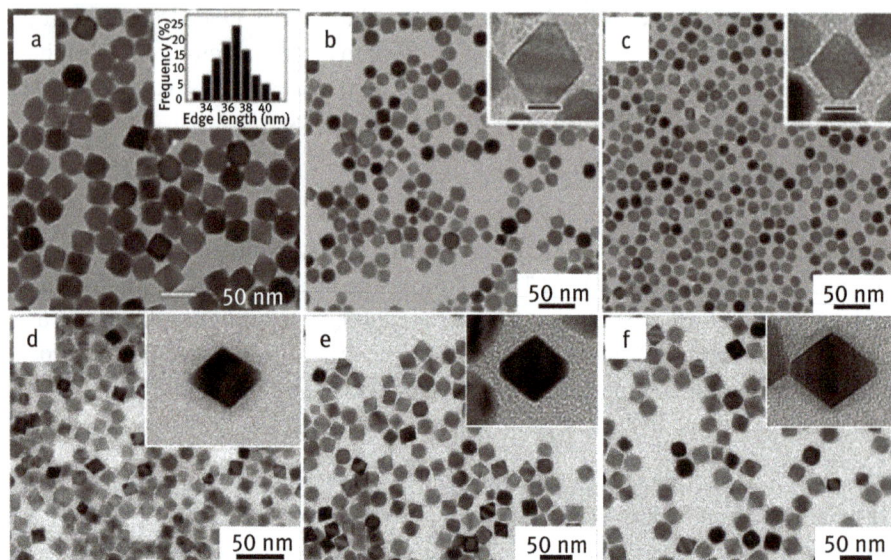

Figure 3.3: (a–c) TEM images of ~37 nm, ~21 nm and ~14 nm Pd octahedra that based on seed-mediate synthesis (Adapted with permission from ref. 76. Copyright 2012, The Royal Society of Chemistry). (d–f) TEM images of ~13 nm, ~18 nm and ~23 nm Pd octahedra that based on etching and regrowth method (Adapted with permission from ref. 77. Copyright 2013, American Chemical Society).

When using ~13 nm Pd nanocubes as starting materials, Pd nanooctahedra with an edge length of ~9 and ~16 nm would be produced when 15 µL (180 µmol) and 5 µL (60 µmol) of HCl was introduced into the reaction solution, respectively. The products were collected by centrifugation at 16,000 rpm for 15 min, and washed with mixture ethanol/acetone (V/V = 1/1) twice, finally dispersed in ultra-pure water.

3.2.1.3 Pd rhombic dodecahedra (RD) NCs with {110} facets

Although the {110} facets is belonged to low-index facets, for *fcc* noble metal NCs, the specific surface energy of {110} facts is higher than {100} and {111} facets, and even higher than some high-index facets, such as {221} facets. So it is much difficult to prepare Pd RD NCs bound by {110} facets.

3.2.1.3.1 Pd RD NCs based on one-pot synthesis [35]

In a typical synthesis, H_2PdCl_4 aqueous solution (10 mmol/L, 0.5 mL) and KI solution (1 mmol/L, 50 µL) were subsequently added into CTAB aqueous solution (25 mmol/L, 10 mL) at 95 °C under stirring at 250 rpm. After 5 min, freshly prepared AA aqueous solution (0.1 mol/L, 80 µL) was quickly added, and the reaction was allowed to proceed for additional 30 min. The size of as-prepared Pd RD NCs was about 45 nm. The products were collected by centrifugation at 10000 rpm for 12 min, and washed with ultra-pure water twice, finally dispersed in ultra-pure water.

3.2.1.3.2 Pd RD NCs with a size about 70 nm by seeded growth [34]

Typically, KI solution (1 mmol/L, 25 µL), CTAB solution (0.1 mol/L, 5 mL), H_2PdCl_4 solution (10 mmol/L, 125 µL) and 40 µL of ~22 nm Pd nanocube seed solution (see Section 3.2.1.1.2) were mixed thoroughly in a water bath at 80 °C. Then, freshly prepared AA solution (0.1 mol/L, 100 µL) was added to initiate reaction. The resulting solution was allowed to be kept at 80 °C for 1 h. The reactions were stopped by centrifugation (6000 rpm, 10 min), and wished twice by using doubly distilled water (Figure 3.4 shows the SEM and TEM images of the products). Finally, the precipitates were redispersed in ultra-pure water.

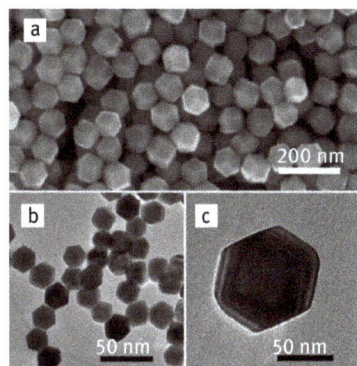

Figure 3.4: (a) SEM, (b, c) TEM images of Pd RD NCs based on seed-mediate growth (Adapted with permission from ref. 34. Copyright 2010, American Chemical Society.).

3.2.1.3.3 Large-scale synthesis of ~71 nm Pd RD NCs [41]

With modified methods of Section 3.2.1.3.2, large-scale (about 60-fold high than Section 3.2.1.3.2) synthesis of Pd RD NCs with similar size can be achieved.

In a typical synthesis, 0.25 mL of 10 mmol/L NaI solution was added into 6.8 mL of deionized water containing 1 g of CTAB, then 3.2 mL of 0.1 mol/L H_2PdCl_4 was mixed together under stirring in a water bath at 30 °C. After that, 10 mL of aged ~11 nm Pd nanocube seeds (see Section 3.2.1.1.2) was added to the growth solution under vigorous stirring, followed by the addition of 5 mL of 1 mol/L AA. After vigorously stirring for 6 h, the products were collected by centrifugation at 10,000 rpm for 5 min to remove excess CTAB and were subsequently redispersed in deionized water.

This protocol can be further modified to yield of up to 70 mg of Pd nanocubes (about 500-fold high than Section 3.2.1.3.2) with size about 45 nm for reaction volumes of 1 L [73].

3.2.2 Synthetic approach for single-crystalline Pd nano-polyhedra with well-defined high-index facets

Compared to NCs with low-index facets, the exposure of high-energy facets is energetically unfavorable, and these facets eventually vanish during crystal growth in order to minimize the total energy in the growth system [26–28, 78–81]. However, these high-index facets usually possess high density of atomic steps, ledges, kinks, and abundant unsaturated coordination sites, so the NCs with high-index surfaces usually exhibit superior performances than those with low-index surfaces. Therefore, the controlled synthesis of Pd NCs with exposed high-index facets is important for both fundamental and application-based research.

3.2.2.1 Concave Pd nanocubes with controllable {hk0} facets based on one-pot synthesis [48]

In a typical synthesis, 5 mL of 0.6 mmol/L H_2PdCl_4 aqueous solution was added into 5 mL aqueous mixed solution containing CTAC (0.01 mol/L) and CTAB (0.01 mol/L) with different volume ratios at 30 °C. Then, 100 µL of freshly prepared 0.1 mol/L aqueous AA solution was quickly added to the resulting solution with gentle shaking and left undisturbed for about 7 h. The exposed high-index {hk0} facets of as-prepared concave Pd nanocubes can be roughly adjusted by the volume ratios between CTAC and CTAB. For example, the concave Pd nanocubes with {730}, {410} and {610} facets would be produced when the volume ratios of CTAC and CTAB were 4:1 (Figure 3.5a), 1:1 and 1:4, respectively. The final sizes were all about 20 nm. The products were collected by centrifugation at 12,000 rpm, for about 10 min, and wished with ultra-pure water twice, then redispersed in ultra-pure water.

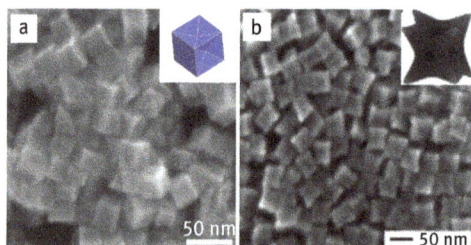

Figure 3.5: (a) SEM of concave Pd nanocubes based on one-pot synthesis (Adapted with permission from ref. 48. Copyright 2011, Wiley-VCH). (b) TEM image concave Pd nanocubes based on seed-mediate growth (Adapted with permission from ref. 47. Copyright 2011, Wiley-VCH).

3.2.2.2 Size controlled concave Pd nanocubes with {730} facets via seed-mediated growth [47]

Typically, 7.7 mL of an aqueous solution containing 105 mg of PVP (MW~55,000), 60 mg of AA and 300 mg of KBr, and 0.3 mL of the Pd seeds (see Section 3.2.1.1.1, 1.8 mg/mL for elemental Pd) were mixed in a glass vial at 60 °C under magnetic stirring. Subsequently, 3.0 mL of an aqueous solution of Na_2PdCl_4 (14.5 mg) was added to the mixture by pipette and the reaction was allowed to proceed at 60 °C for about 3 h. Finally, the product was centrifuged and washed three times with water (Figure 3.5b). The sizes of final concave Pd nanocubes are highly dependent on the sizes of the Pd nanocube seeds used, for example, the seeds with sizes of ~6, ~10, and ~18 nm would produce the concave Pd nanocubes with sizes of ~15, ~27, and ~37 nm, respectively.

3.2.3 Synthetic approach for Pd NCs with CPT structure

In addition the single-crystal NCs, there is a growing interest in synthesizing CPT NCs with twin defects, such as decahedral and icosahedral NCs, due to the fascinating physicochemical properties arising from their unique geometric structures [82–86]. For example, the CPT Pd icosahedral NCs exhibits superior H_2 storage as compared to its single crystalline octahedral counterpart due to the unique crystal lattice expansions exist in two dimensions in CPT icosahedra [87]. Pd icosahedral NCs showed superior catalytic activity than other single-crystal Pd tetrahedral NCs bound by same {111} facets in the electrooxidation of formic acid [88]. Compared to the single-crystalline, the synthesis of CPT nanostructures is less rich, and the approach is not well developed due to their formation mechanism is not fully understood. The difficulty in the synthesis of CPT Pd NCs greatly hampers the exploration of their properties.

3.2.3.1 Solvothermal process [88]

In a typical synthesis, 12.2 mg of Pd(acac)$_2$ and 20 µL of oleylamine were dissolved in 5 mL of toluene, and the solution was vigorously stirred for 10 min. Then, 25 µL of formaldehyde was added, and the mixed solution was continually stirred at room temperature for another 10 min. After that, the resulting solution was transferred to an autoclave and heated at 100 °C for 8 h and then slowly cooled to room temperature. The products were collected using centrifugation and washed with a mixed hexane/ethanol (v/v, 1:3). Interestingly, the shapes of as-prepared Pd NCs were highly related to the amount of oleylamine used, for example, 20 µL for icosahedra, 100 µL for decahedra and tetrahedra, and 1 mL for octahedra and triangular plates (the stirring periods at room temperature were both reduced to 5 min, while keep other experiment condition unchanged). The separation of tetrahedra and triangular plates can be achieved by cooling the products slowly, the products were on the bottom of the autoclave.

3.2.3.2 CPT Pd icosahedra with controllable sizes in the range of 15–42 nm via polyol route [89]

In a typical synthesis, 0.666 g of PVP (MW~360,000, 6 mmol) and 0.3 mL of 1 mol/L NaCl aqueous solution were added to 30 mL of ethylene glycol (EG) in a glass tube under stirring. After about 30 min, 0.3 mL of 0.5 mol/L Na$_2$PdCl$_4$ aqueous solution was introduced under vigorously stirring. The reaction was heated in air at 120 °C for 48 h. The sizes of as-prepared Pd icosahedra can be roughly controlled by the reaction temperature. For example, the final sizes were ~42, ~31, and ~15 nm, when the reaction temperatures were hold at 100, 120, and 150 °C, respectively. The final product was collected with acetone by centrifugation, and then washed repeatedly with ethanol to remove any residual EG or PVP.

3.2.3.3 CPT Pd icosahedra with controllable sizes in the range of 5–17 nm via polyol route [90]

In a typical synthesis, 30 mg of PVP was dissolved in 2 mL of EG in a 20 mL vial and preheated in an oil bath at 160 °C for 20 min under magnetic stirring. 1 mL 50 mmol/L H_2PdCl_4 in EG solution, which was prepared by dissolving $PdCl_2$ in a mixture of EG and 37 % (v/v) HCl, in which the molar ratio of HCl to $PdCl_2$ was 4:1 was then quickly injected into the using a pipette. A specific amount of HCl was also added to achieve a final concentration of 134 mmol/L in the reaction solution. After the reaction proceeded for different time, Pd icosahedra with different sizes would be produced. When the reaction was 10 s, 1 min, 2 min and 3 h, the average size of the Pd icosahedra was 5, 12, 14, and 17 nm, respectively. The reaction was stopped by immersing the vial in an ice-water bath. The product was collected by centrifugation, washed with acetone and ultra-pure water, and finally redispersed in ultra-pure water.

3.2.3.4 CPT Pd decahedra with controllable sizes in the range of 5–15 nm via polyol route [91]

In a standard procedure for the synthesis of Pd decahedra with a size about 12.4 nm, 80 mg of PVP, and 40 mg of Na_2SO_4 were subsequently introduced into 2 mL of diethylene glycol (DEG) in a 20 mL vial, and the mixed solution was preheated in an oil bath at 105 °C for 20 min under magnetic stirring. Then, 1 mL of DEG solution containing 15.5 mg of Na_2PdCl_4 was injected using a pipette. After the reaction had proceeded for 3 h, it was quenched by immersing the vial in an ice-water bath. The sizes can be roughly controlled by stopping the reaction at different time, for example, when the reaction was 0.5, 1, 2, 4, 6 and 24 h, the average size of the Pd icosahedra was 5.1, 6.5, 10, 13.9, 14.6 and 15.2 nm, respectively. The product can be collected by centrifugation and washed once with acetone and then twice with ultra-pure water.

3.2.4 Synthetic approach for monodisperse spherical Pd nanoparticles (NPs)

3.2.4.1 Oleylamine-mediated synthesis of 4.5 nm Pd NPs [32]

In a standard procedure, 75 mg of Pd(acac)$_2$ was mixed with 15 mL of oleylamine under a nitrogen flow. The resulting solution was preheated to 60 °C in 10 min. After that, 3 mL oleylamine containing 300 mg of Borane tributylamine complex (BTB) was quickly injected. Then the reaction temperature was raised to 90 °C at a rate of 3 °C/min. After kept at 90 °C for 60 min, the solution was cooled down to room temperature. 30 mL of ethanol was added to precipitate the products. To obtain sufficiently large electrochemically active surface area, these oleylamine-coated Pd NPs can be readily "cleaned" by acetic acid washing. Finally, the product was dispersed in hexane.

3.2.4.2 Pd NPs with sizes about 2.0–4.8 nm [92]

Synthesis of 2.0 nm Pd NPs: In a 100 mL three-neck flask equipped with a gas-inlet adapter, a silicon septum cap, and a thermometer, 112 mg of Pd(acac)$_2$ (0.5 mmol) and 3.3 mL of oleylamine (10 mmol) were mixed at 50 °C. Then, 3.2 mL oleic acid (10 mmol) was then added to the solution under stirring at 50 °C for 1 h. 20 mL of chloroform was added to the resulting solution and the flask was purged with nitrogen to avoid etching of the Pd nanoparticles. The solution was then transferred to an ice water bath, and the temperature was cooled to approximately 2 °C. Subsequently, a fresh 5 mL of chloroform solution containing 1.54 g of TBAB (6 mmol) was quickly injected into the solution under vigorous stirring, and kept at 2 °C for 1 h, followed by the addition 3.3 mL of OAm (10 mmol). The resulting solution was left for several hours to cool to room temperature in a nitrogen atmosphere. Finally, the Pd nanoparticles were purified with ethanol and centrifugation and were redispersed in a nonpolar solvent.

Generally, the size of NPs is strongly dependent on the nucleation rate. The higher of the nucleation rate is, the smaller of prepared NPs will be. By decreasing the amount surfactants to lower the nucleation rate, and meanwhile elevating the reaction temperature to increase the growth rate, the larger sizes of as-prepared Pd NPs can be obtained.

Synthesis of 3.0 to 4.8 nm Pd NPs: In a 100 mL recovery flask equipped with a gas-inlet adapter, 112 mg of Pd(acac)$_2$ (0.5 mmol) and 3.3 mL of oleylamine (10 mmol) were mixed at 50 °C. Then, 3.2 mL of oleic acid (10 mmol) was added to the solution under stirring at 50 °C for 1 h, resulting in a pale yellow clear solution. And 2 mL chloroform solution containing 386 mg of TBAB (1.5 mmol) was subsequently quickly injected into the solution with vigorous stirring, and allow the reaction proceed at 50 °C for 1 h. The sizes of as-prepared Pd NPs can be roughly tuned by varying the amount of TBAB. With 1.0, 0.6, and 0.3 mmol of TBAB, Pd NPs were obtained with sizes of 3.4 ± 0.3, 4.0 ± 0.3, and 4.8 ± 0.4 nm, respectively. The products can be collected by adding 30 mL ethanol to the solution after it cooled to room temperature. The Pd NPs were purified by centrifugation, and redispersed in a nonpolar solvent.

3.2.4.3 Pd NPs with controllable sizes in the range of 5–10 nm via seed-mediated growth [93]

Synthesis of 5 nm monodisperse Pd NPs: typically, 5 mg of Pd(acac)$_2$ was dissolved in the mixture of 10 mL of n-octylamine and 0.2 mL of formaldehyde. The resulting growth solution was then transferred to a 25-mL Teflon-lined autoclave and maintained at 200 °C for 3 h. After it cooled to room temperature, the product was collected by centrifugation (12,000 rpm for 10 min), and wished with ethanol several times. Finally, the product was dispersed in hexane for further use.

As for the preparation of Pd NPs of about 8 nm, the as-prepared 5 nm Pd NPs solution without separation was employed as the seed solution. 2 mL of seed solution was mixed with the growth solution containing 5 mg of Pd(acac)$_2$, 10 mL of n-octylamine and 0.2 mL of formaldehyde, followed by transferring to a 25-mL Teflon-lined

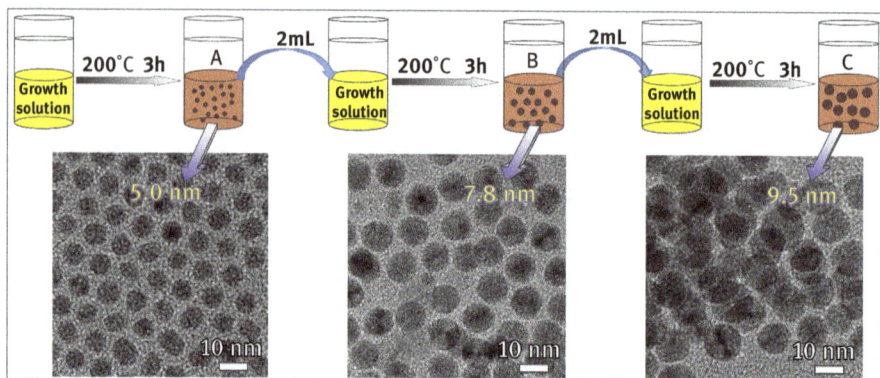

Figure 3.6: Schematic illustration of the synthetic procedure used for the preparation of monodisperse Pd NPs with different sizes by stepwise seed-mediated solvothermal growth. (Adapted with permission from ref. 93. Copyright 2012, Springer.).

autoclave and maintained at 200 °C for 3 h. Similarly, about 10 nm Pd NPs were synthesized using unseparated 8 nm Pd NPs as growth seeds for another growth round at 200 °C for 3 h (Figure 3.6).

3.2.5 Synthetic approach for typical anisotropic Pd NCs

Anisotropy is one of the basic properties of crystals. Engineering NCs with different types of morphologies is very important to explore their novel physical and chemical properties [61, 94–96]. Up to date, some excellent works have been done in the synthesis of Pd NCs with novel shapes. It has been demonstrated that some anisotropic Pd NCs have unexpected physical and chemical properties. For example, Zheng and coworkers reported the synthesis of freestanding hexagonal Pd nanosheets that are less than 10 atomic layers thick [16]. Interestingly, these Pd nanosheets are blue in color, and the surface plasmon resonance peak in the near-infrared region can be well tunable by tailoring their edge lengths. Xia and coworkers reported the synthesis of singly twinned Pd right bipyramids enclosed by {100} facets, which exhibited enhanced catalytic activities toward the formic acid oxidation reaction due to their unique shape [97]. Most recently, Jin and coworkers reported the Pd nanoframes exhibit exceptional catalytic activity and great durability toward formic acid oxidation reaction due to their novel geometry [98]. With an aim to maximize the utilization of noble metal Pd NCs, the exploration of novel anisotropic shapes is very necessary.

3.2.5.1 Freestanding Pd nanoplates with an edge length from 28–160 nm [16]

In a typical synthesis of Pd nanoplates with an edge length about 60 nm (Figure 3.7a, b), 50 mg of Pd(acac)$_2$, 160 mg of PVP (MW = 30000), and 185 mg of CTAB were dissolved in

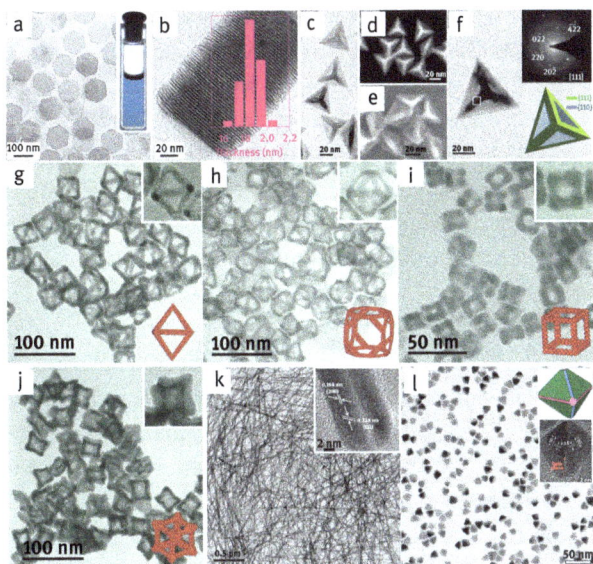

Figure 3.7: (a–b) TEM image of the flat and vertical blue Pd nanosheets (Adapted with permission from ref. 16. Copyright 2011 Nature Publishing Group). (c–f) TEM, HAADF-STEM, SEM, HRTEM images of concave tetrahedral Pd NCs (Adapted with permission from ref. 99. Copyright 2009, American Chemical Society). (g–j) TEM images of the octahedral, cuboctahedral, cubic, and concave cubic Pd nanoframes (Adapted with permission from ref. 98. Copyright 2017, American Chemical Society). (k) TEM and HRTEM images of Pd nanowires (Adapted with permission from ref. 100. Copyright 2009, American Chemical Society). (l) TEM, HRTEM images and model of Pd right bipyramids (Adapted with permission from ref. 97. Copyright 2013, American Chemical Society).

a mixed solution containing 10 mL of N, N-dimethylformamide (DMF) and 2 mL of water. The resulting homogeneous yellow solution was transferred to a glass pressure vessel charged with CO to 1 bar, the reaction was proceeding at 100 °C for 3 h before it was cooled to room temperature. The dark blue products were precipitated by acetone, separated via by centrifugation and further purified by an ethanol–acetone mixture. The sizes can be roughly controlled by reaction time or the solvent. When the DMF and water were used as solvent, the as-prepared Pd nanoplates had edge lengths of about 28, 46 and 60 nm when the reaction times were 0.5, 1.5, and 3 h, respectively. When different solvents used, such as dimethylpropionamide, 2-phenylethanol, and benzyl alcohol, the final average edge lengths for nanosheets produced in 3 h reactions were about 35, 125, and 160 nm, respectively.

3.2.5.2 Pd nanoplates with an edge length from 6–59 nm [17]
Although the above protocol for the synthesis of Pd nanoplates is straightforward, the used CO is toxic. In modified method by Zhang et al., they replaced CO gas with $W(CO)_6$, Pd nanoplates with controlled edge length were also successfully prepared.

In a typical synthesis of Pd nanoplates with an edge length about 36 nm, 16 mg of Pd(acac)$_2$, 10 mg of citric acid, 60 mg of CTAB, and 30 mg of PVP were dissolved in 10 mL of DMF in 25 mL flask under stirring. After 1 h, 100 mg of W(CO)$_6$ were added into the resulting homogeneous solution under an Ar atmosphere. Then, the flask was capped and heated at 80 °C for 1 h. The prepared Pd nanosheets were collected by centrifugation using a sufficient amount of acetone, and then re-dispersed in ethanol. This process was repeated three times. By simply varying the amount of citric acid from 10 to 50, 90, 130, and 170 mg, Pd nanosheets with edge lengths from 36 to 24, 16, 9, and 6 nm were prepared, respectively.

. The larger Pd nanosheets were prepared through seeded growth. For the Pd nanosheets with an edge length about 47 nm, 13.1 mmol of as-preformed Pd nanosheets with an average edge length of 36 nm, 8 mg of Pd(acac)$_2$ and 50 mg of PVP were mixed in 10 mL of DMF in a 25 mL flask. Then, 50 mg of W(CO)$_6$ was added into the flask under an Ar atmosphere. The flask was capped and heated at 60°C for 4 h. Finally, the Pd nanosheets were collected by centrifugation using a sufficient amount of acetone, and then re-dispersed in ethanol. This process was repeated three times. Similarly, Pd nanosheets with an edge length about 59 nm were synthesized by using 47 nm Pd nanosheets as the seeds, with all other experiment conditions were kept the same as in the aforementioned procedure.

3.2.5.3 Concave tetrahedral Pd NCs [99]

In a typical synthesis of concave tetrahedral Pd NCs (Figure 3.7c–f), 50 mg of Pd (acac)$_2$, 160 mg of PVP (MW = 30000), and 0.1 mL formaldehyde solution (40 %) were dissolved in 10 mL benzyl alcohol. The resulting homogeneous yellow solution was transferred to a 20 mL Teflon-lined stainless-steel autoclave, and heated at 100 °C for a desired time (i. e., 1 h, 2 h, 4 h, 8 h and 15 h) before it was cooled to room temperature. Their average side length was 16 nm at 1 h, 24 nm at 2 h, 38 nm at 4 h, and 45 nm at 8 h. The black nanoparticles were precipitated by acetone, separated via centrifugation, and further purified by an ethanol–acetone mixture.

3.2.5.4 Pd nanoframe with different shapes via etching process [98]

The octahedra (Figure 3.7g), cuboctahedra (Figure 3.7h), cubic (Figure 3.7i), and concave cubic Pd nanoframes (Figure 3.7j) were prepared by excavating corresponding solid NCs. These solid were prepared according to the above method. For the synthesis of 18 nm Pd nanocube seed, please refer to see Section 3.2.1.1.1. The final products were redispersed in 11 mL water for further use; for the synthesis of 37 nm Pd nanooctahedra seed, please refer to Section 3.2.1.2.2, the final products were collected by centrifugation, washed two times with water, and finally redispersed in 20 mL of DMF for further use; for the synthesis of Pd cuboctahedra seed, the synthetic method is as same as the synthesis of 37 nm Pd nanooctahedra, but the Na$_2$PdCl$_4$ was reduced to 8.7 mg; for the synthesis of

37 nm concave Pd nanocubes, please refer to Section 2.2.1.2, the final products were collected by centrifugation, washed two times with water, and finally redispersed in 20 mL DMF for further use.

Etching of Pd NCs: in a standard procedure, 1.5 mL of DMF solution containing 20 mg of PVP (MW~55000), 1.5 mg of KI, 1.0 mL of the as-obtained Pd NCs, and 5 μL HCHO (10-fold dilution) were placed in a vial capped with a rubber plug. The reaction system was evacuated, subsequently, and a 10 mL of O_2 was injected with a syringe. The reaction proceeded at 100 °C for 1 h. After collection by centrifugation and washing three times with water, the final product was redispersed in water.

3.2.5.5 Pd nanowires [100]
In a typical synthesis of Pd nanowires (Figure 3.7k), 17.7 mg of $PdCl_2$, 300 mg of NaI, and 800 mg of PVP (MW = 30000) were mixed in 12 mL ultra-pure water. The resulting homogeneous dark red solution was transferred to a 20 mL Teflon-lined stainless-steel autoclave and heated at 200 °C for 2 h before it was cooled to room temperature. The nanowires have a uniform diameter of 9.0 ± 1.0 nm along their entire length, which is in the range of micrometers and can be up to 3 μm. The products were precipitated by isopropanol, separated via centrifugation at 8000 rpm, and further purified twice by an ethanol–isopropanol mixture.

3.2.5.6 Pd nanorods [101]
Firstly, the Pd seeds were prepared according to following protocol. In a round-bottom flask, 10 mL of 2.5×10^{-4} M Na_2PdCl_4 aqueous solution was bubbled with nitrogen for 30 min. Then the solution was injected into a two-neck flask containing 0.3645 g CTAB and stirred until the CTAB powder was dissolved. During stirring, 10 mL of ice-cold 15 mM $NaBH_4$ solution was prepared in another flask. Next, 220 μL of the $NaBH_4$ aqueous solution was quickly injected into the above solution. This solution must be kept under nitrogen to avoid exposure to air.

In a typical synthesis of Pd nanorods with an average length of 308 nm, two 20 mL vials labeled A and B were used. 0.345 g of CTAB was mixed with 10 mL of Na_2PdCl_4 aqueous solution into both vial A and B, and stirred until a pale orange solution formed. Then 78 μL of 0.1 M AA was added as a reducing agent into vial A. Next, 84 μL of the seed solution was injected into vial A with shaking about for 5 s. After 2 h, the solution in vial A turned blackish green. At this time, 100 μL of 4.0×10^{-3} M copper(II) acetate solution and 78 μ L of 0.1 M AA was subsequently added to vial B. Then 84 μ L of the solution in vial A was withdrawn and added to vial B. This vial was shaken for 5 s, and the mixture was left undisturbed at 30 °C for 20 h. The products were at the bottom of the vial, and it was collected by decanting the upper solution, then washed with deionized water twice. The edge length is very sensitive to the amount of copper acetate. When only 50 μL of 4.0×10^{-3} M copper acetate solution used, Pd nanorods with an average length of 200 nm were prepared.

3.2.5.7 Pd right bipyramids [97]

In a typical synthesis of Pd right bipyramids (Figure 3.71), in a 25 mL three-neck flask, 5 mL of EG containing 400 mg PVP and 150 mg NaI was added. The flask was transferred to an oil bath, and pre-heated to 160 °C under magnetic stirring. Then, 1 mL of 15 mg/mL Na_2PdCl_4 EG solution was rapidly injected into the flask using a pipette. After 2 h, the reaction was terminated by quickly placing the flask in an ice-water bath with gentle shake for 2–3 min. The product was collected by centrifugation, and washed with acetone once and DI water four times.

Alternatively, the Pd right bipyramids can be synthetized following [38]: A vial containing 0.15 g of NaI, 0.40 g of PVP, 6.2 mL of water, and 6.0 mL of ethanol was sonicated until got a clear solution. Then, 1.0 mL of 0.1 M Na_2PdCl_4, 0.80 mL of 0.1 M Na_2CO_3, and a dilute 1.0 mL of HCHO solution (0.37–0.40 wt %) were added to the vial. The vial was gently stirred, and the pH value was between 8.0 and 8.5. The resulting homogeneous dark red solution was transferred to a 20 mL Teflon-lined stainless steel autoclave. The sealed vessel was then heated at 140 °C for 4–12 h before it was cooled to room temperature. The products were collected by centrifugation at 10,000 rpm for 15 min and washed three times with ethanol.

3.3 Characterization methodologies and instrumentation techniques

3.3.1 Routine characterization

3.3.1.1 X-ray diffraction (XRD)

XRD is a fast and reliable tool for the identification of the crystal phase of crystal materials, because every crystal has its own unique pattern. For the most of the noble metal NCs that synthetized via wet-chemical methods, they were dissolved in the solution, such as water, ethanol, hexane, one can simply put the droplet of the as-prepared suspension on a glass side, as shown in Figure 3.8. After evaporation of the solution, the formed smooth film can be measured by X-ray powder diffraction.

During the measurement, there will be several characteristic peaks, which provide a unique fingerprint of the as-prepared NCs. For *fcc* Pd, we usually measure from 20° to 90° (2 Theta), and the characteristic peaks are located at 40.1° (111), 46.66° (200), 68.12° (220), 82.10° (311) and 86.62° (222), respectively. By comparison with standard reference patterns (e. g. JCPDS No. 46-1043), the type of NCs can be identified. It should be pointed out that if the diffraction peaks were not perfectly symmetrical, and the barycenter of these peaks had a little left deviation, it may mean two-dimensional lattice expansion exists in the as-prepared nanoparticles, which is a unique characteristic in the XRD pattern of CPT nanostructures [89].

It also should be noted that different diffraction peaks have different intensity. The relative intensity cannot be used to determinate the surface structure of as-prepared well-defined structure. For example, it is not correct to conclude

Figure 3.8: Schematic diagram of the preparation of sample for XRD analysis.

that the as-prepared NCs are bounded by {100} facets, by just said the measured intensity ratio between {100} and {111} of as-prepared NCs is high than that of standard pattern. Nevertheless, there is preferential orientation for special shapes with well-defined surface structures. For example, for well-shaped octahedral Au NCs, only (111) peak appears in the pattern, indicating the {111} planes of Au NCs have a preferential orientation (parallel to the substrate) [102]. Similarly, for Pd nanocubes with exclusive {100} facets, they tend to orient along {100} planes parallel to the substrate, thus giving higher (200) diffraction intensity than that of other peaks [31].

Besides the identification of the crystal phase, XRD can be used to estimate the sizes of as-prepared NCs (size ranges from 1–100 nm) according to the Debye-Scherrer eq. (3.1) by the analysis of the diffraction peak of XRD pattern.

$$D = K\lambda/\beta\cos\theta \tag{3.1}$$

where D is the average crystallite size; λ is the X-ray wavelength in nanometer (nm); K is Scherrer constant related to crystallite shape, it is normally taken as 0.9 when the half maximum intensity used to determinate the peak width (β) of the diffraction peak. The value of β in 2θ axis of diffraction profile should be in radians. It should be noted that, the actual value of β should be after subtracting the instrumental line broadening (also in radians). θ is Bragg angle (in degree).

The size estimated by Scherrer formula is usually smaller than its actual size. This is because there are many other factors can result in the width of the diffraction peak besides the crystalline size, such as stacking faults, twinning defects, and so on.

3.3.1.2 Scanning electron microscopy (SEM)

SEM is a useful technics to directly give us the 3-dimension (3D) images of as-prepared NCs. The preparation of sample is quite simple. Make one droplet of as-prepared suspension of NCs on a silicon wafer, after the evaporation of the solution, the silicon wafer was transferred to SEM holder and for the next analysis.

The low magnification SEM images can give us the information of the uniformity of as-prepared NCs. Usually, for the NCs with various shapes, it is better to provide some high magnification SEM images from different angles. The visualization can help us build the 3D geometrical models of the as-prepared NCs, and get the useful information of the exact exposed facets.

3.3.1.3 Transmission electron microscopy (TEM)
TEM is a useful technics to directly give us the 2-dimension (2D) images of as-prepared NCs. For the preparation of the TEM samples, make one droplet of as-prepared suspension of NCs on a TEM Cu grid, after the evaporation of the solution, the grid was transferred to TEM holder and for the next analysis.

In addition to the characterization of the shapes, it is easy to get the intrinsic crystal information, such as the structural defects, by selected area electron diffraction (SAED) and high resolution TEM (HRTEM). For the NCs with well-defined shapes, the TEM images from different zone axes can help obtaining an ideal 3D geometrical models, which would further be used to estimate the actual surface structures of the NCs [48, 103–105].

If the samples lie flat well on the TEM Cu grid, we can also get the size distribution of the as-prepared NCs by counting lots of samples.

Due to the instrumental intrinsic imperfections of electron lenses, the resolution of TEM is usually effected. The development of spherical aberration-corrected (C_s) TEM help compensate for many of instrumental limitations responsible, such as eliminating the contrast delocalization in images, and it would give us more structural details of the NCs [106, 107].

High-angle-annular-dark-field scanning TEM (HAADF-STEM) can provide the Z-contrast imaging, is widely used to obtain the structural information of the NCs, especially the NCs containing two different noble metals, such as core-shell structures [108, 109]. In addition, by combining HAADF-STEM with a reconstruction algorithm based on a series of images from different orientations, it is possible to obtain the 3D structures of as-prepared NCs, even with atomic resolution [110].

3.3.2 Identification of surface structures of NCs with different shapes

3.3.2.1 Identification of NCs with low-index facets
Pd is belong to *fcc* structure. The basic shapes that with low-index facets are cube with 6 {100} facets, octahedron with 8 {111} facets, and RD with 12 {110} facets. The surface energies of these three kinds of low-index surfaces follow the order of {111} < {100} << {110}. Thus, compared to the frequently prepared cubic and octahedral Pd NCs with low surface energies, the Pd RD NCs are more difficult to be obtained due to their higher surface energies.

The identification of these three kinds of NCs can be done by SEM, TEM, and corresponding SAED. The SEM provides the overall morphologies of as-prepared Pd NCs, the TEM further confirm the shape from certain zone axis (for cube, usually [100] zone axis; for octahedron, usually [111] or [110] zone axis; for RD, usually [110] zone axis), which can be determined by corresponding SAED.

3.3.2.2 Identification of NCs with high-index facets

When additional atomic steps and/or kinks appear on these basic (111, 100), and (110) surface, the high-index facets (which can be expressed by a set of Miller indices {hkl} with at least one index being greater than 1 formed) can be described by a combination of terraces and steps from low-index facets (Figure 3.9) [78, 80]. For example, {hkk} (e. g. (211) facets) surfaces can be described as a combination of (111) terraces and (100) steps; {hhl} (e. g. (221) facets) surfaces are a combination of (111) terraces and (110) steps, and {hk0} (e. g. (310) facets) surfaces are a combination of (100) terraces and (110) steps. In addition to the presence of terraces and steps, the {hkl} surface structures contain atomic kinks. For example, on a (541) plane, both (110) terraces and (111) steps are present, and the kinks (marked by a black zigzag dashed line) also appear in the junctions between the (110) terraces and (111) steps. Compared to those on low-index flat (100) and (111) facets, these high-index facets with additional steps and kinks usually possess higher surface energies.

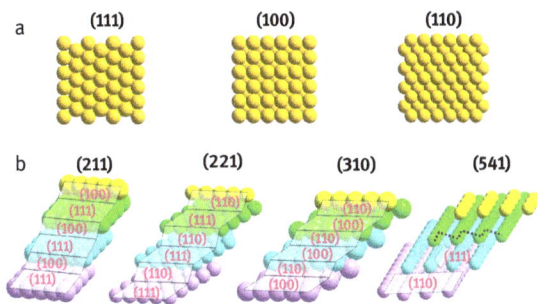

Figure 3.9: Surface atomic arrangements of several typical (a) low-index facets and (b) high-index facets of *fcc* noble metal NC.

Geometrically, the shapes with different {hkl} facets can be constructed from basic three basic shapes with low-index facets (the shapes are summarized in Figure 3.10). For the representative trisoctahedral (TOH) shapes with {hhl} facets (on the [1$\bar{1}$0] zone axis), it can be generated from an octahedron by "pulling out" the centers of the eight triangular {111} facets. For the representative trapezohedral (TPH) shapes with {hkk} (on the [01$\bar{1}$] zone axis), it can be generated from an

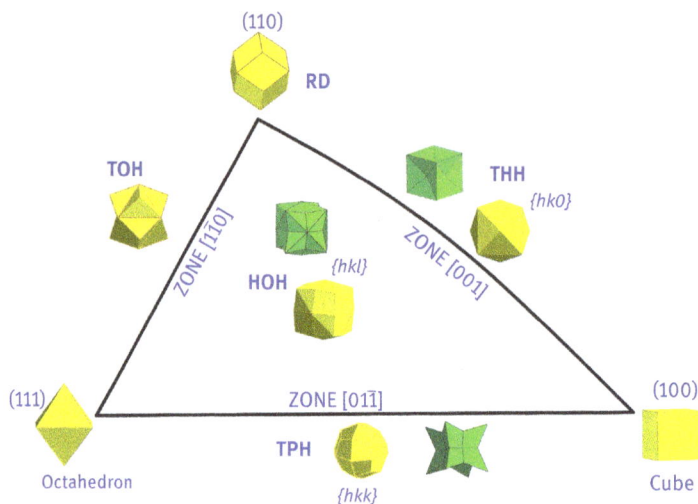

Figure 3.10: Schematic polyhedral models of *fcc* NCs with the various well-defined specific crystal facets. The green 3D model represents the corresponding concave shapes.

octahedron by "pulling out" the centers of the eight triangular {111} facets and the 12 edges simultaneously. For the representative THH shapes with {*hk*0} (on the [001] zone axis), it can be generated from a cube by "pulling out" the centers of the six square {100} facets. For the representative hexoctahedral (HOH) shapes with {*hkl*} facets (h ≠ k ≠ l > 1), it can be obtained from a THH shape after "pulling out" the edge centers in the {100} squares.

For the identification of each type of NCs with high-index facets, the first step is to take SEM images. Then the preliminary 3D model can be constructed, and the model is further optimized according the TEM images along from different zone axes. Finally, the exposed facets can be concluded on the basis of the 3D model. The accuracy of the obtained surface structure can be further confirmed by HRTEM along a specific zone axis. For examples, the stepped facets on THH, TOH can be clearly observed along [001] and [011] zones axis, respectively.

On the other hand, the 3D model of these shapes with high-index facets can be draw mathematically. Thus for a specific type of high-indexed shape, the exposed facets can be calculated from the projection angles along certain zone axis, as shown in Figure 3.11. Consequently, by comparison with the theoretical values, the exposed high-index facets on the as-prepared NCs can be determined by measuring the angles along the certain zone axis.

It is hard to directly image the concave faces for some types of concave NCs, such as concave nanocubes with {hk0} facets. Focused ion beam technique is a useful to cut the NCs make the cross section of the concave nanocube to be exposed, thus the concave face can be directly imaging by HRTEM [111].

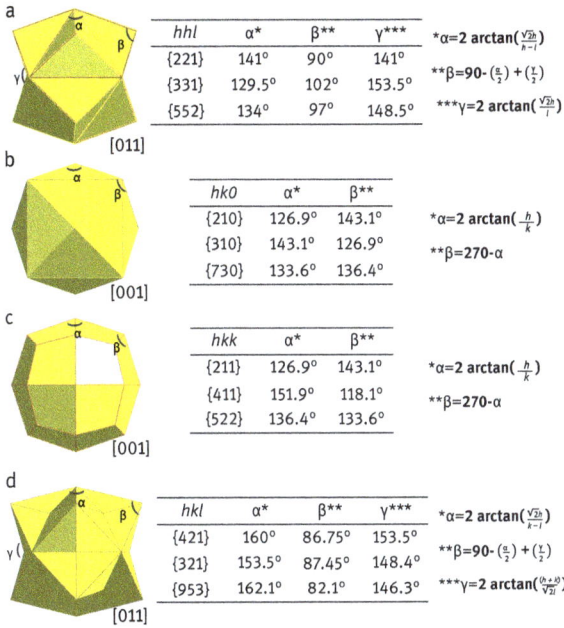

hhl	α*	β**	γ***
{221}	141°	90°	141°
{331}	129.5°	102°	153.5°
{552}	134°	97°	148.5°

$*\alpha = 2\arctan\left(\frac{\sqrt{2}h}{h-l}\right)$

$**\beta = 90 - \left(\frac{\alpha}{2}\right) + \left(\frac{\gamma}{2}\right)$

$***\gamma = 2\arctan\left(\frac{\sqrt{2}h}{l}\right)$

hk0	α*	β**
{210}	126.9°	143.1°
{310}	143.1°	126.9°
{730}	133.6°	136.4°

$*\alpha = 2\arctan\left(\frac{h}{k}\right)$

$**\beta = 270 - \alpha$

hkk	α*	β**
{211}	126.9°	143.1°
{411}	151.9°	118.1°
{522}	136.4°	133.6°

$*\alpha = 2\arctan\left(\frac{h}{k}\right)$

$**\beta = 270 - \alpha$

hkl	α*	β**	γ***
{421}	160°	86.75°	153.5°
{321}	153.5°	87.45°	148.4°
{953}	162.1°	82.1°	146.3°

$*\alpha = 2\arctan\left(\frac{\sqrt{2}h}{k-l}\right)$

$**\beta = 90 - \left(\frac{\alpha}{2}\right) + \left(\frac{\gamma}{2}\right)$

$***\gamma = 2\arctan\left(\frac{(h+k)}{\sqrt{2}l}\right)$

Figure 3.11: Schematic models and corresponding theoretical a, β and γ values of (a) an ideal TOH enclosed by different {hhl} facets projected along the [011] direction; (b) an ideal THH enclosed by different {hk0} facets projected along the [001] direction; (c) an ideal TPH enclosed by different {hkk} facets projected along the [001] direction; and (d) an ideal HOH enclosed by different {hkl} facets along the [011] direction.

3.3.3 Identification of CPT Pd decahedra and icosahedra

CPT NCs (usually in the form of icosahedron and decahedron) are frequently observed in many *fcc* noble metal NCs. The decahedron or icosahedron particle consists of 5 or 20 single-crystal tetrahedral crystallites with (111) faces, respectively, as shown in Figure 3.12. To form a perfect CPT icosahedral or decahedral NC, the

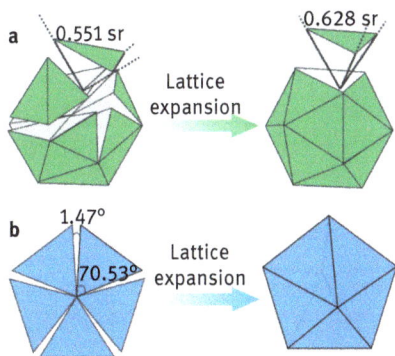

Figure 3.12: Schematic diagram of (a) icosahedron and (b) decahedron with {111} twinned boundaries.

tetrahedral single-crystallites subunits should be expanded. The CPT icosahedron has a pseudo-spherical shape, and contains two-, three-, and fivefold rotational axes. The decahedron has a pentagonal profile, and contains 5-fold rotational axis. These structural characteristics can be well disclosed by TEM images. For example, as shown in Figure 3.13, the HRTEM images clearly show the profiles of an icosahedron from different orientations [86].

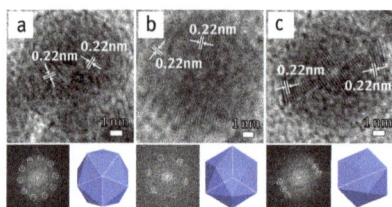

Figure 3.13: HRTEM images of Rh icosahedron oriented along three typical projections (a) 5-, (b) 3-, and (c) 2-fold axis and their corresponding FFT images and geometrical models. (Adapted with permission from ref. 86. Copyright 2017, Tsinghua University Press and Springer-Verlag Berlin Heidelberg.).

3.4 General growth mechanisms of shape and sized-controlled Pd NCs

For the growth of colloidal metal NCs in solution phase, it has been widely accepted that the process can be described by LaMer curve, as shown in Figure 3.14 [95, 112]. The growth process can be divided into three steps. Firstly, the metal atoms (building blocks for a metal NC) formed by decomposition or reduction of metal precursors. With the continually accumulation of atoms, the stable small nuclei form once its

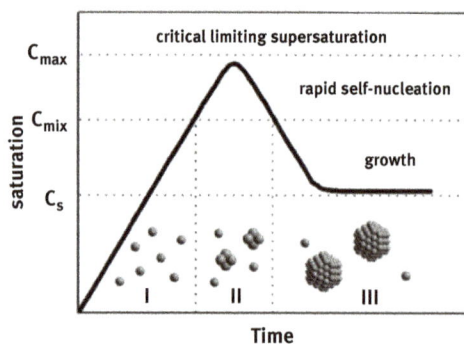

Figure 3.14: LaMer curve illustrating three stages (generation of atoms, nucleation, and subsequent growth) of metal nanocrystal formation in solution system (Adapted with permission from ref. 95. Copyright 2016, American Chemical Society.).

concentration reaches a point of supersaturation (C_{min}). Subsequently, the nuclei continually grow with depleting of newly formed atoms. After the concentration of atoms drops below C_{min}, no additional nucleation will occur. Then the nuclei grow into large NCs through the addition of metal atoms or agglomeration at low supersaturation.

The driving force for this crystallization is the supersaturation of the solute in the growth solution [113]. And the supersaturation can be defined as the difference of the chemical potentials between the infinitely large mother and new phases at the particular value of the temperature. Thus one can describe the nucleation process according to Gibbs free energy, and higher supersaturation usually results in smaller crystallites during the crystallization. Based on this, one can obtain small nanoparticles by increasing the concentration of nucleation via fast reduction of metal precursor. Examples can be found in the synthesis of small Au seeds (about 5 nm) by using strong reducing agents (e. g. $NaBH_4$) [114–117]. In addition, for controlling the NC with uniform size, it is important to make the nucleation occurring in a very short time (step II in Figure 3.14), otherwise polydispersed NCs will form due to that the secondary nucleation happens in the growth stage [66]. For example, in oleylamine-mediated synthesis of Pd nanoparticles (Section 3.2.4.1) [32], the strong reducing agent BTB could greatly boost the reaction rate and thus leading to the formation uniform 4.5 nm Pd nanoparticles. Similarly, the size can also be rough controlled by precisely adjusting the reaction rate. For instance, by intentionally adjusting the nucleation rate through involving different amount TBAB, and the growth rate by changing the reaction temperature, the sizes of Pd nanoparticles can be well controlled from 2 to 4.8 nm (Section 3.2.4.2) [92]. Another effective approach to obtain the uniform NCs is to separate the nucleation and growth steps by using seed-mediate methods [115–119].

For the controlled synthesis of noble metal NCs with various shapes, the types of preformed seeds during the nucleation stage play a key role in the final structures [19]. Except the single-crystalline phase, the *fcc* noble metal NCs usually contain many crystal defects, such as twin defects and stacking faults. The single-crystal seed would result in the single-crystal NCs, and the seed with defects would result in NCs containing twinned defects (e. g. CPT nanoparticles) or stacking faults (e. g. nanoplates). The structures of the prepared shapes of noble metal NCs can be well controlled by precisely tuning the thermodynamic and kinetic factors.

Pd is a classic *fcc* noble metal, the surface energy of exposed facets usually follow the order {111} < {100} < {110} < {*hkl*} (high-index facets). During the growth of NCs, the surface with low surface energy usually exposed in order to minimize the total system energy. The exposure of high-index facets is energetically unfavorable, and they vanish quickly during crystal growth. Theoretically, from the viewpoint of thermodynamics, the surface energy can be well controlled by judiciously choosing capping agents [19, 65, 66, 117]. The selected capping agents can selectively absorb on specific facet, lower its surface energy and finally make the specific facet to be exposed. In this aspect, the theoretical calculation and experimental empirical knowledge on surface science studies may provide guidelines for the rational design of NCs with desired surface structures. Alternatively, the shapes of NCs can be controlled kinetically. During the growth of NCs, the growth tendency is determined thermodynamically, while the crystal growth paths depend on kinetic factors. Therefore, various transition states (including nonconvex shapes, high-energy facets, and so on) that are metastable thermodynamically can be obtained by precisely controlling the growth kinetics [61, 120–122]. During the growth of NCs, both thermodynamic and kinetic factors have effects on the final shapes, they should be considered comprehensively to get clear growth mechanism of different types of NCs. Following, different growth mechanisms are clarified to illustrate the growth behaviors of Pt NCs with various shapes and sizes.

3.4.1 Surface-regulating agent-mediated synthesis

In the wet-chemical methods based synthesis of Pd NCs, the Pd precursor (such as H_2PdCl_4, Na_2PdCl_4, $Pd(acac)_2$) is reduced in the presence of surface-regulating agents. These agents mainly act two roles. On the one hand, they are capping agents that can lower the surface energy of specific facets; on the other hand, at the same time, they are dispersant that can prevent the as-prepared Pd NCs from aggregation. In the past decades, it has been witnessed the great success in employing a large variety of surface-regulating agents (e. g. surfactants, polymeric molecules, small molecule adsorbates and polymers) to prepare Pd NCs bounded with various facets.

For example, in the Section 3.2.1.1.2, CTAB, one of kinds of cationic surfactants was applied to the synthesis of Pd nanocubes [31]. The Br^- ions from CTAB have strong affinity to the {100} facets of Pd NCs, thus greatly lowering the surface energy of {100} facets and giving rise to the formation of Pd nanocubes. If CTAB was replaced with CTAC, polydisperse nanoparticles with rough surfaces were obtained. Similarly, by directly using halides that containing Br^-, including NaBr, KBr, Pd nanocubes can also be obtained. Xia and coworkers found that the high-quality Pd nanocubes with {100} facets can be prepared by applying KBr (serving as a capping agent) and PVP (acting as a dispersant, see Section 3.2.1.1.1 for detail) [33]. Interestingly, some pseudo-halide ion (such as SCN^-) was found to play similar role in stabilizing the facets of Pd NCs. For example, Xu and coworkers reported truncated rhombic dodecahedra enclosed by twelve {110}, eight {111} and six {100} facets were obtained by taking advantage of the capping effect of SCN^- on the {110} facets [123]. Besides, some small molecules have been proven effective surface-regulating agents that can stabilize specific crystal facets of Pd NCs. For instance, as shown in Figure 3.15, Yang et al. found that NO_2 can stabilize Pd {111} facets, thus leading to the formation of octahedral Pt@Pd NCs at high concentration of NO_2 [124]. Zheng et al. reported that

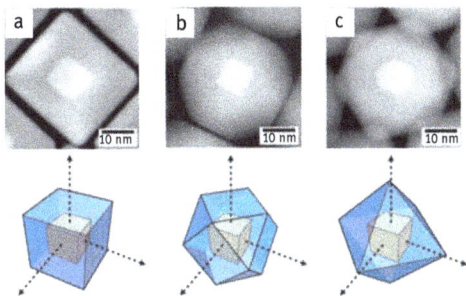

Figure 3.15: HAADF-STEM images of Pt@Pd core-shell nanostructures evolved from cube to cuboctahedra and finally octahedra with gradually increasing the amount of NO_2. (Adapted with permission from ref. 125. Copyright 2007 Nature Publishing Group.).

CO can strongly bind to {111} facets of Pd, resulting in the formation of Pd nanosheets [16]. Zheng et al. found that the addition of formaldehyde can introduce the formation of {110} facets of Pd NCs. Increasing the amount of formaldehyde in the synthesis, the percentage of {110} facets in tetrahedron was gradually increased from 0 to 44 %, and finally unique concave tetrahedron formed (Section 3.2.5.3) [99].

In fact, except the capping role on specific facets, these surface-regulating agents may have significant roles in changing the growth kinetics. For example, the halide species are easy to coordinate with Pd^{2+} ions and form corresponding complex, such as $[PdBr_4]^{2-}$ and $[PdCl_4]^{2-}$ [48, 125, 126]. The Br^- binds Pd^{2+} ions more strongly than Cl^-, greatly reducing the effective concentration of Pd^{2+} ions in the growth solution, thus reducing the rate of reduction of the Pd^{2+} ions. At the same time, the standard reduction potential of $[PdBr_4]^{2-}$ is lower than the $[PdCl_4]^{2-}$, which leading to the slower reduction rate of Pd in presence of Br^- ions than that in the presence of Cl^- ions. As mentioned above, the fast reduction of noble metal ions would result in large number of seeds formed in the initial stage, and the sizes of final NCs would be reduced when the amount of the precursor is kept the same. Therefore, the sizes may be controlled by varying the amount or kinds of additives. For example, in the standard synthesis of Pd nanocubes with an edge length about 18 nm (Section 3.2.1.1.1), the usage of KBr was 600 mg [33]. When the amount was reduced to 300 mg while keeping other condition unchanged, Pd nanocubes with an average size about 10 nm were obtained. In order to further accelerate the reaction rate to obtain smaller Pd nanocubes, certain amount of KBr was replaced with KCl. The nanocubes with size about 7 nm were produced when 75 mg KBr and 141 mg KCl was used. If further reduce the amount of KBr to 5 mg and at the same time increase the KCl to 185 mg, 6 nm Pd nanocubes were prepared. With reducing the amount of KBr, the consumption of Pd precursors was accelerated, which was confirmed by monitoring the change of UV-vis spectroscopy of growth solution [33]. The faster consumption of the Pd precursor in the synthesis of 6-nm Pd nanocubes than in the syntheses of larger Pd nanocubes confirms that the formation of smaller Pd nanocubes results from a faster reduction rate, and vice versa.

3.4.2 Seed-mediated synthesis

As mentioned above, the growth of NCs usually involves two steps, nucleation and subsequent growth. However, up to date, it is still difficult to monitor the nucleation process, make the precisely control of the seeds very challenging, especially in the one-pot synthesis of noble metal NCs. Seed-mediated growth that uses pre-formed small NCs with well-defined shape as seeds skips the complicated nucleation process and provide a powerful approach to control the final structures of noble metal NCs through controlling the thermodynamic and kinetic factors in the growth step. For example, Xu et al. found that the uniform Pd RD can be prepared by adding moderate amount KI in the seeded growth of Pd nanocubes (detailed in Section 3.2.1.3.2). The I⁻ ions may replace the Br⁻ ions on the surface of the Pd NCs, and stabilize the {110} facets thermodynamically, thus favor the formation of RD Pd NCs [34]. By using HCHO as reducing agent, Xia et al. reported that octahedral Pd NCs in high yields can be produced when using the pre-made Pd nanocubes as seeds in the presence of PVP (Figure 3.16) [76]. The reduced Pd atoms preferentially deposited on the {100} facets, resulting the faster growth along <100> directions relative to <111> directions, so the as-prepared Pd NCs evolved from nanocubes into truncated nanocubes, cuboctahedra, truncated octahedra, and octahedra with gradually increasing the amount of Pd precursor.

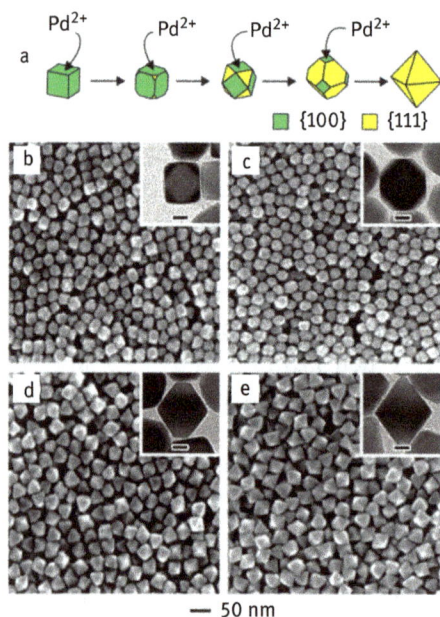

Figure 3.16: (a) Schematic illustrating how continuous growth of Pd nanocubes with {100} planes into an octahedron enclosed by {111} facets. (b-e) SEM images of the products obtained with different amount of Na₂PdCl₄ (b) 5.8 mg, (c) 8.7 mg, (d) 17.4 mg, and (e) 29.0 mg. (Adapted with permission from ref. 76. Copyright 2012. The Royal Society of Chemistry.).

In addition to the fine control the surface structure of NCs, the size of as-prepared NCs can also be tuned by changing the concentration or the dimension of involved seeds. For example, with decreasing the concentration of Pd nanocube seeds, the final Pd nanocubes continually grow to larger one [31]. By taking small Pd nanocubes as seeds, the final octahedral Pd NCs also became small [76].

3.4.3 Supersaturation-controlled methods

During the growth of NCs, the driving force for the crystallization is the supersaturation of the solute in the growth solution (for metal NC, the solute is the zero-valence metal atoms form the reduction of the corresponding ions) [113]. For the growth of NCs at low supersaturation (stage III as shown in Figure 3.14), the shapes of NCs can be determined by the Wulff construction rule due to the growth is under equilibrium growth conditions, thus the exposed crystal surface can be controlled by tuning the surface energy using appropriate surface-regulating agents. However, in most cases, the growth of noble-metal NCs occurs under nonequilibrium conditions. In such cases, supersaturation is an important factor driving crystal growth. Recently, upon careful investigation of NC growth behavior from thermodynamic viewpoint, Xie and coworkers deduced "Thomson-Gibbs"-like eq. (3.2) [28, 127]:

$$\Delta\mu = \mu_1 - \mu_c = \frac{2\sigma v}{h} \tag{3.2}$$

where $\Delta\mu$ is the supersaturation or the difference in the chemical potentials of the solutes in solution (μ_l) and the crystal (μ_c), σ is the specific surface energy of the crystallite, v is the volume of a single building block, and h is the distance from the crystallite's center to its surface (i. e. the size of crystallite).

It indicates that the surface energy of crystal face is in proportion to the supersaturation of crystal growth units during the crystal growth, and higher supersaturation would result in the formation of crystallites with higher surface-energy faces. This theoretic guidance have been successfully applied to guide and or explain the synthesis of various NCs with different surface structures, including noble metal NCs, ionic crystal, organic molecular crystal, metal oxide, and metal-organic frameworks [28, 127, 128].

For Pd NCs, the crystal growth process involves the reduction of Pd^{2+} ions (in most Pd precursors, Pd^{2+} is main form) to Pd atoms (step 1) and the subsequent growth of Pd atoms into crystallites (step 2).

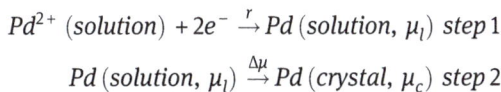

$$Pd^{2+} (solution) + 2e^- \xrightarrow{r} Pd (solution, \mu_l) \; step \, 1$$

$$Pd (solution, \mu_l) \xrightarrow{\Delta\mu} Pd (crystal, \mu_c) \; step \, 2$$

Since supersaturation is the difference between the chemical potentials of the reduced metal atoms and the solid crystal, the supersaturation of the Pd atoms increases as the reduction rate increases. Therefore, accelerating the reduction rate is a straightforward approach to get Pd NCs with high surface energies.

As mentioned above, some surface-regulating agents (such as CTAB, CTAC, halide ions) may change the reduction of metal ions due to the formation of complex. Usually, the reduction rate accelerates with decreasing the concentration of these surface-regulating agents. Therefore, Pd NCs with high index faces could be prepared by simply decreasing or removal of the surface-regulating agents. For example, when the H_2PdCl_4 was reduced by AA without adding any capping agent, the products were dominated with THH Pd NCs with {730} facets, and only Pd nanocubes obtained when CTAC involved (Figure 3.17a, b) [127]. By adjusting the ratio of the two capping agents, CTAC and CTAB, concave Pd nanocubes with high-index {hk0} facets were successfully obtained (Figure 3.17**c**) [48]. The reduction rate of the Pd precursor became slower with increasing the composition of CTAB, leading to the formation of Pd NCs with more flat surface with lower surface energy (Figure 3.17c–e). In addition, by increasing the concentration of reductant is also directly route to increase the reduction rate of metal ions. In a seed-mediate growth of Pd NCs using Pd nanocube as seeds, concave Pd nanocube with high index facets can only be

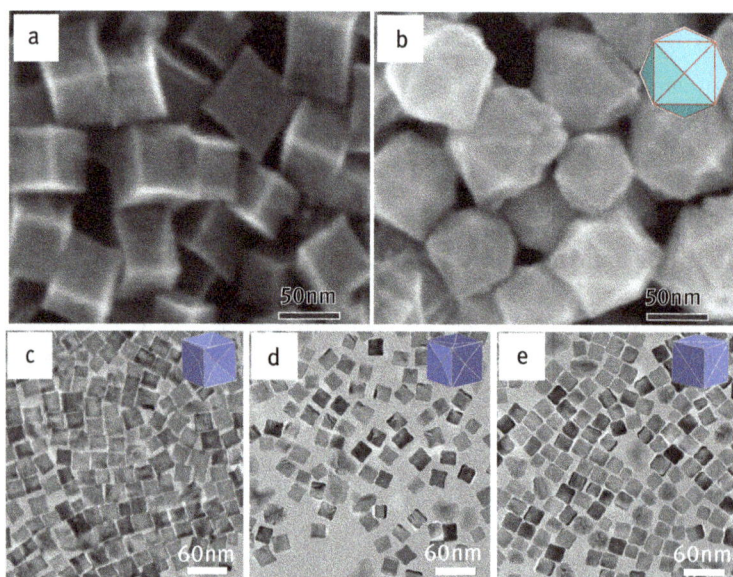

Figure 3.17: SEM images of the cubic Pd NCs (a) produced in the presence of CTAC while Pd THH (b) synthesized without CTAC. (Adapted with permission from ref. 128. Copyright 2013, American Chemical Society.) (c-e) TEM images of the concave cubic Pd NCs became more and more flat with the increasing the ratio of CTAB/CTAC in the growth solution. (Adapted with permission from ref. 48. Copyright 2011, Wiley-VCH.).

achieved via increasing the concentration of AA or reducing the concentration of Br⁻, in contrast, at low reduction rate, Pd cubes was obtained [47]. These results indicate that the supersaturation is the key to determine the surface structures, that faster reduction rate would result in higher supersaturation thus leading to the formation of NCs with higher-surface-energy facets.

3.4.4 Template-directed method

Template-directed method has been demonstrated as a simple and versatile route to fabricate NCs with various surface structures through epitaxial overgrowth second metal on template metal. During the epitaxial growth, three prerequisites should be met: the lattice mismatch of two metals should be below 5%, the electronegativity of the second metal should be lower than the template metal, and the bond energy between second metal atoms should be smaller than that between the second atoms and template atoms [102]. Through epitaxial overgrowth, a secondary metal can be deposited on the well-defined template NCs, and the newly formed shell layer can inherit the surface structure of the template. For example, as shown in Figure 3.18, through employing as-prepared well-shaped Au NCs as templates, Wang et al. successfully prepared Au@Pd NCs with {221} facets and {730} facets by depositing a thin layer of Pd on TOH and enlarged THH Au NCs, respectively [129].

During the construction of the core-shell structures, the lattice mismatch of the core and shell metals, and synthetic condition may affect the shell's final structure. For

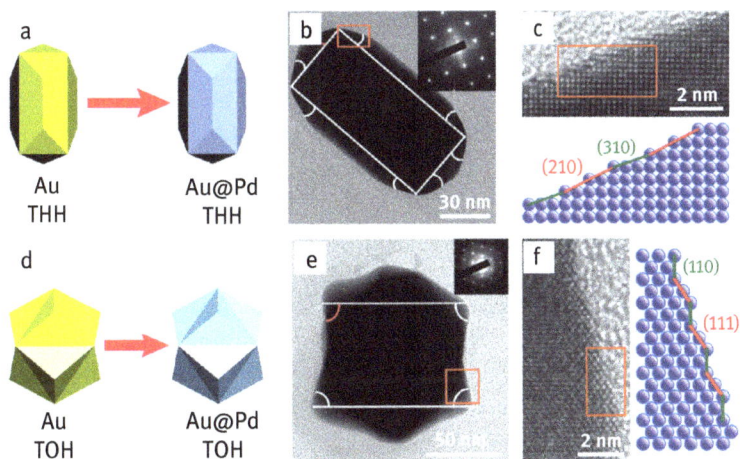

Figure 3.18: (a and d) Schematic of the growth of THH and TOH Au@Pd core-shell NCs from THH and TOH Au NCs, respectively. (b and e) TEM image of a single THH and TOH core-shell NC. (c and f) HRTEM image of the region indicated with the box in (b and e), and corresponding atomic model shows that the as-prepared Au@Pd core-shell NCs adopt same surface structure of the THH and TOH Au seed, respectively. (Adapted with permission from ref. 130. Copyright 2011, American Chemical Society.).

example, Huang and coworker demonstrated that THH Au-Pd core-shell NCs with high index {730} facets can be successfully obtained by virtue of the large lattice mismatch between the Au nanocube seed and Pd shell [44]. The formation of Pd shell with high-energy facets can help releasing the lattice strain caused by lattice mismatch between newly formed Pd shells on Au nanocube templates. Lee and co-workers reported that in addition to concave TOH Au@Pd NCs with high index {hhl} facets, concave hexocta-hedral (HOH) Au@Pd NCs with high index {hkl} facets and THH Au@Pd NCs with {hk0} facets can be prepared by carefully controlling of the growth kinetics when the Pd shell deposited on TOH Au NC seeds [46]. By increasing the reduction rate via decreasing the amount of the Au NC seeds, Huang and coworkers found that core-shell Au@Pd octahedra, truncated octahedra, cuboctahedra, truncated cubes, and concave cubes were produced progressively [130]. As confirmed by time-dependent UV-vis absorption spectra, fast Pd source consumption rates favored the formation of concave nanocubes and truncated nanocubes, while octahedra produced at a much slow reduction rate. The evolution from concave nanocubes with high-energy facets to octahedra with low index facets may also related to the decrease of supersaturation due to the lower reaction rate in the synthesis of octahedra.

3.4.5 Etching-mediated method

Chemical etching of the as-prepared NCs has been regarded as a powerful approach to generate novel and complex architectures such as concave, hollow and framework structures [50, 131–134]. The most used etchants contain organic molecules that can coordinate with metals, halide ions/O_2 pairs, and metal ions with high standard reduction potential that can displace the other metal templates via galvanic replace-ment reaction [133, 134].

Matsui and coworkers develop an ingenious approach to prepared concave Pd nanocubes and hollow Pd nanocages from etching Pd nanocubes (Figure 3.19). In their protocol, reverse micelle CTAB/octanol/H_2O was selected as a reactor [51]. The surface energy gradient on Pd nanocube was constructed by changing the packing density of CTAB on Pd through gradually evaporating the water phase at high reaction temperature under the protection of Ar. When the CTAB on the Pd nano-cubes slowly transformed from bilayer into monolayer, and finally leave the center of {100} facets alone, the Pd atoms were triggered to desorb, resulting in the formation of concave Pd NCs and finally hollow Pd nanocages.

Coupling oxidative etching with co-reduction is another effective methods to pre-pare NCs with controllable shapes. The key for this method is to modulate the equili-brium between etching and regrowth. For example, Pd octahedra with controllable sizes from seeded etching/regrowth of Pd nanocubes can be prepared by con-trolling the ratio of the etching and regrowth rates ($R_{etching}$ and $R_{regrowth}$) simply through varying the amount of HCl [77]. The Pd atoms that etched by HCl/O_2

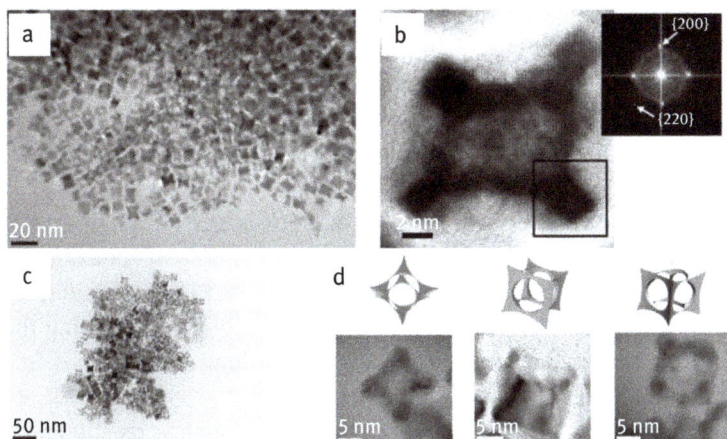

Figure 3.19: (a) TEM and (b) correspond HRTEM image of concave Pd nanocubes. (c) TEM and (d) correspond HRTEM image of hollow Pd nanocages based on etching method. (Adapted with permission from ref. 51. Copyright 2014, Nature Publishing Group.).

were subsequently reduced by solvent (Tri-EG in this case) and deposited back onto the surface of the Pd nanocube seeds, resulting the growth of nanocube to octahedra. Interestingly, in the DMF solution containing PVP, and O_2/KI and HCHO were used as the etchant and reducing agent, respectively, the novel Pd nanoframes can obtained from corresponding solid NCs by precisely controlling the concentrations of etching agent O_2/KI and reducing agent HCHO [98].

To date, the selection of suitable etchants that possess moderate etching power and specific etching direction on noble metal NCs is still a challenge. More studies should be engaged in this interesting and charming field to obtain novel nanostructures with amazing physicochemical properties.

3.4.6 Formation of CPT Pd NCs

Geometrically, crystal lattice of the building block should expand when constructing the perfect *fcc* packing CPT NCs [86, 87, 135]. There should be additional lattice strain in the CPT NCs although they have relatively small surface area compared to their single crystal counterparts. From the thermodynamical point of view, the formation of CPT structure is due to the competition of crystal cohesive energy and surface free energy [88, 135]. It is possible to estimate the crystal structures according to the difference of free energy between the formation of single-crystal (U_{single}) and CPT structure (U_{CPT}) that is given by eq. (3.3) [135]:

$$U_{single} - U_{CPT} = \Delta S\gamma - VW \qquad (3.3)$$

where ΔS is the difference between the surface areas of single crystals and CPT crystals; y, V and W are the specific surface energy, the volume, and the elastic strain energy density for CPT structures, respectively. ΔS is a positive value due to the smaller specific area of CPT structure than that of the single-crystal structure. The W is positive correlated with the crystal cohesive energy E_c, which is constant for the same metal.

So it is reasonable to determine that the transformation from CPT to single crystal structure may be achieved by changing the surface free energy of NCs. For example, in the synthesis of CPT Pd icosahedra, Li et al. found that these CPT nanostructures gradually transformed to single-crystal tetrahedra and octahedra by increasing the amount oleylamine [88]. The oleylamine that can greatly lower the surface free energy of Pd NCs accounts for this transformation.

Besides the thermodynamical factors, the growth kinetics (such as reaction rate) may also have effect on the formation of CPT nanostructures. Recently, Xia and coworkers found the reaction rate can be taken as probe to determinate the twin structure of Pd NCs (Figure 3.20) [136]. In a polyol synthesis of Pd NCs, Na_2PdCl_4 was reduced by EG or diethylene glycol (DEG) in the presence of PVP. The high reaction rate due to the strong reducing power of EG favored the formation of single-crystal Pd NCs, while the slow reaction rate because of weak reducing power of DEG led to the formation of CPT Pd icosahedral NCs.

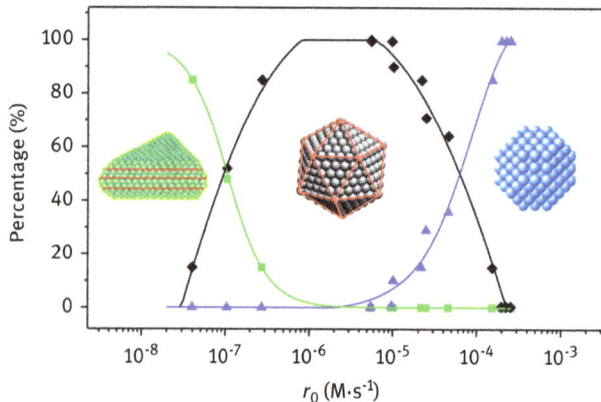

Figure 3.20: Schematic illustration showing the initial reaction rate-dependent crystal structure evolution. (Adapted with permission from ref. 137. Copyright 2015, American Chemical Society.).

For well-controlling the CPT Pd nanostructures, both thermodynamical and kinetic factors should be considered. Especially capping agents are wildly used in the synthesis of NCs, the role of them should be carefully studied because they not only change the surface free energy of NCs thermodynamically, but also may have effect on the growth kinetics due to the complex interaction with the metal precursors. Although the success in elucidating the reaction rate in tuning the twin nanostructures, why it can change the crystal structure is still unclear, and need further exploration.

3.5 Critical safety considerations

In this article, most of the chemicals for the synthesis of Pd NCs are nontoxic in the condition of one can strictly compliance with experimental safety regulations. However, some chemicals and experiment conditions should be paid special attention.

a. For all the synthetic procedures in the lab, safety glasses, lab coat, long pants, and gloves are required.

b. All the experiments should be conducted in fume hood.

c. If using high reaction temperature rather than the room temperature, please make sure the instruments work under proper condition. Additional attention should be paid when operating hot vessels.

d. Special attention should be paid when the organic solvents (e. g. oleylamine, octylamine, DMF, HCHO) are used. These organic solvents may cause serious discomfort and pain if inhalation or in contact with skin, or swallowed.

e. Special attention should be paid when CO gas is used. The CO is toxic, may cause death if inhalation. In addition, $W(CO)_6$ solid is easy to volatilize, and also toxic if inhalation, or in contact with skin, or swallowed.

3.6 Conclusions and future perspective

Over the past decades, great progresses have been made on the synthesis of Pd NCs with various shapes and controllable sizes. It has been demonstrated that the physicochemical properties can be finely tuned by tailoring these structural parameters. In this article, we have summarized the state-of-the-art research progress in the shape and size-controlled synthesis of noble-metal Pd NCs, which is based on the wet-chemical synthesis. The key factors that control the structures, including CPT, single-crystal nanostructures, and size are carefully elucidated. The detailed studies on the growth mechanisms of NCs provide a powerful guideline to the rational design and synthesis of noble-metal NCs with enhanced properties.

However, there are still some problems in both fundamental studies and practical use of Pd NCs. Firstly, it is still great challenge to synthesize reproducibly large quantities uniform monodisperse well-defined Pd NCs with high yield. Secondly, the exact role of varies kinds of surface-regulating agents on the formation of different surface structures is still not clear. The development of *in situ* surface characterization techniques and corresponding theoretical simulations would help improve our understanding of the exact role of surface-regulating agents. Thirdly, the effects of the surface-regulating agents on the performances of the Pd NCs should be intensely studied. For some special applications, the surface cleaned Pd NCs are needed to establish reliable shape-dependent activity and durability. Thus, it is extremely important to develop effective and mild methods to remove surfactants and adsorbates from chemically synthesized Pd NCs without spoiling their surface structures. Besides, it has been found that some specific surface ligands play vital role in improving the catalytic selectivity of Pd NCs. Thus the actual surface/interface structures of the Pd nanocatalysts in the catalytic process should be carefully investigated. In addition, due to the harsh practical environment, the stability of the surface structures should be considered during applications.

Compared to single-crystal Pd nanostructures, the CPT Pd nanostructures contain additional structural defects, which are usually active catalytic sites for lots of catalytic application. However, up to date, the growth mechanism of CPT noble metal nanostructures is still not fully understood, especially how and what the kinetic factors could affect the growth behaviors. In the future, understanding of the growth of NCs at kinetic and microscopic levels should be improved. Very recently, the single atom Pd catalysts attracted lots of attentions because it can maximize the efficiency of Pd and have been demonstrated their outstanding catalytic performance. However, the loading amount and stability of Pd single atoms need to be further improved.

Overall, we believe that comprehensive studies of the growth mechanisms of Pd NCs will help us to rationally design and synthesize functional Pd nanomaterials with exceptional properties, which would help to maximize the efficiency of Pd in applications.

Funding: This work was supported by the National Basic Research Program of China (No. 2015CB932301), National Natural Science Foundation of China (No. 21771151, 21333008 and 21603178), China Postdoctoral Science Foundation (2016M602066, 2017T100468), and National Key Research and Development Program of China (2017YFA0206801).

References

[1]   Narayanan R, El-Sayed MA. Catalysis with transition metal nanoparticles in colloidal solution: nanoparticle shape dependence and stability. J Phys Chem B. 2005;109:12663–76.

[2]   Narayanan R, El-Sayed MA. Effect of catalysis on the stability of metallic nanoparticles: Suzuki reaction catalyzed by PVP-palladium nanoparticles. J Am Chem Soc. 2003;125: 8340–47.

[3]   Wilde M, Fukutani K, Ludwig W, Brandt B, Fischer JH, Schauermann S, et al. Influence of carbon deposition on the hydrogen distribution in Pd nanoparticles and their reactivity in olefin hydrogenation. Angewandte Chemie International Edition. 2008;47:9289–93.

[4]   Marshall ST, O'Brien M, Oetter B, Corpuz A, Richards RM, Schwartz DK, et al. Controlled selectivity for palladium catalysts using self-assembled monolayers. Nat Mater. 2010;9: 853–58.

[5] Friedrich M, Penner S, Heggen M, Armbruster M. High CO_2 selectivity in methanol steam reforming through ZnPd/ZnO teamwork. Angewandte Chemie International Edition. 2013;52:4389–92.

[6] Xue T, Lin Z, Chiu C, Li Y, Ruan L, Wang G, et al. Molecular ligand modulation of palladium nanocatalysts for highly efficient and robust heterogeneous oxidation of cyclohexenone to phenol. Sci Adv. 2017;3:e1600615.

[7] Long R, Rao Z, Mao K, Li Y, Zhang C, Liu Q, et al. Efficient coupling of solar energy to catalytic hydrogenation by using well-designed palladium nanostructures. Angewandte Chemie International Edition. 2015;54:2425–30.

[8] Reetz MT, Breinbauer R, Wanninger K. Suzuki and Heck reactions catalyzed by preformed palladium clusters and bimetallic clusters. Tetrahedron Lett. 1996;37:4499–502.

[9] Suzuki A. Recent advances in the cross-coupling reactions of organoboron derivatives with organic electrophiles. J Organomet Chem. 1999;576:147–68.

[10] Adams BD, Chen A. The role of palladium in a hydrogen economy. Mater Today. 2011;14: 282–89.

[11] Konda SK, Chen A. Palladium based nanomaterials for enhanced hydrogen spillover and storage. Mater Today. 2016;19:100–08.

[12] Bianchini C, Shen PK. Palladium-based electrocatalysts for alcohol oxidation in half cells and in direct alcohol fuel cells. Chem Rev. 2009;109:4183–206.

[13] Jayashree RS, Spendelow JS, Yeom J, Rastogi C, Shannon MA, Kenis PJ. Characterization and application of electrodeposited Pt, Pt/Pd, and Pd catalyst structures for direct formic acid micro fuel cells. Electrochim Acta. 2005;50:4674–82.

[14] Larsen R, Ha S, Zakzeski J, Masel R. Unusually active palladium-based catalysts for the electrooxidation of formic acid. J Power Sources. 2006;157:78–84.

[15] Zhou WP, Lewera A, Larsen R, Masel RI, Bagus PS, Wieckowski A. Size effects in electronic and catalytic properties of unsupported palladium nanoparticles in electrooxidation of formic acid. J Phys Chem B. 2006;110:13393–98.

[16] Huang X, Tang S, Mu X, Dai Y, Chen G, Zhou Z, et al. Freestanding palladium nanosheets with plasmonic and catalytic properties. Nat Nanotechnol. 2011;6:28–32.

[17] Li Y, Yan Y, Li Y, Zhang H, Li D, Yang D. Size-controlled synthesis of Pd nanosheets for tunable plasmonic properties. CrystEngComm. 2015;17:1833–38.

[18] Niu W, Zhang W, Firdoz S, Lu X. Controlled synthesis of palladium concave nanocubes with sub-10-nanometer edges and corners for tunable plasmonic property. Chem Mater. 2014;26:2180–86.

[19] Xia Y, Xiong Y, Lim B, Skrabalak SE. Shape-controlled synthesis of metal nanocrystals: simple chemistry meets complex physics?. Angewandte Chemie International Edition. 2009;48:60–103.

[20] Niu W, Xu G. Crystallographic control of noble metal nanocrystals. Nano Today. 2011;6: 265–85.

[21] Chen M, Wu B, Yang J, Zheng N. Small adsorbate-assisted shape control of pd and pt nanocrystals. Adv Mater. 2012;24:862–79.

[22] Zhang H, Jin M, Xiong Y, Lim B, Xia Y. Shape-controlled synthesis of Pd nanocrystals and their catalytic applications. Acc Chem Res. 2013;46:1783–94.

[23] An K, Somorjai GA. Size and shape control of metal nanoparticles for reaction selectivity in catalysis. ChemCatChem. 2012;4:1512–24.

[24] Somorjai GA, Park JY. The impact of surface science on the commercialization of chemical processes. Catal Lett. 2007;115:87–98.

[25] Somorjai GA, Park JY. Molecular factors of catalytic selectivity. Angewandte Chemie International Edition. 2008;47:9212–28.

[26] Jiang ZY, Kuang Q, Xie ZX, Zheng LS. Syntheses and properties of micro/nanostructured crystallites with high-energy surfaces. Adv Funct Mater. 2010;20:3634–45.

[27] Zhang JW, Kuang Q, Jiang YQ, Xie ZX. Engineering high-energy surfaces of noble metal nanocrystals with enhanced catalytic performances. Nano Today. 2016;11:661–77.

[28] Zhang JW, Chen QL, Cao ZM, Kuang Q, Lin HX, Xie ZX. Surface structure-controlled synthesis of nanocrystals through supersaturation dependent growth strategy. SCIENTIA SINICA Chimica. 2017;47:507–16.

[29] Cao S, Tao FF, Tang Y, Li Y, Yu J. Size- and shape-dependent catalytic performances of oxidation and reduction reactions on nanocatalysts. Chem Soc Rev. 2016;45:4747–65.

[30] Lim B, Xiong Y, Xia Y. A water-based synthesis of octahedral, decahedral, and icosahedral Pd nanocrystals. Angewandte Chemie International Edition. 2007;46:9279–82.

[31] Niu WX, Li ZY, Shi LH, Liu XQ, Li HJ, Han S, et al. Seed-mediated growth of nearly monodisperse palladium nanocubes with controllable sizes. Cryst Growth Des. 2008;8: 4440–44.

[32] Mazumder V, Sun S. Oleylamine-mediated synthesis of Pd nanoparticles for catalytic formic acid oxidation. J Am Chem Soc. 2009;131:4588–89.

[33] Jin MS, Liu H, Zhang H, Xie ZX, Liu J, Xia Y. Synthesis of Pd nanocrystals enclosed by {100} facets and with sizes <10 nm for application in CO oxidation. Nano Res. 2011;4:83–91.

[34] Niu WX, Zhang L, Xu GB. Shape-controlled synthesis of single-crystalline palladium nanocrystals. ACS Nano. 2010;4:1987–96.

[35] Zhang HX, Wang H, Re YS, Cai WB. Palladium nanocrystals bound by {110} or {100} facets: from one pot synthesis to electrochemistry. Chem Commun. 2012;48:8362–64.

[36] Collins G, Schmidt M, O'Dwyer C, Holmes JD, McGlacken GP. The origin of shape sensitivity in palladium-catalyzed suzuki-miyaura cross coupling reactions. Angewandte Chemie International Edition. 2014;53:4142–45.

[37] Kim M, Kim Y, Hong JW, Ahn S, Kim WY, Han SW. The facet-dependent enhanced catalytic activity of Pd nanocrystals. Chem Commun. 2014;50:9454–57.

[38] Lu N, Chen W, Fang G, Chen B, Yang K, Yang Y, et al. 5-fold twinned nanowires and single twinned right bipyramids of Pd: utilizing small organic molecules to tune the etching degree of O_2/Halides. Chem Mater. 2014;26:2453–59.

[39] Yin X, Liu X, Pan YT, Walsh KA, Yang H. Hanoi tower-like multilayered ultrathin palladium nanosheets. Nano Lett. 2014;14:7188–94.

[40] Long R, Wu D, Li YP, Bai Y, Wang CM, Song L, et al. Enhancing the catalytic efficiency of the Heck coupling reaction by forming 5 nm Pd octahedrons using kinetic control. Nano Res. 2015;8:2115–23.

[41] Klinkova A, Larin EM, Prince E, Sargent EH, Kumacheva E. Large-scale synthesis of metal nanocrystals in aqueous suspensions. Chem Mater. 2016;28:3196–202.

[42] Zhang Y, Zhu X, Guo J, Huang XQ. Controlling palladium nanocrystals by solvent-induced strategy for efficient multiple liquid fuels electrooxidation. ACS Appl Mater Interfaces. 2016;8:20642–49.

[43] Tian N, Zhou ZY, Sun SG. Electrochemical preparation of Pd nanorods with high-index facets. Chem Commun. 2009:1502–4.

[44] Lu CL, Prasad KS, Wu HL, Ho JA, Huang MH. Au nanocube-directed fabrication of Au-Pd core-shell nanocrystals with tetrahexahedral, concave octahedral, and octahedral structures and their electrocatalytic activity. J Am Chem Soc. 2010;132:14546–53.

[45] Tian N, Zhou ZY, Yu NF, Wang LY, Sun SG. Direct electrodeposition of tetrahexahedral Pd nanocrystals with high-index facets and high catalytic activity for ethanol electrooxidation. J Am Chem Soc. 2010;132:7580–81.

[46] Yu Y, Zhang Q, Liu B, Lee JY. Synthesis of nanocrystals with variable high-index Pd facets through the controlled heteroepitaxial growth of trisoctahedral Au templates. J Am Chem Soc. 2010;132:18258–65.

[47] Jin MS, Zhang H, Xie ZX, Xia Y. Palladium concave nanocubes with high-index facets and their enhanced catalytic properties. Angewandte Chemie International Edition. 2011;50:7850–54.

[48] Zhang JW, Zhang L, Xie SF, Kuang Q, Han XG, Xie ZX, et al. Synthesis of concave palladium nanocubes with high-index surfaces and high electrocatalytic activities. Chem A Eur J. 2011;17:9915–19.

[49] Xia XH, Xie SF, Liu MC, Peng HC, Lu N, Wang J, et al. On the role of surface diffusion in determining the shape or morphology of noble-metal nanocrystals. Proc Natl Acad Sci U S A. 2013;110:6669–73.

[50] Zhang ZC, Nosheen F, Zhang JC, Yang Y, Wang PP, Zhuang J, et al. Growth of concave polyhedral Pd nanocrystals with 32 facets through in situ facet-selective etching. ChemSusChem. 2013;6:1893–97.

[51] Wei Z, Matsui H. Rational strategy for shaped nanomaterial synthesis in reverse micelle reactors. Nat Commun. 2014;5:3870.

[52] Wei L, Xu CD, Huang L, Zhou ZY, Chen SP, Sun SG. Electrochemically shape-controlled synthesis of Pd concave-disdyakis triacontahedra in deep eutectic solvent. J Phys Chem C. 2016;120:15569–77.

[53] Gallon BJ, Kojima RW, Kaner RB, Diaconescu PL. Palladium nanoparticles supported on polyaniline nanofibers as a semi-heterogeneous catalyst in water. Angewandte Chemie International Edition. 2007;46:7251–54.

[54] Li B, Long R, Zhong X, Bai Y, Zhu Z, Zhang X, et al. Investigation of size-dependent plasmonic and catalytic properties of metallic nanocrystals enabled by size control with HCl oxidative etching. Small. 2012;8:1710–16.

[55] Zhang JF, Feng C, Deng YD, Liu L, Wu YT, Shen B, et al. Shape-controlled synthesis of palladium single-crystalline nanoparticles: the effect of HCl oxidative etching and facet-dependent catalytic properties. Chem Mater. 2014;26:1213–18.

[56] Chanda K, Rej S, Liu SY, Huang MH. Facet-dependent catalytic activity of palladium nanocrystals in Tsuji-Trost allylic amination reactions with product selectivity. Chem Cat Chem. 2015;7:1813–17.

[57] Gao D, Zhou H, Wang J, Miao S, Yang F, Wang G, et al. Size-dependent electrocatalytic reduction of CO_2 over Pd nanoparticles. J Am Chem Soc. 2015;137:4288–91.

[58] Wilson NM, Flaherty DW. Mechanism for the direct synthesis of H_2O_2 on Pd clusters: heterolytic reaction pathways at the liquid-solid Interface. J Am Chem Soc. 2016;138:574–86.

[59] Wang H, Jiang B, Zhao TT, Jiang K, Yang YY, Zhang JW, et al. Electrocatalysis of ethylene glycol oxidation on bare and bi-modified Pd concave nanocubes in alkaline solution: an interfacial infrared spectroscopic investigation. ACS Catal. 2017;7:2033–41.

[60] Zhang JW, Zhang L, Jia YY, Chen GX, Wang X, Kuang Q, et al. Synthesis of spatially uniform metal alloys nanocrystals via a diffusion controlled growth strategy: the case of Au-Pd alloy trisoctahedral nanocrystals with tunable composition. Nano Res. 2012;5:618–29.

[61] Chen QL, Jia YY, Xie SF, Xie ZX. Well-faceted noble-metal nanocrystals with nonconvex polyhedral shapes. Chem Soc Rev. 2016;45:3207–20.

[62] Jiang YQ, Su JY, Yang YN, Jia YY, Chen QL, Xie ZX, et al. A facile surfactant-free synthesis of Rh flower-like nanostructures constructed from ultrathin nanosheets and their enhanced catalytic properties. Nano Res. 2016;9:849–56.

[63] Wang X, Peng Q, Li Y. Interface-mediated growth of monodispersed nanostructures. Acc Chem Res. 2007;40:635–43.

[64] Wiley B, Sun Y, Xia Y. Synthesis of silver nanostructures with controlled shapes and properties. Acc Chem Res. 2007;40:1067–76.

[65] Tao AR, Habas S, Yang P. Shape control of colloidal metal nanocrystals. Small. 2008;4:310–25.

[66] Peng ZM, Yang H. Designer platinum nanoparticles: control of shape, composition in alloy, nanostructure and electrocatalytic property. Nano Today. 2009;4:143–64.

[67] Hao R, Xing R, Xu Z, Hou Y, Gao S, Sun S. Synthesis, functionalization, and biomedical applications of multifunctional magnetic nanoparticles. Adv Mater. 2010;22:2729–42.

[68] Mahmoud MA, Narayanan R, El-Sayed MA. Enhancing colloidal metallic nanocatalysis: sharp edges and corners for solid nanoparticles and cage effect for hollow ones. Acc Chem Res. 2013;46:1795–805.

[69] Personick ML, Mirkin CA. Making sense of the mayhem behind shape control in the synthesis of gold nanoparticles. J Am Chem Soc. 2013;135:18238–47.

[70] Watt J, Cheong S, Tilley RD. How to control the shape of metal nanostructures in organic solution phase synthesis for plasmonics and catalysis. Nano Today. 2013;8:198–215.

[71] Long R, Li Y, Song L, Xiong Y. Coupling solar energy into reactions: materials design for surface plasmon-mediated catalysis. Small. 2015;11:3873–89.

[72] Wu Y, Wang D, Li Y. Understanding of the major reactions in solution synthesis of functional nanomaterials. Sci China Mater. 2016;59:938–96.

[73] Klinkova A, Cherepanov PV, Ryabinkin IG, Ho M, Ashokkumar M, Izmaylov AF, et al. Shape-dependent interactions of palladium nanocrystals with hydrogen. Small. 2016;12:2450–58.

[74] Shao M, Yu T, Odell JH, Jin MS, Xia Y. Structural dependence of oxygen reduction reaction on palladium nanocrystals. Chem Commun. 2011;47:6566–68.

[75] Hong JW, Kim D, Lee YW, Kim M, Kang SW, Han SW. Atomic-distribution-dependent electro-catalytic activity of Au-Pd bimetallic nanocrystals. Angewandte Chemie International Edition. 2011;50:8876–80.

[76] Jin MS, Zhang H, Xie ZX, Xia Y. Palladium nanocrystals enclosed by {100} and {111} facets in controlled proportions and their catalytic activities for formic acid oxidation. Energy Environ Sci. 2012;5:6352–57.

[77] Liu M, Zheng Y, Zhang L, Guo L, Xia Y. Transformation of Pd nanocubes into octahedra with controlled sizes by maneuvering the rates of etching and regrowth. J Am Chem Soc. 2013;135:11752–55.

[78] Tian N, Zhou ZY, Sun SG. Platinum metal catalysts of high-index surfaces: from single-crystal planes to electrochemically shape-controlled nanoparticles. J Phys Chem C. 2008;112:19801–17.

[79] Zhang L, Niu WX, Xu GB. Synthesis and applications of noble metal nanocrystals with high-energy facets. Nano Today. 2012;7:586–605.

[80] Quan Z, Wang Y, Fang J. High-index faceted noble metal nanocrystals. Acc Chem Res. 2013;46:191–202.

[81] Kuang Q, Wang X, Jiang ZY, Xie ZX, Zheng LS. High-energy-surface engineered metal oxide micro- and nanocrystallites and their applications. Acc Chem Res. 2014;47:308–18.

[82] Wu J, Li P, Pan YT, Warren S, Yin X, Yang H. Surface lattice-engineered bimetallic nanoparticles and their catalytic properties. Chem Soc Rev. 2012;41:8066–74.

[83] Wu J, Qi L, You H, Gross A, Li J, Yang H. Icosahedral platinum alloy nanocrystals with enhanced electrocatalytic activities. J Am Chem Soc. 2012;134:11880–83.

[84] Yin AX, Min XQ, Zhu W, Wu HS, Zhang YW, Yan CH. Multiply twinned Pt-Pd nanoicosahedrons as highly active electrocatalysts for methanol oxidation. Chem Commun. 2012;48: 543–45.

[85] Xiong Y, Shan H, Zhou Z, Yan Y, Chen W, Yang Y, et al. Tuning surface structure and strain in Pd-Pt core-shell nanocrystals for enhanced electrocatalytic oxygen reduction. Small. 2017;13:1603423.

[86] Yang YN, Zhang JW, Wei YJ, Chen QL, Cao ZM, Li HQ, et al. Solvent-dependent evolution of cyclic penta-twinned rhodium icosahedral nanocrystals and their enhanced catalytic properties. Nano Res. 2017. DOI: 10.1007/s12274-017-1672-6

[87] Huang H, Bao SX, Chen QL, Yang YN, Jiang ZY, Kuang Q, et al. Novel hydrogen storage properties of palladium nanocrystals activated by a pentagonal cyclic twinned structure. Nano Res. 2015;8:2698–705.

[88] Niu Z, Peng Q, Gong M, Rong H, Li Y. Oleylamine-mediated shape evolution of palladium nanocrystals. Angewandte Chemie International Edition. 2011;50:6315–19.

[89] Li C, Sato R, Kanehara M, Zeng H, Bando Y, Teranishi T. Controllable polyol synthesis of uniform palladium icosahedra: effect of twinned structure on deformation of crystalline lattices. Angewandte Chemie International Edition. 2009;48:6883–87.

[90] Lv T, Wang Y, Choi SI, Chi M, Tao J, Pan L, et al. Controlled synthesis of nanosized palladium icosahedra and their catalytic activity towards formic-acid oxidation. ChemSusChem. 2013;6:1923–30.

[91] Huang H, Wang Y, Ruditskiy A, Peng HC, Zhao X, Zhang L, et al. Polyol syntheses of palladium decahedra and icosahedra as pure samples by maneuvering the reaction kinetics with additives. ACS Nano. 2014;8:7041–50.

[92] Sato R, Kanehara M, Teranishi T. Homoepitaxial size control and large-scale synthesis of highly monodisperse amine-protected palladium nanoparticles. Small. 2011;7:469–73.

[93] Zhang L, Wang L, Jiang ZY, Xie ZX. Synthesis of size-controlled monodisperse Pd nanoparticles via a non-aqueous seed-mediated growth. Nanoscale Res Lett. 2012;7:312.

[94] Fang Z, Wang Y, Liu C, Chen S, Sang W, Wang C, et al. Rational design of metal nanoframes for catalysis and plasmonics. Small. 2015;11:2593–605.

[95] Nasilowski M, Mahler B, Lhuillier E, Ithurria S, Dubertret B. Two-dimensional colloidal nano-crystals. Chem Rev. 2016;116:10934–82.

[96] Chen X, Shi S, Wei J, Chen M, Zheng N. Two-dimensional Pd-based nanomaterials for bioap-plications. Sci Bull. 2017;62:579–88.

[97] Xia X, Choi SI, Herron JA, Lu N, Scaranto J, Peng HC, et al. Facile synthesis of palladium right bipyramids and their use as seeds for overgrowth and as catalysts for formic acid oxidation. J Am Chem Soc. 2013;135:15706–09.

[98] Wang Z, Wang H, Zhang Z, Yang G, He T, Yin Y, et al. Synthesis of Pd nanoframes by excavating solid nanocrystals for enhanced catalytic properties. ACS Nano. 2017;11:163–70.

[99] Huang XQ, Tang S, Zhang H, Zhou Z, Zheng N. Controlled formation of concave tetrahedral/trigonal bipyramidal palladium nanocrystals. J Am Chem Soc. 2009;131:13916–17.

[100] Huang XQ, Zheng N. One-pot, high-yield synthesis of 5-fold twinned Pd nanowires and nanor-ods. J Am Chem Soc. 2009;131:4602–03.

[101] Chen YH, Hung HH, Huang MH. Seed-mediated synthesis of palladium nanorods and branched nanocrystals and their use as recyclable suzuki coupling reaction catalysts. J Am Chem Soc. 2009;131:9114–21.

[102] Fan FR, Liu DY, Wu YF, Duan S, Xie ZX, Jiang ZY, et al. Epitaxial growth of heterogeneous metal nanocrystals: from gold nano-octahedra to palladium and silver nanocubes. J Am Chem Soc. 2008;130:6949–51.

[103] Zhang H, Jin MS, Liu H, Wang J, Kim MJ, Yang D, et al. Facile synthesis of Pd-Pt alloy nanocages and their enhanced performance for preferential oxidation of CO in excess hydrogen. ACS Nano. 2011;5:8212–22.

[104] Tian N, Zhou ZY, Sun SG, Ding Y, Wang ZL. Synthesis of tetrahexahedral platinum nanocrystals with high-index facets and high electro-oxidation activity. Science. 2007;316:732–35.

[105] Huang XQ, Zhao Z, Fan J, Tan Y, Zheng N. Amine-assisted synthesis of concave polyhedral platinum nanocrystals having {411} high-index facets. J Am Chem Soc. 2011;133:4718–21.

[106] Lu N, Wang J, Xie S, Brink J, McIlwrath K, Xia Y, et al. Aberration corrected electron microscopy study of bimetallic Pd-Pt nanocrystal: core–shell cubic and core-frame concave structures. J Phys Chem C. 2014;118:28876–82.

[107] Bu L, Zhang N, Guo S, Zhang X, Li J, Yao J, et al. Biaxially strained PtPb/Pt core/shell nanoplate boosts oxygen reduction catalysis. Science. 2016;354:1410–14.

[108] Xie SF, Choi SI, Lu N, Roling LT, Herron JA, Zhang L, et al. Atomic layer-by-layer deposition of Pt on Pd nanocubes for catalysts with enhanced activity and durability toward oxygen reduction. Nano Lett. 2014;14:3570–76.

[109] Wang X, Choi SI, Roling LT, Luo M, Ma C, Zhang L, et al. Palladium-platinum core-shell icosahedra with substantially enhanced activity and durability towards oxygen reduction. Nat Commun. 2015;6:7594.

[110] Goris B, Polavarapu L, Bals S, Van Tendeloo G, Liz-Marzan LM. Monitoring galvanic replacement through three-dimensional morphological and chemical mapping. Nano Lett. 2014;14:3220–26.

[111] Zhang J, Langille MR, Personick ML, Zhang K, Li S, Mirkin CA. Concave cubic gold nanocrystals with high-index facets. J Am Chem Soc. 2010;132:14012–14.

[112] LaMer V, Dinegar R. Theory, production and mechanism of formation of monodispersed hydrosols. J Am Chem. 1950;72:4847–54.

[113] Markov I. Crystal Growth for Beginners: fundamentals of nucleation, crystal growth, and epitaxy. Singapore: World Scientific Publishing Company; 2003.

[114] Jana NR, Gearheart L, Murphy CJ. Seed-mediated growth approach for shape-controlled synthesis of spheroidal and rod-like gold nanoparticles using a surfactant template. Adv Mater. 2001;13:1389–93.

[115] Nikoobakht B, El-Sayed MA. Preparation and growth mechanism of gold nanorods (NRs) using seed-mediated growth method. Chem Mater. 2003;15:1957–62.

[116] Rodriguez-Fernandez J, Perez-Juste J, de Abajo FJG, Liz-Marzan LM. Seeded growth of submicron Au colloids with quadrupole plasmon resonance modes. Langmuir. 2006;22:7007–10.

[117] Grzelczak M, Perez-Juste J, Mulvaney P, Liz-Marzan LM. Shape control in gold nanoparticle synthesis. Chem Soc Rev. 2008;37:1783–91.

[118] Xia Y, Gilroy KD, Peng HC, Xia X. Seed-mediated growth of colloidal metal nanocrystals. Angewandte Chemie International Edition. 2017;56:60–95.

[119] Yu H, Gibbons PC, Kelton KF, Buhro WE. Heterogeneous seeded growth: a potentially general synthesis of monodisperse metallic nanoparticles. J Am Chem Soc. 2001;123:9198–99.

[120] Lai J, Niu WX, Luque R, Xu GB. Solvothermal synthesis of metal nanocrystals and their applications. Nano Today. 2015;10:240–67.

[121] Xia Y, Xia X, Peng HC. Shape-controlled synthesis of colloidal metal nanocrystals: thermodynamic versus kinetic products. J Am Chem Soc. 2015;137:7947–66.

[122] Marks LD, Peng L. Nanoparticle shape, thermodynamics and kinetics. J Phys Condens Matter. 2016;28:053001.

[123] Zhang L, Niu W, Xu G. Seed-mediated growth of palladium nanocrystals: the effect of pseudo-halide thiocyanate ions. Nanoscale. 2011;3:678–82.

[124] Habas SE, Lee H, Radmilovic V, Somorjai GA, Yang P. Shaping binary metal nanocrystals through epitaxial seeded growth. Nat Mater. 2007;6:692–97.

[125] Veisz B, Király Z. Size-selective synthesis of cubooctahedral palladium particles mediated by metallomicelles. Langmuir. 2003;19:4817–24.

[126] Xiong Y, Chen J, Wiley B, Xia Y, Aloni S, Yin Y. Understanding the role of oxidative etching in the polyol synthesis of Pd nanoparticles with uniform shape and size. J Am Chem Soc. 2005;127:7332–33.

[127] Lin HX, Lei ZC, Jiang ZY, Hou CP, Liu DY, Xu MM, et al. Supersaturation-dependent surface structure evolution: from ionic, molecular to metallic micro/nanocrystals. J Am Chem Soc. 2013;135:9311–14.

[128] Ouyang J, Pei J, Kuang Q, Xie ZX, Zheng LS. Supersaturation-controlled shape evolution of alpha-Fe_2O_3 nanocrystals and their facet-dependent catalytic and sensing properties. ACS Appl Mater Interfaces. 2014;6:12505–14.

[129] Wang F, Li C, Sun LD, Wu H, Ming T, Wang J, et al. Heteroepitaxial growth of high-index-faceted palladium nanoshells and their catalytic performance. J Am Chem Soc. 2011;133:1106–11.

[130] Yang CW, Chanda K, Lin PH, Wang YN, Liao CW, Huang MH. Fabrication of Au-Pd core-shell heterostructures with systematic shape evolution using octahedral nanocrystal cores and their catalytic activity. J Am Chem Soc. 2011;133:19993–20000.

[131] Ruditskiy A, Xia Y. The science and art of carving metal nanocrystals. ACS Nano. 2017;11:23–27.

[132] Chen C, Kang Y, Huo Z, Zhu Z, Huang W, Xin HL, et al. Highly crystalline multimetallic nanoframes with three-dimensional electrocatalytic surfaces. Science. 2014;343:1339–43.

[133] Xia X, Wang Y, Ruditskiy A, Xia Y. 25th anniversary article: galvanic replacement: a simple and versatile route to hollow nanostructures with tunable and well-controlled properties. Adv Mater. 2013;25:6313–33.

[134] Wu Y, Wang D, Niu Z, Chen P, Zhou G, Li Y. A strategy for designing a concave Pt-Ni alloy through controllable chemical etching. Angewandte Chemie International Edition. 2012;51:12524–28.

[135] Bao SX, Zhang JW, Jiang ZY, Zhou X, Xie ZX. Understanding the formation of pentagonal cyclic twinned crystal from the solvent dependent assembly of Au nanocrystals into their colloidal crystals. J Phys Chem Lett. 2013;4:3440–44.

[136] Wang Y, Peng HC, Liu J, Huang CZ, Xia Y. Use of reduction rate as a quantitative knob for controlling the twin structure and shape of palladium nanocrystals. Nano Lett. 2015;15:1445–50.

Bionotes

Jiawei Zhang received his B.S degree from Jinan University (Shandong Province, China) in 2009. Then he joined department of chemistry at Xiamen University as a graduate and obtained his Ph.D. degree under the supervision of Prof. Zhaoxiong Xie in 2015. He studied as a visiting graduate student in Prof. Dong Qin's group at Georgia Institute of Technology from 2013 to 2015. He is currently working as a postdoctoral research fellow in Professor Zhaoxiong Xie's group. His research focuses on the controlled synthesis of noble metal nanocrystals, and their applications in catalysis and fuel cells.

Huiqi Li received her BS degree from department of chemistry, Heilongjiang University in 2015. Then she continued pursuing her PhD degree at Xiamen University under the supervision of Prof. Zhaoxiong Xie. She is interested in wet chemical synthesis of noble metal-based nanocrystals with excavated structures and well-defined facets, and exploring their fantastic applications in catalysis.

Zhiyuan Jiang received his B.S. degree from Peking University in 1990 and Ph.D. degrees from department of chemistry, Xiamen University respectively in 2002 under the supervision of Prof. Lansun Zheng. Since 2013, he holds the position of Professor of physical chemistry at Xiamen University. His current research focuses on the syntheses and application of functional inorganic nanomaterials.

Zhaoxiong Xie received his B.S., M. S. and Ph.D. degrees from department of chemistry, Xiamen University in 1987, 1990 and 1995, respectively. During 1997–1998, he was a postdoc in Centre d'Etudes de Saclay, France, and Ulm University, Germany. Since 2002, he holds the position of Professor of physical chemistry at Xiamen University. He won the National Distinguished Young Scientist Fund of China in 2007 and Chang Jiang Chair Professorship in 2014. So far, he has published more than 200 peer-reviewed research journal publications. His current research interests are focused on surface/interfacial chemistry of functional inorganic nanomaterials.

Linlin Xu and Jun Yang

4 Size and shape-controlled synthesis of Ru nanocrystals

Abstract: Mastery over the size/shape of nanocrystals (NCs) enables control of their properties and enhancement of their usefulness for a given application. Within the past decades, the development of wet-chemistry methods leads to the blossom of research in noble metal nanomaterials with tunable sizes and shapes. We herein would prefer to devote this chapter to introduce the solution-based methods for size and shape-controlled synthesis of ruthenium (Ru) NCs, which can be summarized into five categories: (i) Synthesis of spherical Ru NCs; (ii) synthesis of one-dimensional (1D) Ru NCs, e.g. wires and rods; (iii) synthesis of two-dimensional (2D) Ru NCs, e.g. nanoplates; (iv) synthesis of Ru NCs with hollow interiors and (v) synthesis of Ru NCs with other morphologies, e.g. chains, dendrites and branches. We aim at highlighting the synthetic approaches and growth mechanisms of these types of Ru NCs. We also introduce the detailed characterization tools for analysis of Ru NCs with different sizes/shapes. With respect to the creation of great opportunities and tremendous challenges due to the accumulation in noble metal nanomaterials, we briefly make some perspectives for the future development of Ru NCs so as to provide the readers a systematic and coherent picture of this promising field. We hope this reviewing effort can provide for technical bases for effectively designing and producing Ru NCs with enhanced physical/chemical properties.

Graphical Abstract:

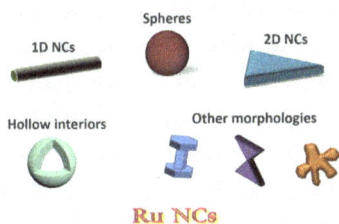

The solution-based methods for size and shape-controlled synthesis of ruthenium nanocrystals as well as the mechanisms behind them are extensively reviewed.

This article has previously been published in the journal *Physical Sciences Reviews*. Please cite as: Yang, J., Xu, L. Size and Shape-Controlled Synthesis of Ru Nanocrystals. *Physical Sciences Reviews* [Online] **2018**, 3. DOI: 10.1515/psr-2017-0080

https://doi.org/10.1515/9783110636666-004

Keywords: ruthenium, nanocrystal, wet-chemistry, controlled synthesis, growth mechanism, characterization

4.1 Introduction

Noble metal nanomaterials with controlled sizes/shapes have garnered sustained research interest due to their immense potential for catalytic or photonic applications [1–18]. One typical example to show the size influence on catalytic behavior of noble metal nanocrystals (NCs) is the reduction of *p*-nitrophenol to *p*-aminophenol over CTAB-stabilized gold (Au) nanoparticles with size in the range of 3.5–56 nm. It has been shown that the activity of CTAB-stabilized Au nanoparticles is neither very efficient for the smallest particles (3.5 nm) nor for the larger ones (28 and 56 nm). Instead, it turned out that CTAB-stabilized Au nanoparticles of an intermediate size (13 nm) are the most active ones for the catalytic reduction of *p*-nitrophenol to *p*-aminophenol [19]. In addition, for noble metal NCs with different shapes, they could display different activities for the same catalytic reaction due to their different crystallographic surfaces [20, 21]. In other words, noble metal NCs with different shapes often display quite different catalytic behaviors [12, 13]. For example, Wang *et al.* have demonstrated the monodispersed Pt NCs with controlled sizes of 3–7 nm and shapes of polyhedron, truncated cube or cube are active catalysts for the oxygen reduction reaction (ORR) in acidic medium. However, the measured current density for 7-nm Pt nanocubes is four times that of 3-nm polyhedral (or 5 nm truncated cubic) Pt NCs, suggesting a dominant effect of particle shape on the oxygen reduction [9].

We are currently witnessing the impressive successes in synthesizing noble metal NCs. Over the past decades, numerous wet-chemistry approaches, including the reduction of metal precursors in solution-phase [14, 22–27], in microemulsions [28], or in sol-gel process [29] have been developed to obtain metallic NCs with well-defined sizes and shapes. Further, the size/shape control of metal NCs could also be achieved through control of the nucleation and growth by varying the synthesis parameters, including the activity of the reducing agents, the type and concentration of the metal precursors, the nature and amount of surfactants or protective reagents [26, 30–35]. Another common strategy used to generate noble metal NCs with controlled sizes/shapes is the seed-mediated growth method. The core particle in this case is overlaid with a single shell of another material to realize the preparation of metal NCs with desired sizes/shapes [36–39]. The core of the metal NCs could be subsequently removed by calcination or with a solvent for further tailoring of the particle structure [40, 41].

Among the noble metal NCs reported in literatures, the ruthenium (Ru) NCs are of particular interest. Ru NCs play an important role in many catalytic reactions, including hydrogenation [42–47], ammonia synthesis [48, 49], *p*-nitrophenol reduction and ammonia borane dehydrogenation [50], CO oxidation [51–55], CO_2 methanation [56], degradation of congo red [57], hydrogen production [58], methane partial oxidation [59], and the Fischer–Tropsch reaction [60, 61]. In addition, Ru-based catalysts are the only ones that can be used for partial hydrogenation of benzene to cyclohexene [62–64], and especially heterogeneous Ru catalysts are preferable from both industrial and environmental perspectives [65]. The synthesis of Ru NCs could be dated back to 30 years or more. An early example is the preparation of Ru colloid by adding solid metal chloride ($RuCl_3$) to SiH-containing liquid, where SiH-containing compound such as $(EtO)_3SiH$ or $Me_2(EtO)SiH$ serves as reducing agent [66]. However, the Ru NCs thus obtained do not have uniform morphology and size distribution due to lack of surface capping. For the growth of noble metal NCs including Ru in wet-chemistry approaches, both size and shape can be well controlled by manipulating the reaction thermodynamic and kinetic parameters. The reaction thermodynamics determine the growth trends, while the reaction kinetics governs the synthetic pathways, which are associated with the energetic barriers of different growth processes. Therefore, a comprehensive understanding of the fundamental growth mechanisms of Ru NCs is of high significance. An outstanding short review article has been published on the shape-controlled synthesis and catalytic applications of Ru NCs [67]. As a result, we devote this chapter to mainly summarize the solution-based methods for size- and shape-controlled synthesis of Ru NCs. We start with the summary of wet-chemistry approaches for the synthesis of Ru NCs with different sizes and shapes, and then, we introduce the detailed characterization tools for analyzing Ru NCs with different sizes/shapes, followed by elucidating the mechanisms accounting for the formation of Ru NCs with special sizes/shapes. Finally, we briefly discuss the perspectives for the future development of Ru NCs so as to provide the readers a systematic and coherent picture of this promising field. The chapter features in detailed summarization of the synthetic approaches and in-depth discussion of the synthetic mechanisms for Ru NCs, and we aim at providing a handbook for readers working in different fields associated with Ru NCs.

4.2 Preparation methods

The solution-based method is powerful and versatile toward the synthesis of different kinds of nanomaterials [14, 15, 68–72]. In a solution-based synthetic system, the nucleation and growth of NCs can be easily controlled by adjusting the reaction parameters including the concentration of reactants, the nature of reducing agent, the mole ratio between precursors and surfactants, pH of reaction medium, and the reaction temperature and time. In this section, we summarize the wet-chemistry approaches for the synthesis of Ru NCs with controllable sizes/shapes, which is precisely classified into five categories including (i) synthesis of spherical Ru NCs, (ii) synthesis of one-dimensional (1D) Ru NCs, (iii) synthesis of two-dimensional (2D) Ru NCs, (iv) synthesis of Ru NCs with hollow interiors and (v) synthesis of Ru NCs with other morphologies.

4.2.1 Synthetic approaches for spherical Ru NCs

Ru NCs with spherical morphology are the most common products formed using solution-based methods. The spherical Ru NCs are desired model to investigate the size effect on their catalytic properties. In addition, by controlling the size of spherical Ru NCs, the specific surface areas as well as the catalytic performance in term of activity, stability and selectivity could be finely tuned.

4.2.1.1 Spherical Ru stabilized by polymers

The physical and chemical properties of metal NCs depend on their stability. In early studies, the most common tools used for stabilizing Ru NCs are polymers, although they may not be usable in selected chemical or physical applications [73, 74]. For a typical synthesis, in a 500-mL flask, polyvinylpyrrolidone (PVP, 1.04 g, 9.36×10^{-3} mol as monomeric unit) was dissolved in mixed solvents of methanol (200 mL) and H_2O (160 mL). Then, ruthenium(III) chloride hydrate (RuCl$_3 \cdot n$H$_2$O, 0.115 g, 4.68×10^{-4} mol) was put into it. After RuCl$_3 \cdot n$H$_2$O was dissolved, 40 mL of aqueous NaBH$_4$ solution (0.177 g, 4.68×10^{-3} mol) was poured into the vigorously stirred solution. The mixture was continuously stirred for 6 h to obtain the PVP-stabilized Ru colloids. This polymer stabilizing approach can be used to prepare Ru NCs with size range of 1.3–1.8 nm, and different Ru colloids can be conveniently achieved by changing the amount of PVP and NaBH$_4$ (Table 4.1). The Ru NCs in the colloidal solutions could be recovered by rotatory evaporating the solvent at 310 K, and redispersed into 200 mL of methanol.

Table 4.1: Preparation conditions of PVP-stabilized Ru colloids (Adapted with permission from ref. 73. Copyright 1998, Elsevier B. V.)[a].

Ru Colloid	PVP	NaBH$_4$	Average diameter (nm)	Standard deviation (nm)
Ru 1	5	10	1.8	0.72
Ru 2	10	10	1.4	0.61
Ru 3	20	10	1.4	0.51
Ru 4	50	10	1.3	0.43
Ru 5	20	5	1.4	0.47
Ru 6	20	20	1.3	0.47
Ru 7	20	50	1.3	0.49

[a] The numbers in the columns of PVP and NaBH$_4$ are the molar ratio of them to Ru.

4.2.1.2 Spherical Ru NCs by decomposing organometallic precursors

The organometallic derivatives have the potential to provide a general route for the synthesis of monodisperse metal nanoparticles, as illustrated by Sun *et al* [75]. One typical synthetic procedure using organometallic precursors for Ru NCs is as following [76]: Ru(COD)(COT)(158 mg, 0.50 mmol, COD = 1,5-cyclooctadiene, COT = 1,3,5-cyclooctatriene) was introduced in a Fischer–Porter bottle and left in vacuo for

30 min. A solution of 1 g of PVP in 60 mL of THF, degassed by freeze-pump cycles, was then added using a transfer tubing. The resulting yellow solution was stirred for 30 min at room temperature, after which the bottle was pressurized under 3 bar dihydrogen and the solution was allowed to react for 68 h, during which time a black precipitate formed. After elimination of excess dihydrogen, the solution was filtrated and the black precipitate dried in vacuo. The precipitate was then redissolved in 30 mL of methanol, and the resulting solution was filtrated and reduced to 15 mL, after which addition of 30 mL of pentane led to a dark brown precipitate. It was filtered, washed with pentane, and dried in vacuo. At a mass ratio of 5% for Ru/ PVP, crystalline hexagonal close-packed (hcp) Ru NCs with an average size of 1.1 nm were obtained (Figure 4.1).

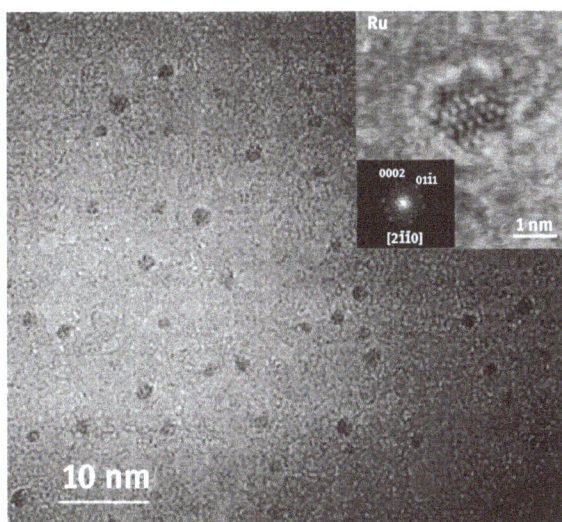

Figure 4.1: Transmission electron micrographs of Ru NCs prepared by decomposing Ru(COD)(COT) in the presence of PVP. Insert are the high-resolution electron micrograph and electron diffraction pattern of a single particle (Adapted with permission from ref. 76. Copyright 2001, American Chemical Society).

By replacing PVP with cellulose acetate (CAC) and also at the mass ratio of 5% for Ru/ CAC, the average size for the as-prepared spherical Ru NCs is 1.7 nm. The same reaction, carried out using various concentrations relative to ruthenium of alkylamines or alkylthiols as stabilizers (L = $C_8H_{17}NH_2$, $C_{12}H_{25}NH_2$, $C_{16}H_{33}NH_2$, $C_8H_{17}SH$, $C_{12}H_{25}SH$, or $C_{16}H_{33}SH$), leads to agglomerated particles (L = thiol) or Ru NCs dispersed in the solution (L = amine), both displaying a mean size near 2–3 nm and a hexagonal close-packed structure. In the case of amine ligands, the Ru NCs are generally elongated and display a tendency to form worm- or rod-like morphologies at high amine concentration.

4.2.1.3 Spherical Ru NCs by polyol reduction

Owing to relatively low reduction potential of $RuCl_3$ to zerovalent Ru ($\varphi^0 = 0.3862$ V), polymer- or ligand-stabilized Ru NCs cannot be synthesized by simply refluxing the $RuCl_3$ precursors in an alcohol–water mixed solvent, or even by reducing Ru^{3+} ion precursors with commonly used reducing agent such as hydroxylamine (NH_2OH) or hydrazine (NH_2NH_2). Instead, reduction of metal slats in liquid polyols has been proven to be a suitable method for the synthesis of monodisperse metal NCs in nanometer size range [77, 78]. The detailed synthetic approaches for spherical Ru NCs using polyol reduction are described in following subsections. This method is facile, and can be easily modified for the synthesis of Ru NCs with controlled sizes.

4.2.1.3.1 PVP-stabilized Ru NCs by polyol reduction

In a typical synthesis [79], to a 250-mL flask, PVP (0.8325 g, 7.5×10^{-3} mol) and $RuCl_3 \cdot nH_2O$ (0.0371 g, 1.5×10^{-4} mol) were dissolved in polyol solvent (150 mL) under stirring to form a dark red solution, which was then heated to reflux (471 K) or to a given temperature (433 or 453 K) with vigorous stirring. The stirring at refluxed state was maintained for 3 h to obtain homogeneous Ru colloidal solution with transparent dark brown color. The as-prepared Ru NCs in polyols can be precipitated using anhydrous acetone, and re-dispersed into methanol for further characterizations. By varying the $PVP/RuCl_3$ molar ratio, the reaction temperature, and polyol solvents used (ethylene glycol [EG], diethylene glycol, or triethylene glycol [TEG]), a series of PVP-stabilized Ru colloids can be obtained, in which the average diameters for Ru NCs can be controlled in the range of 1.4–7.4 nm with relative standard deviations of less than 0.3 (Figure 4.2).

In a modified approach, by using ruthenium nitrosyl nitrate ($Ru(NO)(NO_3)_3$) as a substitute for $RuCl_3$, the average diameters of Ru NCs could be controlled in the range of 1.2–6.5 nm with narrow size distributions by changing the synthesis procedure and reaction conditions. Further, by using the as-prepared 2.6-nm Ru NCs as seeds and L-ascorbic acid as additional reducing agent, in a seed-mediated process, the size of Ru NCs could be precisely controlled in the range of 3.8–7.3 nm by varying the seed to Ru^{3+} ion ratios[80].

4.2.1.3.2 NaOH-assisted synthesis of Ru NCs in polyol

With assistance of NaOH, Wang et al. reported an effective preparation method for the stable "unprotected" Ru NCs with small size and narrow size distribution in EG [81]. "Unprotected" does not mean that the surface of metal NCs are truly bare, but mean that their surfaces are not protected by conventional stabilizes, e.g. surfactants, polymers, or organic ligands. Typically, a total of 5 mL of aqueous solution of $RuCl_3 \cdot 6H_2O$ (95.0 mg, 0.30 mmol) was added into 100 mL of EG. Then 5 mL of aqueous NaOH solution (0.5 M) was added with stirring. The mixture was then heated at 160°C for 3 h with an argon (Ar) flow passing through the reaction system to take away water and organic byproducts. A transparent dark colloidal solution of Ru NCs was obtained without any precipitates. The Ru NCs prepared this way have the

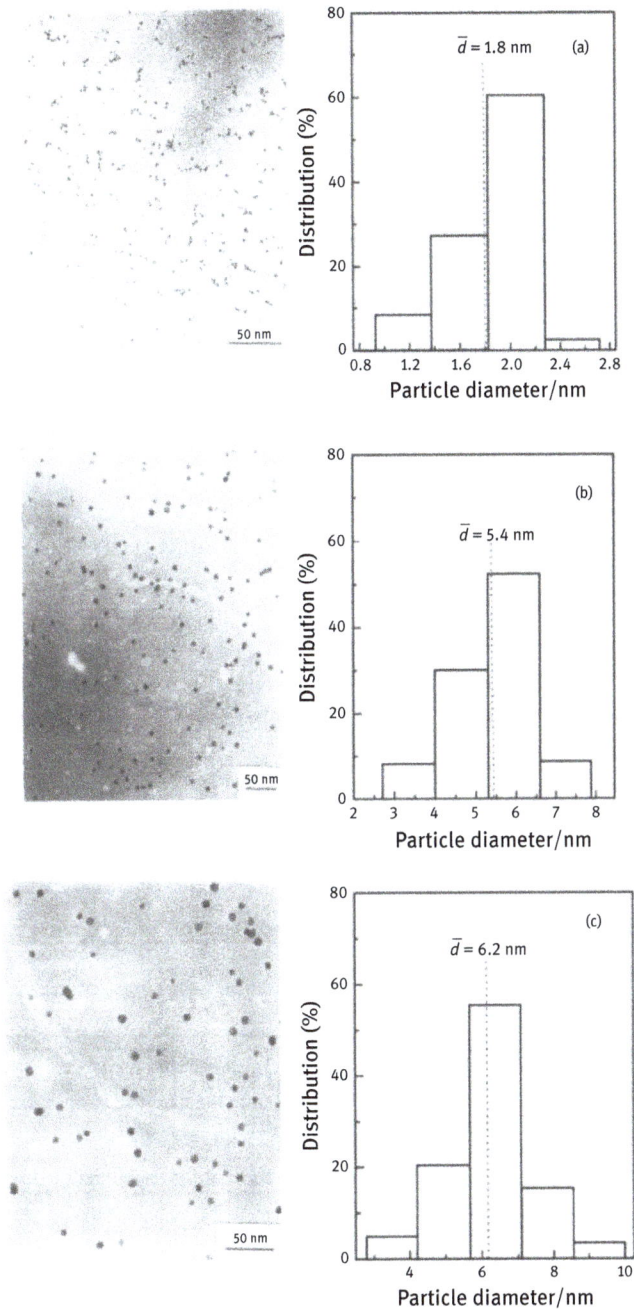

Figure 4.2: TEM photographs (left) and the corresponding particle size distribution histograms (right) of PVP-stabilized Ru NCs prepared in triethylene glycol at 285 °C (a), in ethylene glycol at 198 °C (b), and in ethylene glycol at 180 °C (c), respectively (Adapted with permission from ref. 79. Copyright 2001, Royal Society of Chemistry).

average diameter of 1.1 nm with a size distribution from 0.7 to 2.2 nm. By adjusting pH, the "unprotected" Ru NCs could be easily precipitated from EG, and can be further dissolved conveniently in many organic solvents to form transparent Ru colloidal solutions with high concentration in the absence of usual stabilizing agents.

4.2.1.3.3 Microwave-assisted synthesis of Ru NCs in polyol

Microwave irradiation can heat a substance uniformly through a glass or plastic reaction container, leading to a more homogeneous nucleation and shorter crystallization time compared with those for conventional heating. This is beneficial to the formation of uniform metal colloids [82–93]. A typical experimental apparatus used for microwave heating was shown in Figure 4.3 [85]. A commercial microwave oven was modified by installing a condenser and thermocouple through holes in the top and a magnetic stirrer plate coated with Teflon in the bottom of the oven. A glass flask is placed in the microwave oven and connected to a condenser, into which a mixture of metallic salt, surfactant and, if necessary, reducing agent, even a small amount of nucleation reagent is added. The reagent solution is irradiated by microwave in a continuous wave mode or a pulse mode. The pulse mode is more useful for the temperature control of the heating media.

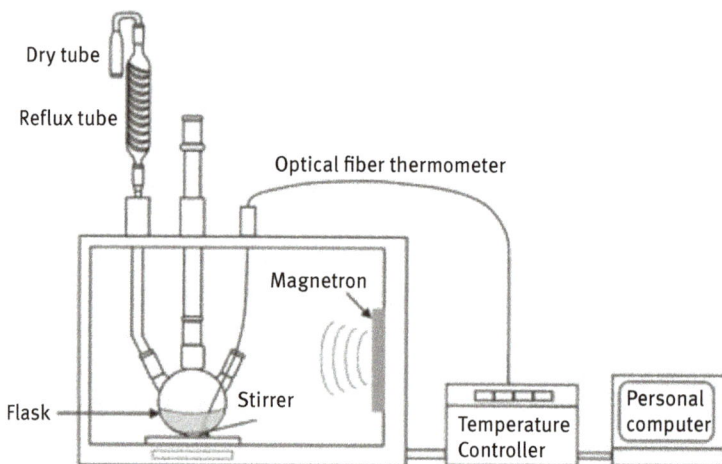

Figure 4.3: Apparatus used for the microwave-assisted synthesis of metallic nanocrystals (Adapted with permission from ref. 85. Copyright 2005, Wiley-VCH Verlag).

In an early study, by combining microwave irradiation with polyol reduction, a typical protocol for synthesize Ru NCs was developed [83]. In detail, $RuCl_3 \cdot 3H_2O$ (2×10^{-5} mol) was mixed with 0.044 g of PVP in 20 mL of EG. The mixture was stirred and heated in a microwave oven (Galanz WP750) operated at 2450 MHz for 30 s, whereupon a dark brown colloidal Ru NCs was obtained, in which the average size of

Ru NCs was determined by transmission electron microscopy (TEM) to be 1.4 nm with standard deviation of 0.3 nm (Figure 4.4). The colloidal Ru NCs in EG could be precipitated by acetone re-dispersed in methanol. By varying the PVP/$RuCl_3$ molar ratio, reaction temperature, reaction time, irradiation power of microwave oven, Ru NCs with different sizes and size distributions could be conveniently obtained [93].

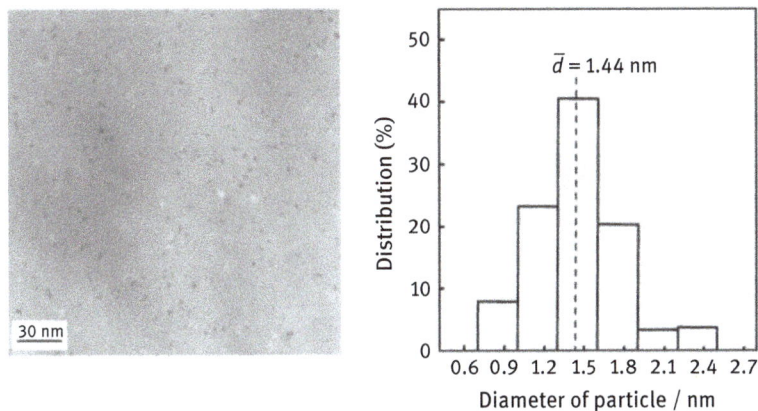

Figure 4.4: TEM photograph (left) and the corresponding particle size distribution histogram (right) of PVP-stabilized Ru NCs as-prepared by microwave irradiation in ethylene glycol (Adapted with permission from ref. 83. Copyright 2000, Royal Society of Chemistry).

4.2.1.3.4 Acatate-stabilized Ru NCs by polyol reduction

Uniform Ru NCs with tunable size from 1 to 6 nm (Figure 4.5) could be synthesized by the polyol reduction of $RuCl_3$ in the presence of sodium acetate trihydrate, and the particle size could be tuned by varying the reaction temperature and concentration of acetate ions [94]. In a typical synthetic procedure, $RuCl_3$ and sodium acetate trihydrate ($CH_3CO_2Na \cdot 3H_2O$) were dissolved in 200 mL of diol (1,2-propane diol, 1,2-ethane diol, or bis(2-hydroxyethyl) ether). The $RuCl_3$ concentration was 3.25×10^{-3} M in all experiments. The sodium acetate concentration was varied from 0.01 to 0.1 M. The polyol solution was heated to a temperature of 150°C for propane diol, 170°C for ethane diol and 180°C for bis(2-hydroxyethyl) ether, respectively, and kept that temperature for 10 min with stirring. At corresponding temperature, the solution turned from intense red to pale green and finally to brown, indicating the reduction. The Ru NCs were separated from polyol by centrifugation. Reactions are considered to be quantitative when the recovered polyol is colorless (a pale green polyol indicates the presence of unreduced Ru^{3+} species).

4.2.1.3.5 Thiol-stabilized Ru NCs by polyol reduction

The surface of acatate-stabilized Ru NCs synthesized by polyol reduction could be further modified by alkanethiols to form thiol-capped Ru NCs, which usually self-assemble into regular patterns on TEM grids due to appropriate van der Waals

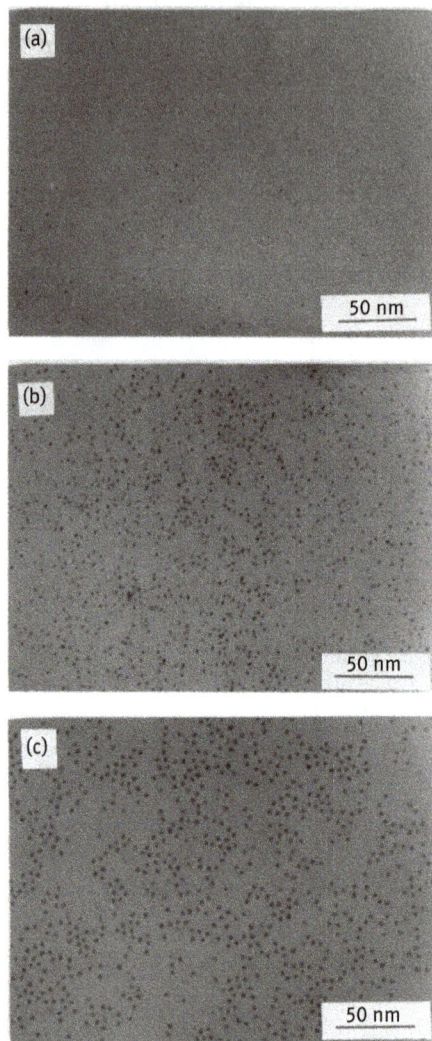

Figure 4.5: Transmission electron microscopy image of Ru NCs prepared in sodium acetate solution in 1,2-propane diol: (a) d_m = 1.4 nm; (b) d_m = 2.5 nm; and (c) d_m = 3.5 nm (Adapted with permission from ref. 94. Copyright 2003, American Chemical Society).

attraction energy between particles (Figure 4.6) [94, 95]. The surface capping of Ru NCs by thiol was realized through a facile extraction process [94]. In principle, polyol and toluene are immiscible at room temperature, and in a mixture of Ru polyol colloid and toluene solution of alkanethiol, the migration of the Ru NCs toward the hydrophobic phase is complete provided that the thiol/Ru molar ratio is high enough, namely higher than 0.1. To favor separation of the two phases, distilled water was added to the polyol. The organization of particles coated with

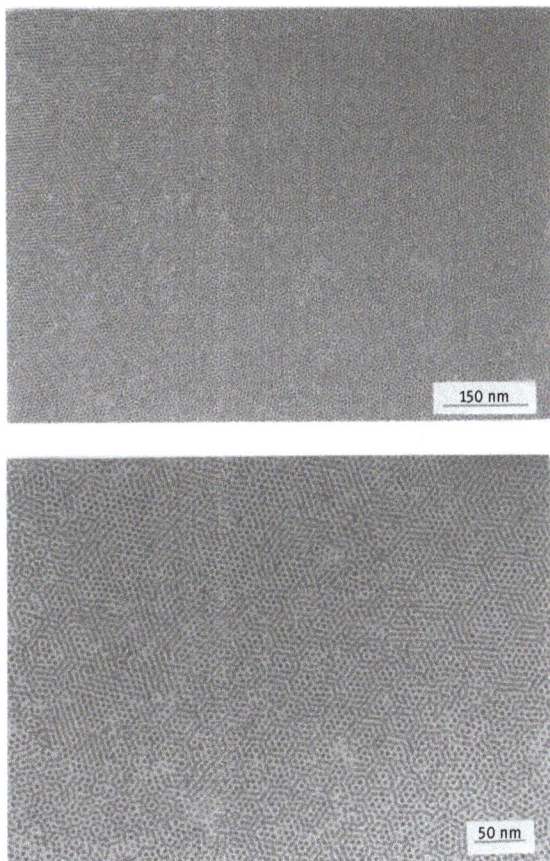

Figure 4.6: TEM images of 4-nm-sized dodecanethiol-stabilized Ru nanocrystal bilayer showing regular self-assemblies (Adapted with permission from ref. 94. Copyright 2003, American Chemical Society).

dodecane thiol on the carbon membrane of the TEM grid was found to depend on the thiol concentration in the solution.

In 2005, Tsukatani and Fujihara firstly reported the synthesis of amphiphilic thiol-stabilized Ru NCs, which can be soluble and stable both in water and organic solvents [96]. In a detailed run for the preparation of Ru NCs using 1-octadeca-nethiol as a stabilizer (ODT-Ru), a solution of $RuCl_3$ (103 mg, 0.5 mmol) and 1-octadecanethiol (286 mg, 1.0 mmol) in 10 mL of anhydrous tetrahydrofuran (THF) was stirred for 1 h at room temperature under an Ar atmosphere. To this vigorously stirred solution was added lithium triethylborohydride (1.0 M, 10 mL) at 0°C, and the resulting solution was stirred for 2 h at room temperature. To the reaction mixture was added anhydrous ethanol (40 mL) at 0°C. After stirring, the organic phase was evaporated to 5 mL in vacuo and mixed with ethanol (300 mL). The resulting precipitate was collected by filtration and washed serially with ethanol. The NCs were redissolved in THF for purification and precipitated with ethanol, and then the Ru NCs were isolated by filtration. These processes were repeated until no free thiol remained.

Other thiol-stabilized Ru NCs (OT-Ru, BNT-Ru and EOT-Ru) were prepared using 1-octanethiol (OT), binaphthyldithiol (BNT) and oligoethyleneoxythiol (EOT) as stabilizers for Ru NCs. In the case of the purification of EOT-Ru NCs, the thiol-stabilized Ru NCs was redissolved in CH_2Cl_2 for purification and precipitated with diethyl ether, and then the particles were isolated by filtration. The OT-Ru, ODT-Ru and BNT-Ru as-prepared are soluble in organic solvents such as n-hexane, benzene, toluene, CH_2Cl_2, $CHCl_3$, THF and diethyl ether. Significantly, the oligoether-functionalized Ru NCs (EOT-Ru) are soluble both in water and organic solvents such as benzene, toluene, CH_2Cl_2, $CHCl_3$, THF, ethyl acetate, methanol, acetone and CH_3CN.

4.2.1.4 Synthesis of water-soluble Ru NCs with spherical morphologies

Stable dispersions of metal NCs in aqueous media are important to many applications. For example, colloidal catalysis in water addresses the important issues of developing efficient chemical transformations under environmental benign conditions. Also, the assembly of metal NCs using biological molecules as templates and molecular design tools necessitates the use of metal nanoparticles dispersible in water. In the following, a number of typical approaches were described for the synthesis of water-soluble Ru NCs with spherical morphology.

4.2.1.4.1 Synthesis of ethylenediamine-protected Ru NCs

In searching for simple procedures for preparing water-based dispersions of Ru NCs, Lee et al. discovered ethylenediamine as a capable capping agent suitable for deployment in aqueous media [97]. In detail, 0.1 mL of ethylenediamine was added dropwise to 20 mL of 2 mM stirred aqueous $RuCl_3$ solution. Stirring was continued for five more minutes after the end of addition to form black precipitates. The black solid was spun down in an ultracentrifuge and the supernatant liquid was discarded. The black precipitate thus obtained was dissolved in 30 mL of ethanol. To this ethanolic solution, 1 mL of freshly prepared 0.1 M aqueous $NaBH_4$ solution was added dropwise under constant stirring. Stirring was continued for three more minutes at the end of addition, thereafter the Ru NCs salted out as black precipitates. The Ru NCs were centrifuged and rinsed thrice with a 30% (v/v) water/ethanol mixture to remove any unbound ethylenediamine and inorganic (Na, Cl, B) impurities. The cleansed precipitates were then dried under vacuum. Ru NCs thus obtained have average diameter of 2.1 nm, and are readily dispersible in water.

4.2.1.4.2 Synthesis of positively charged Ru NCs

In aqueous metal colloids, the stabilization of metal NCs might be achieved by the adsorption of hydronium ions and other hydrated protons on their surfaces, which has been investigated both experimentally and theoretically [98–100]. In case of hydronium ions or other hydrated protons-stabilized Ru NCs, the typical synthetic experiment was depicted below [101]. A 0.1 M aqueous solution of sodium borohydride ($NaBH_4$) was introduced dropwise into 10 mL of 2 mM aqueous $RuCl_3$ solution under vigorous stirring. The volume of the $NaBH_4$ solution was carefully controlled to maintain the pH value of the reaction system always lower than 4.9. The pH value of the final Ru hydrosol was most important to determine the stability of the Ru nanosystem. The Ru hydrosol was very stable when the pH value was lower than 4.9, as no precipitation ever took place even after storage for 3 months.

The Ru NCs prepared this way have an average diameter of 1.8 nm with a standard deviation of 0.4 nm (Figure 4.7a). The hydronium ions and hydrated protons would impart a positive charge to the Ru NCs and stabilize them against agglomeration by electrostatic repulsion.

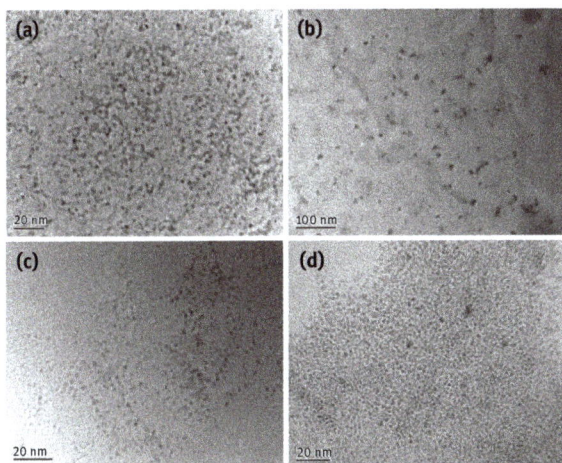

Figure 4.7: TEM images of (a) positively-charged, (b) PVP-stabilized, (c) ethylenediamine-stabilized, and (d) dodecylaime-stabilized Ru NCs (Adapted with permission from ref. 101. Copyright 2004, Elsevier B. V.).

The positively charged Ru NCs could be used as the starting material for further functionalization by PVP, ethylenediamine and dodecylamine [101]. For PVP functionalization, 10 mL of the Ru hydrosol was mixed with an aqueous PVP solution containing 40 mg of PVP (MW = 55,000) in 10 mL of distilled H_2O. Then 120 µL of 1 M aqueous NaOH solution was introduced into the mixture to neutralize the hydrated protons. The procedure for ethylenediamine functionalization is almost identical to the PVP modification besides the PVP solution is replaced by 10 mL aqueous ethylenediamine solution (0.5 mL ethyldiamine in 10 mL of water). For dodecylamine functionalization, 10 mL of Ru hydrosol was mixed with 10 mL of toluene containing

100 µL of dodecylamine. This was followed by the addition of 120 µL of 1 M aqueous NaOH solution. The mixture was stirred for ca. 5 min to allow the replacement of hydronium ions and hydrated protons by dodecylamine on the surface of Ru NCs. TEM examinations (Figure 4.7(b–d)) show that the surface modifications do not induce particle aggregation. However, slight growth in the average particle size was observed, most probably caused by particle agglomeration that occurred during the neutralization of the hydrated protons by sodium hydroxide.

4.2.1.4.3 Synthesis of sulfonated diphosphine-stabilized Ru NCs

Highly stable water-soluble Ru NCs with an average diameter of ca. 1.6 nm were synthesized using a sulfonated diphosphine as stabilizer [102–104]. In a typical synthesis [103], 150 mg of [Ru(COD)(COT)] (0.476 mmol) were introduced in a Fischer–Porter bottle and left in vacuum during 0.5 h and cooled to 193 K. One hundred and fifty milliliter of THF, containing 41 mg of sulfonated diphosphines (1,4-bis[(di-m-sulfonatophenyl) phos-phino]butane (**L1**), 1,4-bis[(dim-sulfonatophenyl)phosphino]propane (**L2**) or 1,4-bis[(di-m-sulfonatophenyl)phosphino]ethane (**L3**), 0.048 mmol, $[L_n]/[Ru] = 0.1$) were then added. The Fischer–Porter bottle was heated to 298 K and then pressurized with 3 bars of dihydrogen. After 18 h, a homogenous brown colloidal solution is obtained. The volume of the solution was reduced to approximately 10 mL by solvent evaporation before its transfer onto a solution of deoxygenated pentane (100 mL). A brown precipi-tate formed which was filtered and dried in vacuum, giving rise to the Ru NCs as a dark brown powder, which could be easily re-dispersed into water leading to stable aqueous Ru colloidal solutions for at least up to 3 months. The transfer into water did not induce any relevant change in dispersion and in their mean diameters as confirmed by TEM analysis (Figure 4.8).

4.2.1.4.4 Synthesis of β-cyclodextrin polymer-stabilized Ru NCs

The strategy for synthesizing aqueous soluble Ru NCs protected by polymer was also developed [105]. In detail, as schematically demonstrated by Figure 4.9, 100 mg of cyclodextrin polymer denoted as poly(CTR-β-CD) (molecular weight = 25,000, 4.05 mmol g^{-1} COOH acidic sites) were dissolved in 5 mL of deionized water in the presence of a controlled amount of $NaHCO_3$ (molar ratio of $NaHCO_3$ to COOH = 1) and the mixture was kept under vigorous stirring for 3 h. Then 269 mg of $Ru(NO)(NO_3)_3$ solution (40 µmol, 1 equiv.) were dispersed in 3 mL of deionized water. The both solutions were mixed together under vigorous stirring for 30 min. Then, 15.2 mg of $NaBH_4$ (400 µmol, 10 equiv.) previously dissolved in 4 mL of water were quickly added to the mixture. The resulting Ru colloidal suspension was kept under vigorous stirring for 24 h in order to check that there is no sedimentation of the metallic particles.

4.2.1.5 Ethanol-mediated phase transfer for the synthesis of Ru NCs

Phase transfer is an important approach for synthesizing organic colloids of noble metals since the methods for direct synthesis of noble metal NCs in nonpolar organic

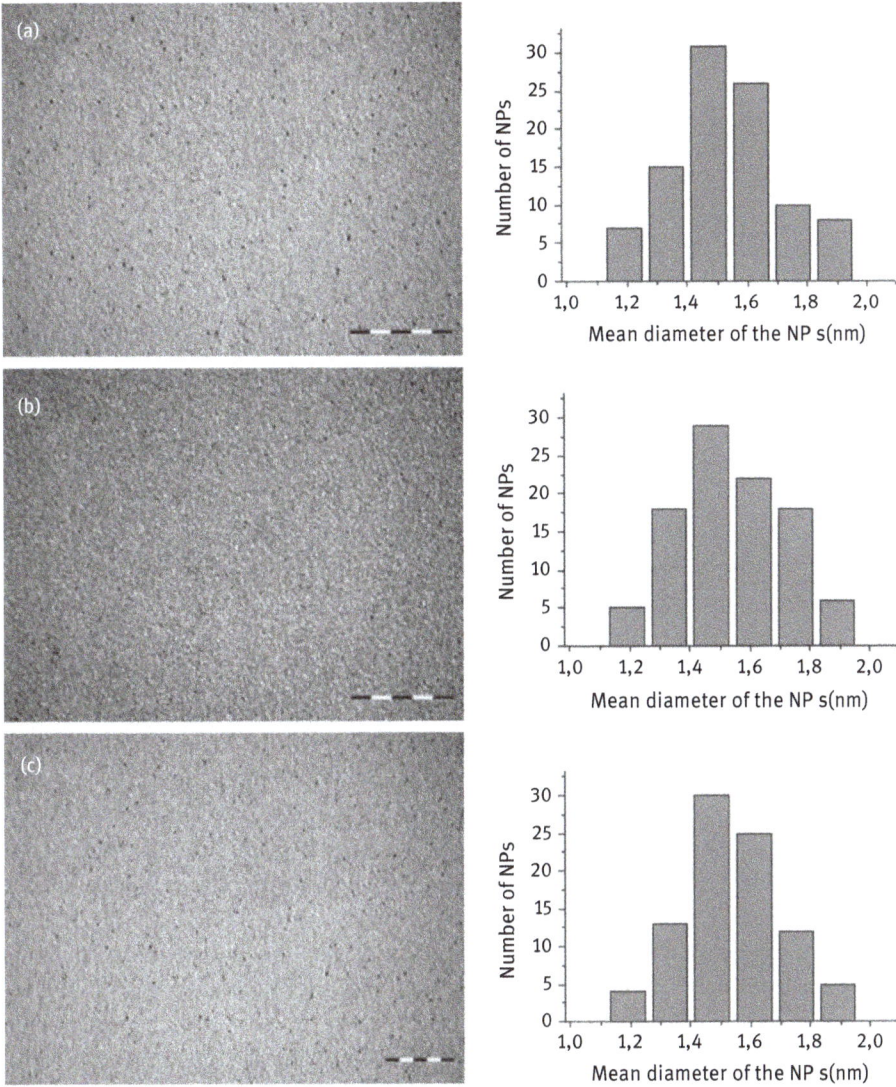

Figure 4.8: TEM images of Ru NCs stabilized with **L1** (a), **L2** (b) and **L3** (c) in the [Lₙ]/[Ru] molar ratio of 0.1 from aqueous colloidal solutions (Adapted with permission from ref. 103. Copyright 2012, Elsevier B. V.).

media, e.g. toluene, hexane, chloroform and benzene, are limited due to poor solubility of the corresponding metal ion precursors in these solvents. In addition, phase transfer is also an effective way to functionalize the noble metal NCs with suitable ligands to create close-packed, ordered arrays and wide-area nanoparticle thin films. The following sections summarize the application of a phase transfer technique for synthesizing Ru NCs capped with alkylamine ligands [106–108].

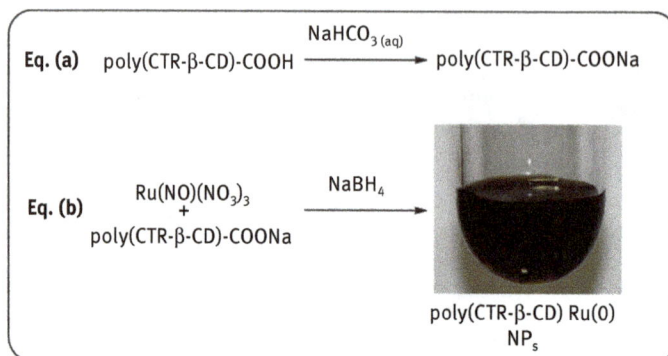

Figure 4.9: Schematic illustration to show the synthesis of aqueous soluble Ru NCs protected by poly(CTR-β-CD) (Adapted with permission from ref. 105. Copyright 2015, Royal Society of Chemistry).

4.2.1.5.1 Ethanol-mediated phase transfer of Ru NCs already formed

In a typical experiment [109], 0.2 mL of 38.8 mM aqueous sodium citrate solution was added to 10 mL of 2 mM aqueous $RuCl_3$ solution. Under vigorous stirring, 1.5 mL of 112 mM aqueous $NaBH_4$ solution was introduced dropwise to prepare a Ru hydrosol in which sodium citrate serves as the stabilizers. The molar ratio of $NaBH_4$ to $RuCl_3$ was kept above 8 to ensure the reduction of Ru^{3+} ions to the zerovalent state. The Ru hydrosol was also left to stand for 4 h after the addition to complete the reduction reaction. The Ru hydrosol was then mixed with 10 mL of ethanol containing 100 μL of dodecylamine and the mixture was stirred for 2 min. Subsequently, 5 mL of toluene was added and stirring continued for three more minutes. Dodecylamine-stabilized Ru NCs were extracted into the toluene layer rapidly, leaving behind a colorless aqueous solution.

In comparison with the Ru particles initially formed in their hydrosol (2.2 nm), the average diameter of Ru NCs after phase transfer has slight increase (3.45 nm), as indicated by the TEM image in Figure 4.10. The slight increase in Ru particle size is most probably caused by particle agglomeration, which occurs during the stabilizer exchange, where citrate was progressively displaced by dodecylamine to form the dodecylamine-stabilized Ru NCs.

4.2.1.5.2 Ethanol-mediated phase transfer of Ru precursors followed by reduction

The ethanol-mediated phase transfer method for metal particles has been extended to transfer the metal ions from aqueous phase to a hydrocarbon layer, in which the metal ions could then be applied toward the synthesis of a variety of noble metal nanoparticles, nanocomposites and nanostructures [110, 111]. In case of Ru NCs, the detailed procedure was depicted below: 10 mL of a 1 mM aqueous $RuCl_3$ solution was mixed with 10 mL of ethanol containing 200 μL of dodecylamine. After vigorous stirring for 3 min, 10 mL of toluene was introduced and the mixture was stirred for 1

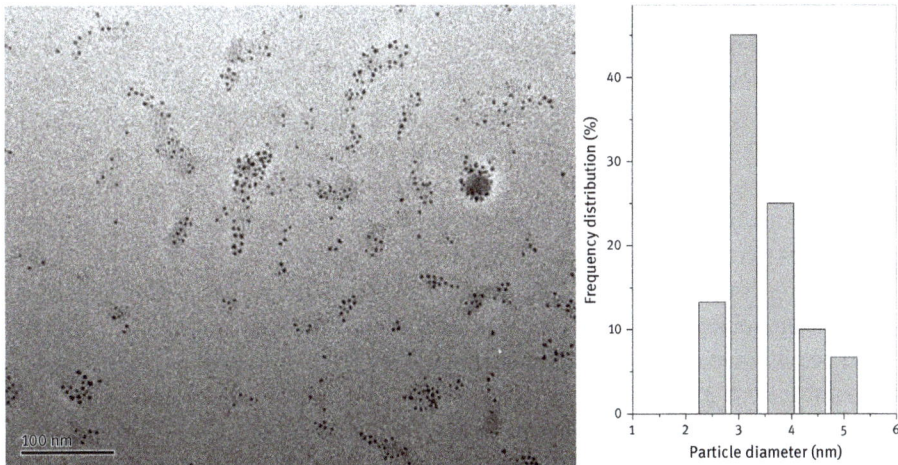

Figure 4.10: TEM image and size distribution of dodecylamine-stabilized Ru NCs prepared by ethanol-mediated phase transfer, $d = 3.45$ nm, $\sigma = 0.68$ nm (Adapted with permission from ref. 109. Copyright 2004, Elsevier B. V.).

more min before the transfer to a 50 mL separation funnel. Phase transfer of Ru^{3+} ions from water to toluene would then occur quickly and completely, as evident by the complete color bleaching of the aqueous phase. Then, the toluene layer was recovered and 1 mL of 100 mM of aqueous $NaBH_4$ or toluene solution of tetrabutylammonium borohydride (TBAB) was added at room temperature or at 100°C, and the mixture was agitated for several min. A black Ru colloidal solution was obtained, in which the average diameter of Ru NCs is 4.1 nm for $NaBH_4$ reduction and 2.48 nm for TBAB reduction, respectively (Figure 4.11). The colloidal Ru solution as prepared this way is highly stable, with no sign of agglomeration even after storage weeks of storage.

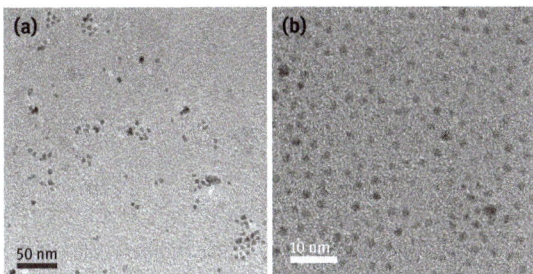

Figure 4.11: TEM image of dodecylamine-stabilized Ru NCs derived by (a) $NaBH_4$ reduction, and (b) by TBAB reduction of $RuCl_3$ in toluene (Adapted with permission from ref. 110 and 111. Copyright 2004, Elsevier B. V. and Copyright 2009, Nature Publishing Group, respectively).

4.2.1.6 Seed-mediated synthesis of Ru NCs with spherical morphology

Among different wet-chemistry strategies, the NC synthesis in oleylamine is of particular interest. Oleylamine is a common reagent in the chemical synthesis of nanoparticles [112]. It can function both as a reducing agent and as a capping ligand to stabilize the surface of NCs. In addition, oleylamine has very high boiling point, and can be a suitable solvent for the chemical reaction. This means amine-based NC synthesis can be conducted in a single-phase reaction system, posing significant advantages to understand the particle growth kinetics.

For direct synthesis of Ru NCs in oleylamine, e.g. reducing $RuCl_3$ in oleylamine at elevated temperature, quasi-spherical Ru NCs with an average diameter of ca. 5.8 nm could be obtained at a temperature as high as 350 °C [113]. Instead, at relatively low temperature, e.g. 320°C, worm- or rod-like Ru NCs with the average diameter of ca. 1.8 nm are the dominant products [114]. However, spherical Ru NCs could be obtained by a seed-mediated method at relatively low temperature [113]. In this method, silver (Ag) nanoparticles with spherical morphology were firstly synthesized in oleylamine as seeds for the subsequent growth of Ru metal. In principle, in the presence of seed particles, the epitaxial growth of the second metal on the previously formed seeds might occur and hence the final morphology of the second metal might be determined by the shape of the seed particles.

In a typical run, 85 mg of $AgNO_3$ was dissolved in 20 mL of oleylamine placed in a three-necked flask equipped with a condenser and stir bar. The solution was heated to 150°C under flowing N_2 and kept at this condition for 1 h for the reduction of Ag^+ ions by oleylamine. Subsequently, a certain amount of $RuCl_3$ was added swiftly, followed by heating and keeping at 320°C for 1 h under flowing N_2 to reduce Ru^{3+} ion precursors. After reaction, the products were purified by precipitation with methanol, centrifugation, washing with methanol and re-dispersed in 20 mL of chloroform. At a Ag/Ru molar ratio of 1/1, as evinced by Figure 4.12(a and b) for TEM and high-resolution TEM (HRTEM) images, the core-shell Ag-Ru NCs formed by this Ag seed-mediated method were nearly spherical in shape and have an average size of 13.7 nm. With the increase of the Ag/Ru molar ratios (2/1 and 4/1), the thickness of in Ru shell decreases, while the nearly spherical morphology of the core-shell Ag-Ru nanoparticles is maintained, as shown in Figure 4.12(c)–12(f).

4.2.1.7 Synthesis of spherical Ru NCs in ionic liquids

Ionic liquids (ILs) are particularly important in synthesis of metal NCs due to the possibility to equip the nano-products with new properties that may not be attainable otherwise [115–119]. Because of their many distinct advantages such as good thermal stability, high ionic conductivity, broad electrochemical potential windows, high synthetic flexibility and environmental benefits deriving from the negligible vapor pressures [120, 121], the use of ILs for the synthesis and tailoring of metal particles represents a burgeoning direction in materials chemistry. In

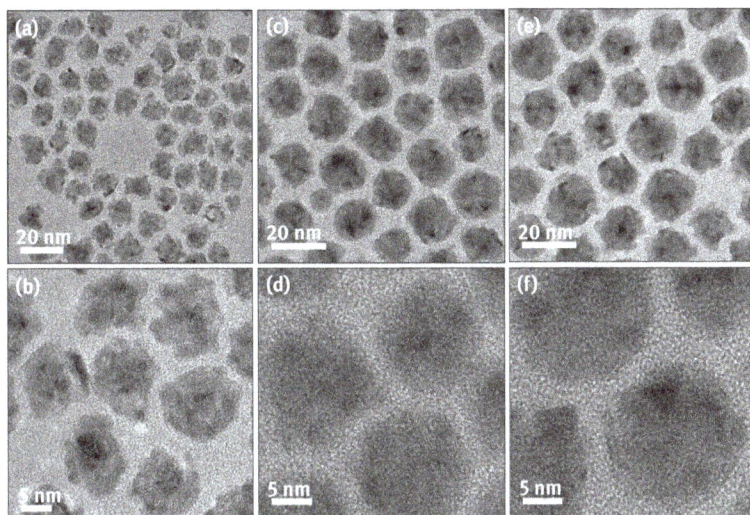

Figure 4.12: TEM (a,c,e) and HRTEM images (b,d,f) of core-shell Ag-Ru NCs as-prepared in oleylamine at Ag/Ru molar ratios of 1/1 (a,b), 2/1 (c,d), and 4/1 (e,f), respectively (Adapted with permission from ref. 113. Copyright 2013, Royal Society of Chemistry).

comparison with wet-chemistry methods employing conventional aqueous or organic solvents, which usually additional surfactants to stabilize the metal NCs formed in them, the ILs demonstrate dual functions in the synthesis, acting both as dispersion media and stabilizing agents [122]. The recent advances in this field have been comprehensively reviewed in a number of publications [116, 119, 122–124]. In this section, the typical routes for the synthesis of Ru NCs in ILs were described.

The central point for synthesizing Ru NCs in ILs is the decomposition of Ru-containing organometallic precursors, e.g. Ru(COD)(2-methylallyl)$_2$ [125–128], Ru (COD)(COT) [129–133] and Ru$_3$(CO)$_{12}$ [134, 135], in ILs under dihydrogen atmosphere with or without additional chemical agents. In a typical synthesis without additional chemical agents [125], a Fischer–Porter bottle was loaded with [Ru (COD)(2-methylallyl)$_2$] (15 mg, 0.047 mmol) in the glovebox. Then 1 mL of the ILs (BMI·NTf$_2$, DMI·NTf$_2$, BMI·BF$_4$ and DMI·BF$_4$; BMI = 1-n-butyl-3-methylimidazolium, DMI = 1-n-decyl-3-methylimidazolium, NTf$_2$ = N-bis(trifluoromethanesulfonyl)imidates, BF$_4$ = tetrafluoroborates) was added via syringe under an Ar flow. The mixture was stirred at room temperature for 60 min, resulting in a turbid dispersion. The system was heated to 50°C, and hydrogen (4 bar) was admitted to the system. After stirring for 18 h, a black suspension was obtained. The Fischer–Porter bottle was then kept under reduced pressure to eliminate the cyclooctane and isobutene formed. The Ru NCs prepared this way are in the range of 2.1–2.9 nm, determined by TEM (Table 4.2).

Table 4.2: Particle size (determined by TEM) of Ru NCs in different ILs (Adapted with permission from ref. 125. Copyright 2008, American Chemical Society).

Ru/IL	Particle size (nm)
Ru/BMI·NTf$_2$	2.1 ± 0.5
Ru/BMI·BF$_4$	2.9 ± 0.5
Ru/DMI·BF$_4$	2.7 ± 0.5
Ru/DMI·NTf$_2$	2.1 ± 0.5

For a typical synthesis of Ru NCs in ILs in the presence of other chemical agents [132], [Ru(COD)(COT)] (66.2 mg, 0.21 mmol) was placed in a Fischer–Porter bottle, on which 5 mL of the IL 1-alkyl-3-methyl-imidazolium bis(trifluoromethanesulfonyl) imide [C$_1$C$_n$ImNTf$_2$] (C$_n$ = butyl, hexyl, octyl, decyl) and 4.2×10^{-2} mmol of the ligands 1-octylamine (OA) or 1-hexadecylamine (HDA) were added. The mixture was a yellow suspension that was stirred for 3 h. The mixture was frozen in a liquid nitrogen bath, and placed in *vacuo*. The system was thermostated to 30°C, filled with H$_2$ (3 bar), and stirred for 20 h. After that time, the remaining hydrogen and volatile alkanes were evacuated under vacuum. The final mixture was a black colloidal suspension that remained stable for months. Selected TEM micrographs with corresponding size histograms were presented in Figure 4.13, which show the Ru NCs prepared this way have average sizes of ca. 1.2 ± 0.1 nm.

4.2.1.8 Supported Ru NCs with spherical morphology

Although high activity in catalysis, it is universally assumed that most important drawback of metal NCs is their tendency to aggregate, which leads to the loss of their main characteristics. Besides capping these particles with suitable ligands or surfactants, another effective way to solve this problem is to stabilize them by immobilizing them on solid substrates, e.g. inorganic oxides, porous carbons or insoluble polymers [136–140]. In addition, the possible interactions between the substrates and their supported metal NCs may have positive contribution to the catalytic reactions [141–143]. The conventional supported Ru NCs are prepared by impregnating an appropriate substrate with an aqueous or organic solution of Ru metal precursors, followed by dihydrogen reduction at elevated temperature. An alternative method to obtain supported metal NCs with well-defined metal particles is the preparation of supported of supported metal NCs from their colloids. The colloid method has the advantage of leveraging upon many existing methods for preparing the metal nanoparticles with well-controlled size, shape and dispersity. There are also several special approaches such as polygonal barrel-sputtering method and plasma electrochemical reduction method for producing supported Ru NCs, and all of them will be introduced in this section.

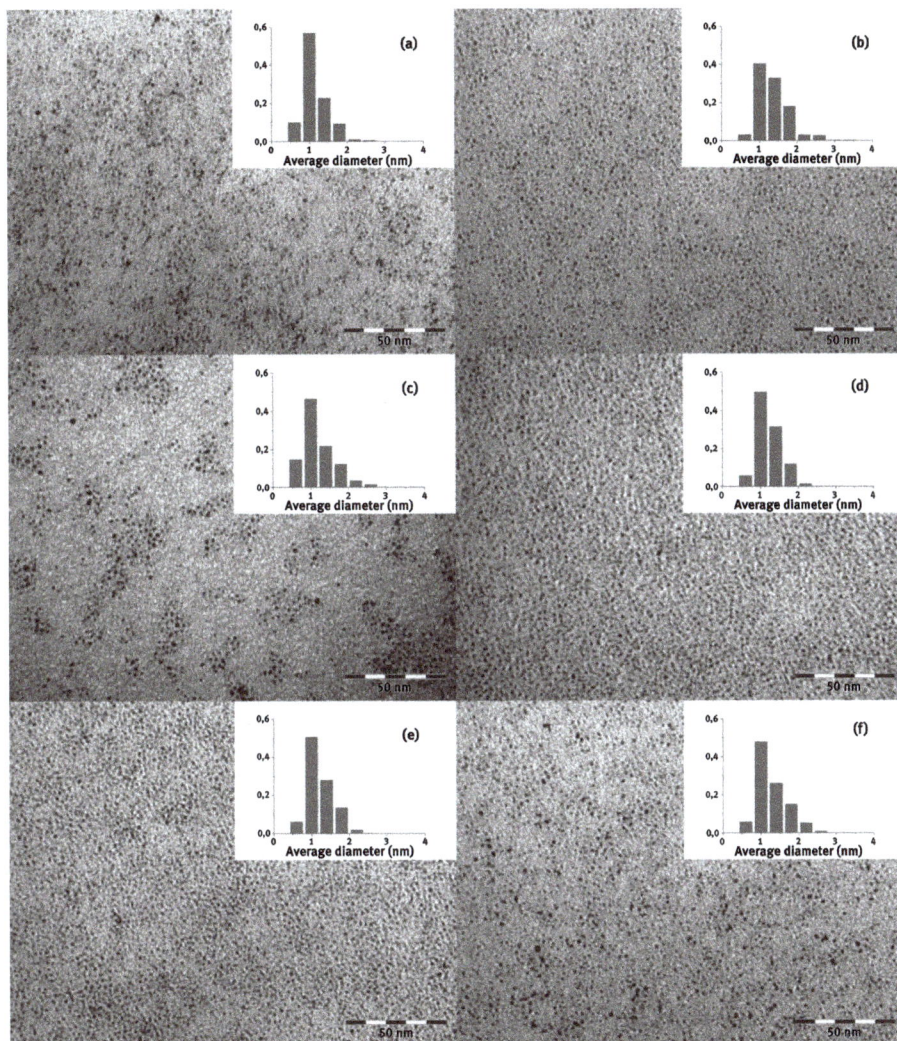

Figure 4.13: A–f Left to right and up to down: Ru colloids in ILs of $C_4 \cdot OA$, $C_4 \cdot HDA$, $C_4 \cdot OA$ no stirring, $C_6 \cdot OA$, $C_8 \cdot OA$, and $C_{10} \cdot OA$ (Adapted with permission from ref. 132. Copyright 2011, Royal Society of Chemistry).

4.2.1.8.1 Spherical Ru NCs supported on carbon substrates

Typically [92], appropriate amount of $RuCl_3$ were dissolved into 100 mL of EG in order to reach a concentration of metal of 0.375 g L^{-1}. Then, pH of the solution was adjusted at 10 by adding a solution of NaOH (1 M) in EG dropwise. Carbon Vulcan XC-72R (150 mg) thermally treated for 4 h at 400°C under nitrogen was then added to the solution for obtaining a nominal metal loading of 20 wt% on carbon, and the mixture was ultra-sonically homogenized for 5 min. The reactor equipped with a condenser was put inside a MARS oven. The synthesis of carbon-supported Ru NCs was performed under continuous microwave irradiation at a power of 1600 W until reaching the desired reaction temperature (130°C), and then microwave pulses were applied to maintain it. After reaction, the carbon-supported Ru powders were washed with acetone and ultrapure water, and filtered. The average sizes of these carbon-supported Ru NCs determined by TEM are smaller than 1.5 nm.

4.2.1.8.2 Carbon-supported Ru NCs from Ru colloids

In a typical experiment, the Ru organosol in toluene was prepared by the ethanol-mediated phase transfer of Ru NCs already formed (also see the description in Section 3.1.5.1) [109]. Then to the Ru organosol, a calculated amount of XC-72R carbon substrates were introduced for obtaining a nominal Ru loading of 20 wt% on carbon. The mixture was stirred vigorously for more than 8 h, and the carbon-supported Ru NCs were collected by centrifugation, washed thrice with methanol, and then dried at room temperature in vacuum. Analysis by inductively coupled plasma-atomic emission spectrophotometry (ICP-AES) indicates that more than 97% Ru NCs have been loaded on carbon substrates by this way.

4.2.1.8.3 Carbon nanotube (CNT)-supported Ru NCs

In a typical synthesis [144], carbon nanotubes (CNTs, 1 g) with outer diameters and lengths of 60–80 nm and 1–12 µm, respectively, and a suitable amount of $RuCl_3 \cdot H_2O$ dissolved in water (5 mL) were loaded into a stainless steel autoclave (15 mL). After being flushed with nitrogen gas for 5 min, the autoclave was sealed and maintained at desired temperature (400°C or 450°C) for 2 h and then cooled down to room temperature naturally. The dark precipitates were separated from the solution by centrifugation. The product was washed with sufficient distilled water and then vacuum dried at 60°C for 6 h. The TEM observation (Figure 4.14(a)) confirms Ru NCs with an average diameter of ca. 5 nm are randomly distributed on the surface of CNT substrates at an initial $RuCl_3$/CNT weight ratio of 1/1. Both increases in $RuCl_3$/CNT weight ratio and in synthetic temperature would lead to the increase in Ru particle size for the CNT-supported Ru NCs (Figure 4.14(b–d)).

4.2.1.8.4 *Graphene-supported Ru NCs in ionic liquids*

In a typical experiment [145], chemically derived graphene (CDG, 4.8 mg, 0.2 wt% related to 2.4 g IL) was dissolved/suspended in the dried and degassed (desoxygenated) IL 1-butyl-3-methylimidazolium tetrafluoroborate ($BMImBF_4$, 2.0 mL, 2.4 g, $\rho = 1.2$ g mL^{-1}) at room temperature with magnetic stirring for 20 h in a microwave reaction vial. Then 50.6 mg of $Ru_3(CO)_{12}$ (0.230 mmol Ru) was added to the CDG slurry in $BMImBF_4$ (1 wt% metal, related to 2.4 g BMImBF4) and suspended with magnetic stirring for 18 h under Ar atmosphere. Subsequently, the stirring bars were removed and the mixture was subjected to microwave irradiation (6 min, 20 W) under Ar atmosphere. Afterwards, the slurry was degassed from CO in vacuo. Distilled water (6 mL) was added to remove the IL from the grapheme-supported Ru

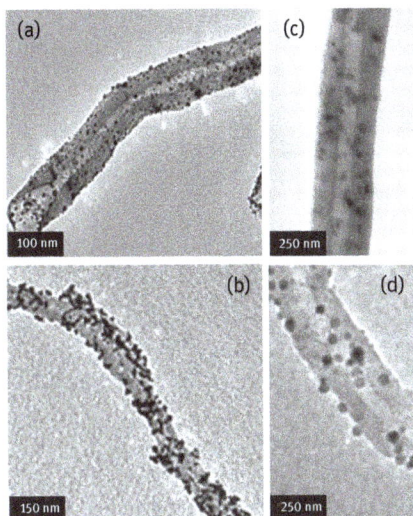

Figure 4.14: TEM images of CNT-supported Ru NCs synthesized at (a) 400 °C and initial RuCl$_3$/CNT weight ratio of 1/1, (b) 400 °C and initial RuCl$_3$/CNT weight ratio of 2/1, (c) 450 °C and initial RuCl$_3$/CNT weight ratio of 1/2, and (d) 450 °C and initial RuCl$_3$/CNT weight ratio of 1/2, respectively (Adapted with permission from ref. 144. Copyright 2005, Wiley-VCH Verlag).

system (Ru/CDG). The black slurry was centrifuged (2 × 15 min, 2000 rpm) and the supernatant liquid H$_2$O/IL phase decanted and discarded. The addition of H$_2$O, centrifugation and decantation was repeated three times. At last, the residue was again dispersed in water, filtered and dried under vacuum. The dry black-greyish residue formed flakes which could easily be removed from the filter to yield 25.0 mg (83 %) Ru/CDG in which the Ru NCs with an average size of 2.2 nm are uniformly dispersed on the grapheme substrates (Figure 4.15).

4.2.1.8.5 Ru NCs entrapped in porous substrates
The impregnation methods have also been used to entrap the Ru NCs in porous substrates, e.g. porous carbon [65], porous silica [42, 46, 144, 145], periodic mesopolymer [146], and montmorillonite (MMT, a type of naturally occurring clay) [147], to gain high performance materials for diverse catalytic applications. As schematically illustrated in Figure 4.16(a and b) for schematic illustrations, in a typical procedure for entrapping Ru NCs in porous carbon and silica [65], first, a porous hard template (either H-form zeolite Y, HY, or mesoporous SBA-15 silica) was impregnated with a RuCl$_3$ solution by sonication for 0.5 h. The concentration of the RuCl$_3$ solution used was 0.1 M and the ratio of solution volume to hard-template mass was 4 mL:1 g. Second, the suspension was evaporated under stirring at 100°C, and then dried in air at 200°C for 3 h. Third, the Ru-impregnated solid (ca. 0.5 g) was placed in a quartz tube and heated to 900°C under a pure N$_2$ flow (30 cm^3 min^{-1}). Subsequently, infiltration of carbon was conducted with benzene as the carbon precursor using

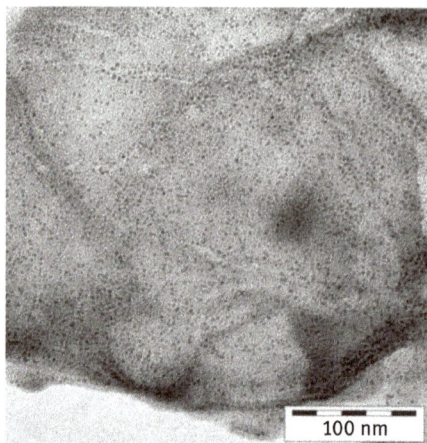

Figure 4.15: TEM image of grapheme-supported Ru NCs prepared from microwave irradiation of Ru_3 $(CO)_{12}$ in ionic liquid BMImBF$_4$ (Adapted with permission from ref. 145. Copyright 2011, Elsevier B. V.).

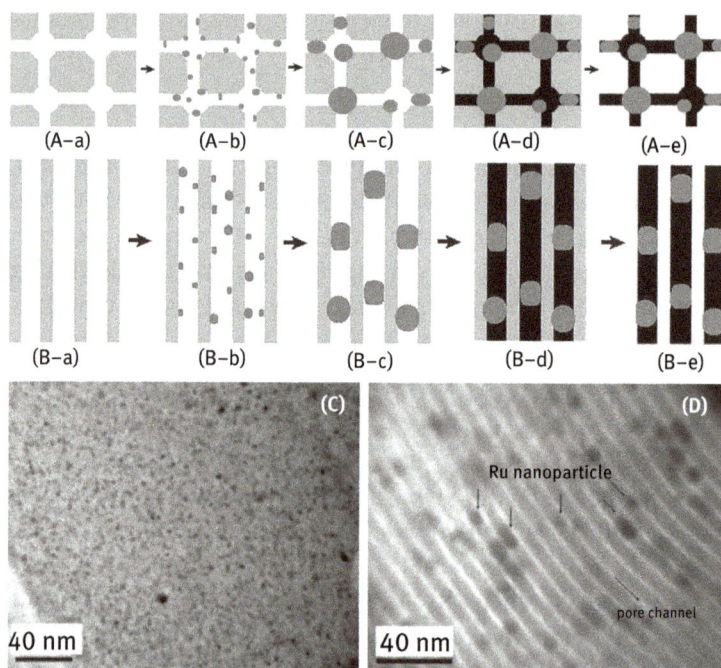

Figure 4.16: A schematic illustration of the synthetic procedure using zeolite (A) HY and (B) meso-porous silica SBA-15 as templates: (a) hard template; (b) Ru NCs confined in the pores of the template by the impregnation method; (c) Ru NCs grow at high temperature in nitrogen gas; (d) infiltration of carbon into the pores using the CVD method; (e) Ru NCs sandwiched in the carbon matrix upon removal of the template; TEM images of catalysts (C) RuC$_1$ and (D) RuC$_2$ samples (Adapted with permission from ref. 65. Copyright 2007, Wiley-VCH Verlag).

the chemical vapor deposition (CVD) method at 900°C for 3 h. Finally, the black sample was treated with a 20% HF solution to remove the template, washed with deionized water, and dried at 12°C overnight. The Ru catalysts thus obtained using HY and SBA-15 as hard templates were designated as RuC_1 and RuC_2, respectively. As confirmed by Figure 4.16(c) for TEM image, the Ru particle sizes range from 1 to 2 nm, comparable to that of the channel and cage sizes of zeolite Y, while when using SBA-15 silica as template, the Ru NCs of 7–8 nm in diameter were obtained (Figure 4.16(d)), agreeing well with the pore-channel size of SBA-15.

4.2.1.8.6 Ru NCs supported on γ-alumina substrates

γ-Alumina ($γ-Al_2O_3$)-supported Ru NCs are the commonly used catalysts for ammonia synthesis [48, 49]. The loading of Ru NCs on $γ-Al_2O_3$ substrates is also based on impregnation methods by (i) deposition of the Ru colloid, which was already formed, onto alumina; and (ii) reduction of $RuCl_3$ or other Ru metal precursors in the presence of alumina. In a typical synthesis for the former case [148], 0.2 g of $RuCl_3$ was dissolved in 50 mL of EG and then 1.5 g of $γ-Al_2O_3$ was added to form a suspension. The mixture was stirred for 15 min at room temperature and a microwave-solvother-mal reaction was carried out under mild conditions at 170–180°C, pressure 5 bar and the hold time of 10 min. After reaction the vessel was rapidly cooled down in an ice-water bath. A gray solid was filtered off, washed with $NaNO_3$ aqueous solution, next with distilled water to remove the sodium and chloride ions, and then dried under vacuum at room temperature.

While for the latter case, an appropriate amount of $RuCl_3$ was dissolved in 1.5 mL of water/g support to bring the samples to incipient wetness. After impregnation, the mixture was left for 2 h at room temperature, next dried overnight at 110°C and reduced in H_2 flow (heating rate of 5°C min^{-1}) at 500°C for 20 h. TEM analyses (Figure 4.17) manifest that the average sizes of Ru NCs on the $γ-Al_2O_3$ substrates are 2.2 nm for former case and 2.4 nm for latter case, respectively. With slight modification, e.g. by impregnation of gallium nitride (GaN) nanowires with a solution of Ru_3 $(CO)_{12}$ in dry tetrahydrofuran, GaN-supported Ru NCs with high Ru dispersion could be obtained for catalyzing nitrogen photofixation[149].

4.2.1.8.7 Ru NCs supported on TiO₂ substrates by a polygonal barrel-sputtering method

The wet-chemistry methods for loading metal NCs on various substrates usually need heating process, which often induces growth of the metal NCs. Instead, Abe et al. developed a "dry" technique, named as "barrel-sputtering" to allow the fabrication of highly dispersed naked metal NCs on a powdery substrates without any heating [56, 150, 151]

Figure 4.18 is a schematic representation of the main components of the hexagonal-barrel sputtering system [56], [150], [151], The system basically consists of a vacuum chamber, diffusion pump, turbo-molecular pump, gas introduction system, high frequency power supply and target stage. The vacuum chamber (200 mm in

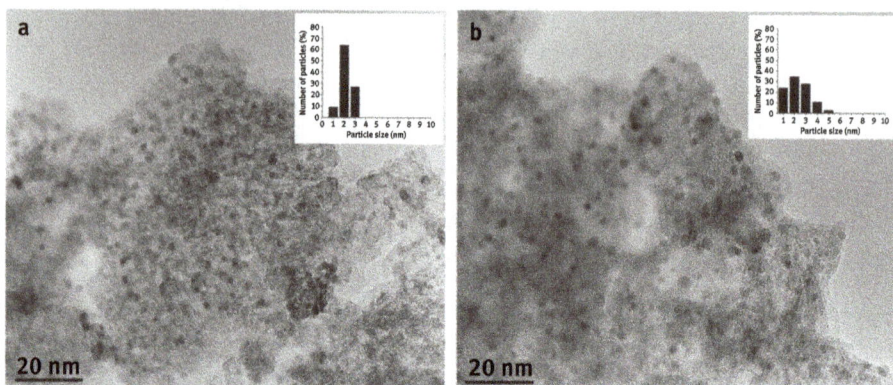

Figure 4.17: Representative TEM micrographs and particle size histograms (insets) of the Ru/γ-Al$_2$O$_3$ catalysts: (a) by impregnation with "'as prepared'" Ru colloids and (b) by impregnation with RuCl$_3$ solution followed by H$_2$ reduction at 500 °C (Adapted with permission from ref. 150. Copyright 2007, Elsevier B. V.).

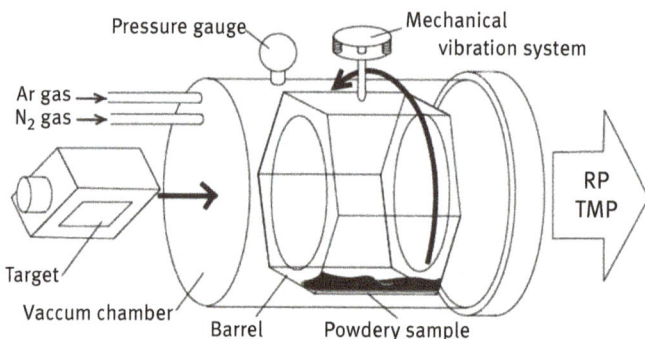

Figure 4.18: Schematic diagram of the hexagonal-barrel equipped RF sputtering system (Adapted with permission from ref. 56. Copyright 2009, Royal Society of Chemistry).

diameter, 200 mm in length) is equipped with a hexagonal-barrel to hold samples. The barrel is equipped with a rotation mechanism. The microparticles are lifted up to a certain level along the wall of the barrel with barrel rotation and then falls to the bottom of the barrel under its own weight. The powder sample is then agitated. The powder particles are prevented from coagulating with each other by repeating this motion during the stirring in this manner. A mechanical mechanism to be vibrated the barrel is mounted on the system that makes the microparticles adhere to the wall of the barrel and prevents the particles from coagulating with each other. The target stage (50 mm in width, 100 mm in length) is cooled by water and the angle to the sample powder could be freely changed to enhance under a given rotation speed.

In a typical synthesis using a polygonal barrel-sputtering method for TiO$_2$-supported Ru NCs [56], 3.0 g of titanium dioxide (TiO$_2$, anatase) powder, which was used as a support after being dried overnight in an oven at 180°C, was loaded into the barrel. Metallic Ru plates (purity: 99.9%, 50 × 100 mm^2) were used as the sputtering target. The vacuum chamber was carefully evacuated up to 8.0 × 10^{-4} Pa, and then high-purity Ar gas (purity: 99.995%) was gradually introduced into the chamber until the pressure increased up to 0.8 Pa. Radio-frequency (13.56 MHz) magnetron sputtering was performed for 25 min with an input power of 100 W, leading to a deposition of Ru on each particle surface in the powder. The amount of Ru loaded onto a support was adjusted to be approximately 0.8 wt% by controlling the sputtering duration. During sputtering, a swinging motion of ±75° at a speed of 4.2 rpm was provided to the hexagonal barrel. After sputtering, N$_2$ gas (99.9998%) was gradually introduced to increase the pressure up to 1 atmosphere, whereupon the prepared samples could be extracted.

4.2.1.8.8 Ru NCs supported on substrates by plasma electrochemical reduction method

The plasma electrochemical reduction method is a non-lithographic, dry approach to patterning films of metal nanoparticles, featured in both low cost and high through-put. Plasma electrochemical reduction is high purity and versatile, allowing a wide-range of noble metal NCs to be formed [152–154].

In a typical strategy developed by Lee *et al* [155], the plasma source was operated without hydrogen gas, confined to micrometer dimensions, and continuously operated at atmospheric pressure, making it possible to write microscale patterns of metal nanoparticles including Ru at ambient conditions. In the course of developing and optimizing patterning technique, two approaches were explored to plasma electroche-mical reduction of metal precursors: (i) solutions of metal salt and polymer were directly reduced by a microplasma to form colloidal (i.e. suspended) metal nanoparti-cles or, alternatively, (ii) the same solutions of metal salt and polymer were spin coated onto a substrate and dried to obtain a thin film, then reduced by a scanning micro-plasma to form patterns of metal nanoparticles. A direct-current, atmospheric-pressure microplasma jet was formed between a stainless-steel capillary tube (I.D. = 180 μm) and a grounded counter electrode (Si substrate) using a negatively biased high voltage power supply. (Ar gas flow was coupled to the tube electrode and controlled by a mass flow controller (MFC). To ignite the microplasma, a high voltage (ca. 2 kV) was applied. The plasma was then stabilized at constant current by a ballast resistor (R = 160 kΩ) and adjusting the power supply voltage. Patterns in films were generated by rastering the microplasma; a moving stage controlled by a pair of numerically controlled stepping motors was used to move the substrate in the x and y directions. The step size for the scanning was 1 mm and the scanning rate could be adjusted between 3 and 30 mm min^{-1} by changing the delay time between steps. The z direction was also controlled by a third stepping motor which fixed the distance between the tube and counter electrode.

4.2.2 Synthetic approaches for one-dimensional (1D) Ru NCs

One-dimensional (1D) structures, such as nanowires, nanorods, nanobelts and nano-tubes (NTs), have become the focus of intensive research owing their unique applica-tions in catalysis and nanoscale devices [156–159]. 1D nanostructures have a number of inherent advantages, such as the preferential exposure of smooth, low-energy crystal-line planes, lower defect site densities and beneficial size-dependent electronic proper-ties, thereby collectively rendering them more active and durable toward a given chemical reaction, e.g. ORR and alcohol electroxidation [160–164]. Unfortunately, in comparison with the abundant literatures in 1D nanostructure of other noble metals, e.g. Au [165–168], Ag [169–173], Pt [174–178] and Pd [161, 179–181], the reports on 1D Ru NCs remain scarce.

In a typical synthesis of 1D Ru NCs, which is based on a solvothermal procedure [182], 0.05 g (0.126 mmol) of ruthenium (III) acetylacetonate (Ru(acac)$_3$) was dis-solved in 6 mL of the hydrocarbon (decalin or toluene) or n-octylamine. The resulting solution was sealed in a swagelok autoclave of 20 mL volume capacity and heated at 320°C for 6 h. The autoclave was then allowed to cool down to room temperature. A black-colored insoluble product was formed which was thoroughly washed with acetone. The Ru NCs as-prepared have rod shapes with a length of ca. 50 nm and an average diameter of ca. 15 nm. However, the uniformity of the rod shape of the Ru NCs needs to be further improved.

Templated-directed synthesis represents a straightforward route to 1D nanoma-terials. In this approach, the templates serve as a scaffold, within or around which another material is generated *in situ* and shaped into nanostructures with their morphology complementary to that of the template [156]. Ponrouch *et al.* reported a strategy to prepare well-aligned Ru NTs, which is based on electrochemical deposi-tion through a porous template [183, 184]. With the assistance of anodized aluminum oxide (AAO) membranes as hard templates, micrometer-long Ru NTs with an outer diameter of ca. 200 nm are successfully fabricated. In brief, the working electrode (WE) is made of a Ti substrate, an AAO (200 nm pore diameter) membrane and a porous glass plate. These three parts are held together with two clamps. The Ru electroplating solution was [HCl] = 0.01 M, [KCl] = 0.1 M and [RuCl$_3$ · 6H$_2$O] = 0.005 M. All Ru deposits were performed using the pulse potentiostatic mode. After electro-deposition, the AAO membrane was dissolved by immersion in 1 M NaOH for 2 h at room temperature.

In a recent study, high-quality, crystalline Ru 1D nanowires were synthesized using a simple, ambient, template-based method (Figure 4.19) [185]. In detail, a saturated precursor solution, consisting of 25 mM potassium hexachlororuthenate (K$_2$RuCl$_6$), was prepared by dissolving the appropriate amount of solid powder (98 mg) into 10 mL of distilled water. A reducing agent solution consisting of 50 mM NaBH$_4$ was also generated by dissolving the NaBH$_4$ powder (19 mg) in 10 mL of distilled water. Prior to assembling the U-tube, wherein a wide variety of precursor

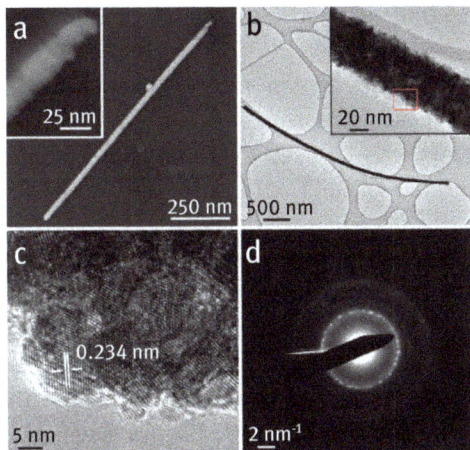

Figure 4.19: (a) Scanning electron microscopy (SEM) and (b) TEM images of individual isolated Ru NWs. Higher-magnification SEM and TEM images are shown as insets. HRTEM (c) and SAED (d) patterns were obtained from a representative area of a central region of a typical wire, as designated by the red box (Adapted with permission from ref. 187. Copyright 2013, American Chemical Society).

solutions can react within the nanosized pores of a membrane, often composed of anodic alumina or polycarbonate (PC) [158, 161], a commercially available PC membrane, serving as the growth template, was immersed in distilled water and sonicated in order to saturate the pores. The reaction was carried out by securing the PC membrane between the two glass half-cells of the U-tube device, and subsequently the precursor and reducing agent solutions are simultaneously added to the two half-cells separated by the template.

The precursor and reducing agent solution diffuse into the template pores and react, leading to the nucleation and growth of the Ru nanowires within the 1D pore channels of the membrane. Crystalline Ru nanowires can be achieved after a reaction time of 60 min. Ru nanowires with average diameters of 44 ± 4, 131 ± 14, and 280 ± 20 nm could be reproducibly prepared from templates with nominal corresponding pore sizes of 15, 50, and 200 nm, respectively.

4.2.3 Synthetic approaches for two-dimensional (2D) Ru NCs

Two-dimensional (2D) noble metal nanostructures, e.g. nanoplates and nanosheets, exhibit fascinating properties, and have shown various potential applications in electronics, optics, biomedicine, sensors and energy [186–195]. In special, 2D nanostructures with relatively high surface area-to-volume ratio and high density of exposed atoms on their surface are promising for catalysis reactions [196–198]. For instance, the ultrathin Pd nanosheets display much higher electrocatalytic activity compared to the tetrahedral and concave tetrahedral Pd NCs as well as the commercial Pd catalyst for formic acid oxidation [199].

In case of 2D Ru NCs, the synthetic approaches are less reported. However, the general concept for fabricating 2D noble metal nanostructures, e.g. by choosing appropriate facet-selective-adsorbing agents, can be used to obtain 2D Ru nanoplates. In a typical synthesis [200], 0.06 mmol of $RuCl_3 \cdot xH_2O$ and 100 mg of PVP were dissolved in 10 mL of water. Then, 0.4 mL of HCHO (40 wt%) was added, and the total volume of the solution was adjusted to 15 mL with water. The homogeneous black solution was transferred to a 25 mL Teflon-lined stainless steel autoclave and sealed. The autoclave was then heated at 160°C for 4 h before it was cooled down to room temperature. The black nanoparticles were centrifuged at 7500 rpm for 10 min with importing 45 mL of acetone and washed with water/acetone three times. The Ru NCs prepared this way are triangle nanoplates (Figure 4.20). By increasing the amounts of $RuCl_3 \cdot xH_2O$ and PVP to 0.45 mmol and 200 mg, respectively, Ru nanoplates with irregular shapes could be obtained as dominant products.

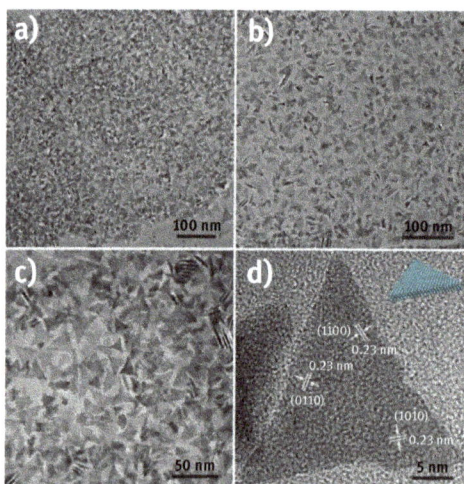

Figure 4.20: TEM (a – c) and HRTEM (d) images of triangle Ru nanoplates (Adapted with permission from ref. 202. Copyright 2012, American Chemical Society).

Template-based strategy was also used to produce 2D Ru nanostructures. For example, by using a seed-mediated growth method, ultrathin Pd nanosheets decorated with a Ru submonolayer, referred to as Pd@Ru NSs, could be prepared (Figure 4.21) [201]. In detail, for the synthesis of ultrathin Pd nanosheets as seeds, PVP with molecular weight of 29,000 (30 mg), critic acid (CA, 170 mg), cetyltrimethylammonium bromide (CTAB, 60 mg), and Palladium (II) acetylacetonate ($Pd(acac)_2$, 16 mg) were dissolved in N,N-dimethylformamide (DMF, 10 mL) and magnetically stirred for 1 h. The resulting homogeneous orange-red solution was transferred into a 50 mL flask. After $W(CO)_6$

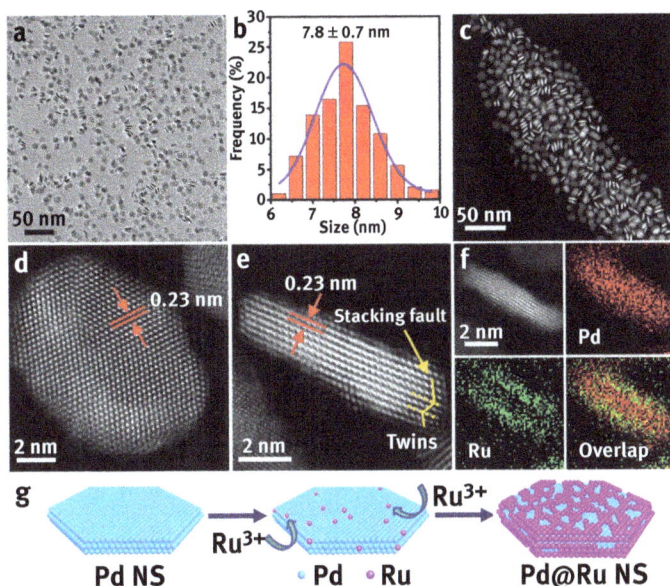

Figure 4.21: (a) TEM, (c) HAADF-STEM, and (d,e) aberration-corrected HAADF-STEM images of Pd@Ru NSs. (b) Statistical analysis of lateral size of Pd@Ru NSs measured in (a). (f) STEM-EDS elemental mapping of a typical Pd@Ru NS. (g) Schematic illustration of the synthetic process of Pd@Ru NSs (Adapted with permission from ref. 203. Copyright 2016, Wiley-VCH Verlag).

(100 mg) was added into the flask under N_2 atmosphere, the flask was capped and heated at 80°C for 1 h in an oil bath. Finally, the resulting product was precipitated by acetone (30 mL), separated by centrifugation at 6000 rpm for 5 min. After being washed with the mixture of ethanol (5 mL) and acetone (40 mL), it was centrifuged at 12,000 rpm for 10 min. The precipitate was re-dispersed in EG (10 mL).

Subsequently, The PVP (50 mg) and ascorbic acid (AA, 50 mg) were dissolved in EG (5 mL), and then mixed with the suspension of the synthesized Pd nanosheets in EG (0.55 mg mL^{-1}, 5 mL). After the mixture was heated to 160°C in an oil bath under magnetically stirring, the solution of Ru(acac)$_3$ in EG (0.8 mg mL^{-1}, 5 mL) was syringed into the aforementioned mixture at the rate of 5 mL h^{-1} under magnetically stirring. After the solution was maintained at 160°C for 1 h under magnetically stirring, it was cooled down to room temperature naturally. Finally, the product was precipitated by acetone (40 mL), separated *via* centrifugation at 12,000 rpm for 10 min, and further washed three times with a mixture of ethanol (5 mL) and acetone (40 mL).

4.2.4 Synthetic approaches for Ru NCs with hollow interiors

Hollow nanostructures made of noble metals are of great interest because of their immense potentials in catalysis [202–206], optics [207–212] and therapeutics [213–217].

Currently, template-assisted selective etching of core-shell particles is the principle methods to generate hollow nanostructures of noble metals. In this section, a number of highly reproducible approaches for the synthesis of Ru NCs with hollow interiors are depicted, although the literatures reported on them are still very few.

4.2.4.1 Synthesis of Ru nano-inorganic cages using Cu_2S nanocrystals as seeds

The seed-mediated synthesis of Ru nanocages includes three steps [220]: (1) Preparation of Cu_2S seeds, (2) Preparation of Ru nano-inorganic caged $Cu_{1.96}S$ nanoparticles (Ru NICed $Cu_{1.96}S$) and (3) Liberation of the Ru nano-inorganic cages (Ru NICs). In detail, copper (II) acetylacetonate (265 mg, 1.0 mmol) was suspended in 25 mL of dodeca-nethiol. The mixture was bubbled with Ar for 30 min and then heated quickly to 200°C for 65 min. Isopropanol was added and the mixture stirred briefly to settle the particles and the supernatant was removed. This was repeated again with isopropanol and then twice with chloroform. The products were suspended in 20 mL of chloroform.

Then, while maintaining an inert atmosphere, a suspension of Cu_2S seeds in chloroform (0.70 mL, 75 µmol of copper) was added to 2.0 g of octadecylamine. The chloroform was removed *in vacuo* and the solution was then heated at 200–210°C for 30 min under Ar. A solution of ruthenium (III) acetylacetonate (1.7 mg, 4.3 µmol) in octyl ether (1 mL) at 200–210°C was added and the entire solution stirred at 205–210°C for 1 h. Particles were isolated by repetitive dissolution of the excess surfactant in warm isopropanol (60–70°C) followed by centrifugation at 4000 rpm for 3 min, yielding a black precipitate and a green supernatant. The precipitate was suspended in chloro-form and centrifuged at 4000 rpm for 2 min. The brown-yellow supernatant was collected as a solution of the desired Ru NICed $Cu_{1.96}S$ nanoparticles.

Subsequently, a chloroform solution of the Ru NICed $Cu_{1.96}S$ nanoparticles was prepared (Optical Density [OD] = 0.5 @ 400 nm). Neocuproine (10 mg, 48 µmol) was added and the solution was stirred for 4 d. The solvent was removed by Ar flush and the precipitate washed three times with ethanol. Empty Ru NICs thus obtained have an average diameter of 14.4 nm with standard deviation of 1.3 nm (Figure 4.22).

4.2.4.2 Synthesis of hollow Ru NCs from Ni@Ru core@shell structures

In a typical synthesis for the Ni seeds, nickel (II) acetylacetonate ($Ni(acac)_2$, 64 mg) were dissolved in 5 mL of diphenyl ether containing 0.75 mL of oleic acid. The solution was heated to 120°C and kept at this temperature for 20 min to remove humidity and oxygen. Next, the solution was cooled down to 90°C, and 1 mL of Superhydride® was quickly injected into the solution. After 1 min, the as-prepared mixture was transferred into a flask containing 15 mL of oleylamine which was preheated to 300°C. The mixture was kept at 250°C for 15 min and then cooled down to 100°C for the Ru deposition [221, 222].

Then to prepare Ni@Ru core-shell nanostructures with complete Ru shell, 4 mL of oleylamine containing 53.3 mg of $Ru_3(CO)_{12}$ was injected into a colloidal suspension of

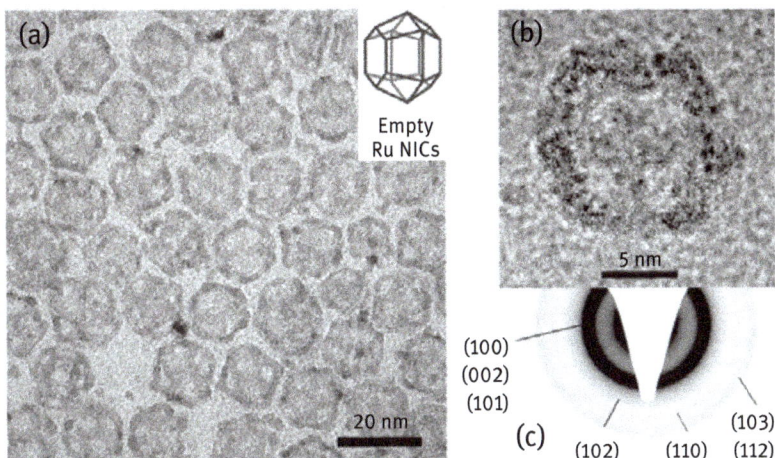

Figure 4.22: (a) TEM image of empty Ru NICs (Inset is a 3D model of an individual Ru NIC), (b) HRTEM image of empty Ru NICs showing very small, typically 1.5–3.5 nm, crystalline domains, (c) selected area electron diffraction (SAED) with broad reflections indexed to hexagonal ruthenium of empty Ru NICs (Adapted with permission from ref. 220. Copyright 2010, Nature Publishing Group).

Ni seeds at 100°C. The mixture was heated to 160°C (2°C min^{-1}) under vigorous stirring, and then slowly raised to 200°C (1°C min^{-1}). After refluxing for 60 min, the reaction was quenched by removing the reaction flask off the mantle. Forty milliliters of ethanol was added, and the product was separated by centrifugation at 9000 rpm for 10 min. After three cycles, the product was dispersed in 5 mL of hexane.

Finally, nitric acid (HNO$_3$) was used to obtain the hollow Ru NCs. Five milliliters of HNO$_3$ (1 M) were added to the Ni@Ru dispersion in hexane. Vigorous shaking (30 min) was necessary to ensure effective contact between Ni core and HNO$_3$ for the complete removal of Ni. Then, the mixture was decanted until two layers of nonpolar (hexane) and polar phase (aqueous HNO$_3$) separate. Next, the aqueous layer was carefully removed using a syringe, followed by addition of an excess of ethanol. Finally, after three cycles of washing with ethanol and centrifugation, the hollow Ru products, which have an average size of 14 nm and apparent brightness contrast between the core and shell regions (Figure 4.23), were dispersed in hexane.

4.2.4.3 Synthesis of hollow Ru NCs from Ag@Ru core@shell structures

In a typical synthesis [223], 85 mg AgNO$_3$ was dissolved in 20 mL of oleylamine placed in a three-necked flask equipped with a condenser and stir bar. The solution was heated to 150°C under flowing Ar or N$_2$ and kept at this condition for 1 h for the reduction of Ag$^+$ ions by oleylamine. Then 104 mg of RuCl$_3$ was added swiftly, followed by heating and keeping at 320°C for 1 h under flowing Ar or N$_2$ for the reduction of the Ru metal precursors. After reaction, the core@shell Ag@Ru

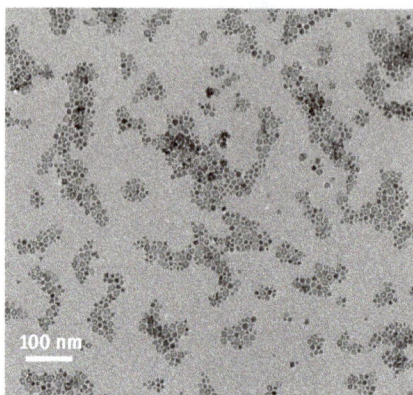

Figure 4.23: TEM image of as-prepared hollow Ru NCs from Ni@Ru core@shell nanostructures (Adapted with permission from ref. 222. Copyright 2012, Royal Society of Chemistry).

nanostructures were purified by precipitation with methanol, centrifugation, washing with methanol and re-dispersed in 20 mL of toluene. Subsequently, for the synthesis of hollow Ru NCs, 4 mL of core-shell Ag-Ru colloidal solution was diluted with toluene to 100 mL. An aqueous bis(p-sulfonatophenyl)phenylphosphane dihydrate dipotassium salt (BSPP) solution was prepared by dissolving 500 mg of BSPP in 100 mL of water. The diluted core-shell Ag-Ru colloidal solution in toluene and the aqueous BSPP solution were then mixed together. The mixture was aged for 48 h under vigorous stirring at room temperature to remove Ag from the core-shell Ag-Ru nanostructures. The mixture was left to stand, and the upper toluene phase was collected, which contains hollow Ru NCs with clear image contrast between the core and shell regions (Figure 4.24).

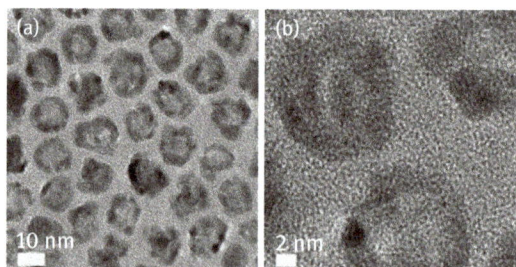

Figure 4.24: (a) TEM image and (b) HRTEM image of as-prepared hollow Ru NCs from Ag@Ru core@shell nanostructures (Adapted with permission from ref. 223. Copyright 2012, American Chemical Society).

4.2.4.4 Synthesis of carbon-supported hollow Ru NCs

The core@shell Ag@Ru-based synthetic approach can be easily extended to produce hollow nanostructures of other noble metals as well as their alloys, e.g. Pt, Rh, Ir, Os, PtRu, PtRh, PtIr, PtOs and PtRuRh, etc [218]. Unfortunately, the expensive nature of

BSPP is a major obstacle to the application of this protocol for large-scale production of hollow structured metal nanomaterials. Instead, a generic, cost-effective, and high-yield approach to the fabrication of carbon-supported hollow structured noble metal nanoparticles, including Ru, was developed [219]. In detail, a calculated amount of carbon powders was added to the core@shell Ag@Ru colloidal solution in toluene. After stirring the mixture for 24 h, the carbon-supported core@shell particles (20% Ru on carbon support) were collected by centrifugation, which were then re-dispersed in 20 mL of acetic acid by ultrasonication. The resulting mixture was subsequently refluxed in acetic acid for 3 h at 120°C to remove the oleylamine from the particle surface. After refluxing, the carbon-supported core@shell Ag@Ru particles were collected by centrifugation, washed thrice with water, and then dried at room temperature in vacuum. Then, to remove the Ag component from the core@shell Ag@Ru nanostructures, 20 mg of carbon-supported Ag@Ru particles after refluxing in acetic acid was added into 20 mL of saturated Na_2S or NaCl solution in a three-necked flask fitted with a stir bar. The mixture was aged for 24 h under vigorous stirring at room temperature. Then the resulting carbon-supported hollow Ru NCs were collected by centrifugation, washed thrice with water, and then dried at room temperature in vacuum.

4.2.4.5 Synthesis hollow Ru NCs by a structural evolution on a solid substrate

The synthetic approach for hollow Ru NCs on a solid substrate *via* structural evolution is very simple [225]. By aging the carbon-supported core@shell Ag@Ru nanostructures for 3 weeks under air at room temperature, the Ag cores are diffused from the core-shell particles, resulting in carbon-supported hollow Ru NCs with intact sizes, as displayed by the TEM images in Figure 4.25.

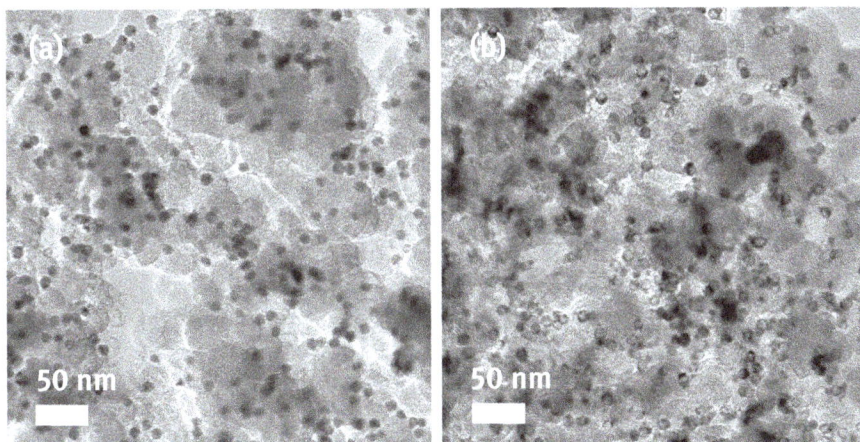

Figure 4.25: TEM images of carbon-supported Ag-Ru nanostructures (a) before and (b) after aging for 3 weeks under air at room temperature (Adapted with permission from ref. 225. Copyright 2015, Springer).

4.2.4.6 Synthesis of face-centered cubic (fcc) Ru nanoframes

The typical synthesis on the basis of seed-mediated method includes three steps [50], as depicted below in detail:

Synthesis of Pd truncated octahedra. In a typical synthesis, 2 mL of an EG solution containing 40 mg of PVP was hosted in a vial and preheated to 160°C in an oil bath under magnetic stirring for 10 min. One milliliter of an EG solution containing 16 mg of Na_2PdCl_4 was then quickly injected into the reaction solution using a pipette. The reaction was allowed to continue at 160°C for 3 h. After being washed with acetone once and ethanol twice *via* centrifugation, the final product was re-dispersed in 1 mL of EG.

Overgrowth of Ru on Pd truncated octahedral seeds. In a standard procedure, 8 mL of an EG solution containing 100 mg of PVP and 50 mg of L-ascorbic acid was hosted in a 50-mL three-neck flask and preheated to 200 °C in an oil bath under magnetic stirring for 5 min. Then, 0.5 mL of the truncated octahedral Pd seeds was added to the flask using a pipette. After 5 min, 8.0 mL of $RuCl_3 \cdot xH_2O$ solution (0.5 mg mL^{-1} in EG) was injected to the flask at a rate of 2.0 mL h^{-1} using a syringe pump. The reaction was allowed to proceed for an additional 10 min after the $RuCl_3 \cdot xH_2O$ precursor had been completed injected. The products (i.e. Pd-Ru core-frame octahedra) were collected by centrifugation, washed once with acetone, two times with water, and finally redispersed in 0.5 mL of deionized water.

Removal of Pd cores from Pd-Ru core-frame octahedra by chemical etching. KBr (150 mg), PVP (25 mg), $FeCl_3$ (25 mg), HCl (0.15 mL, 37%), deionized water (2.85 mL), and an aqueous suspension of the as-prepared Pd-Ru core-frame octahedra (0.5 mL) were mixed together in a 20-mL glass vial under magnetic stirring at room temperature for 10 min. Then, the solution was heated to 80°C in an oil bath under magnetic stirring. After 3 h, the solution was cooled down with a water bath to room temperature and the product was collected by centrifugation, washed once with ethanol, twice with water, and finally re-dispersed in 0.5 mL of deionized water. The microscopic graphs confirm that the Pd cores were successfully eliminated, leaving only Ru nanoframes with an overall octahedral morphology (Figure 4.26).

4.2.5 Synthetic approaches for Ru NCs with other morphologies

Apart from the shapes mentioned above, there are also a number of Ru NCs with other morphologies, e.g. branches, wires, chains and dendrities, reported in literature, which are detailedly introduced in this section.

4.2.5.1 Synthetic approach for Ru NCs with branching morphologies

In a typical synthesis [226], a 25 or 50 mL three-neck round-bottom flask fitted with a water-cooled reflux condenser and thermocouple was purged with N_2. The flask was filled with 4 mL of tri(*n*-octyl)amine, and heated to 175°C. Separately, a vial was

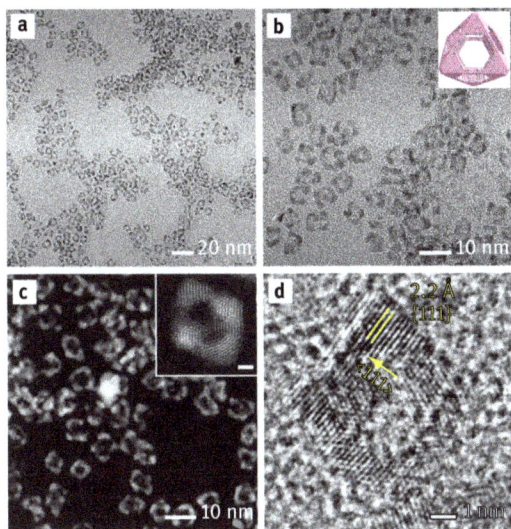

Figure 4.26: (a) Low-magnification TEM image showing the good uniformity. (b) TEM image at a higher magnification showing the overall octahedral shape. Inset shows an ideal atomic model of the sample. (c) High-angle annular dark-field scanning TEM (HAADF-STEM) image. Inset shows a high-resolution HAADF-STEM image of an individual Ru nanoframe (scale bar = 1 nm). (d) HRTEM image of an individual Ru nanoframe orientated along ⟨110⟩ direction (Adapted with permission from ref. 50. Copyright 2016, American Chemical Society).

charged with $Ru_3(CO)_{12}$ (0.03 g, 0.047 mmol) and HDA (1.6 mL), and heated to produce a homogeneous solution. The $Ru_3(CO)_{12}$ solution was injected slowly, drop-wise (1 drop 30 s^{-1}) in the hot tri(n-octyl)amine. The contents were kept at 175°C for a total of 60 min (including the time of addition), at which point 4 mL of toluene was added to quench the reaction. Ethanol was added to the dark brown mixture to induce precipitation of a dark brown solid. The contents were centrifuged at 6500 rpm for 5 min. The supernatant was discarded, and the dark brown precipitate was re-dispersed in toluene. The toluene dispersion was centrifuged (6500 rpm, 5 min) to discard the trace solids. The dark brown supernatant was subject to two additional precipitation/re-dispersion cycles for purification. TEM observation (Figure 4.27(a)) shows that the particles each have a different number of irregular-shaped branches (crystals) with random directionality, and the HRTEM data (Figure 4.27(b)) indicate that the branching structures are polycrystalline, which indicates an attachment mechanism and not anisotropic growth.

4.2.5.2 Synthetic approach for Ru NCs with chain-like morphologies

Chain-like Ru NCs were prepared by reducing the complex of Ru ions with PVP [43]. $RuCl_3$ (0.1 mmol) and PVP of various average molecular weight (90,000, 30,000 or 15,000) (the amount of PVP in each case is 1.1, 0.22, 0.11, 0.04, 0.02 and 0.01 g) were

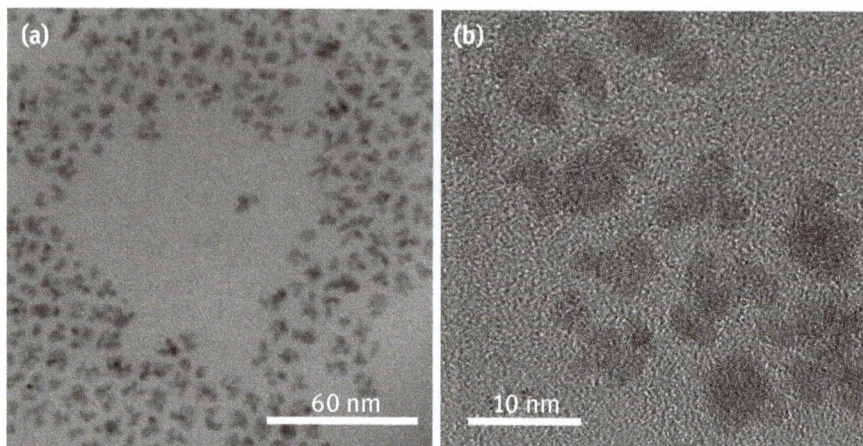

Figure 4.27: TEM (a) and HRTEM image (b) of the as-prepared Ru NCs with branching morphologies (Adapted with permission from ref. 226. Copyright 2008, American Chemical Society).

both dissolved in 60 mL of distilled water with constant stirring, at 353 K, and stirring was allowed to continue for 2 h. A dark-brown transparent solution was obtained where Ru^{3+} ions were coordinated with PVP. After PVP–Ru^{3+} aqueous solution was treated with 4.0 MPa H_2 (or 1 MPa) in a 500-mL stainless steel autoclave with a Teflon liner at 353 K for 2 h, chain-like Ru nanoparticle NCs were obtained.

4.2.5.3 Synthetic approach for Ru nanostars and nanourchins

The Ru nanostars (Figure 4.28(a) and (b)) were synthesized in a one-step process involving the decomposition of $Ru_3(CO)_{12}$ in 10 mL of toluene at 160°C under hydrogen atmosphere (3 bar) in the presence of HDA (HDA/Ru molar ratio = 3) and hexadecanoic acid (PA, PA/Ru molar ratio = 1) for 6 h [227]. To synthesize the nanostars, all compounds were added in a Fischer–Porter bottle at room temperature. The solution was then pressurized under hydrogen, heated to 160°C at a rate of 10°C min^{-1}, and maintained at this temperature for 6 h under hydrogen atmosphere.

The Ru nanourchins (Figure 4.28(c) and (d)) were synthesized from the decomposition of $Ru_3(CO)_{12}$ in 10 mL of toluene at 160°C under hydrogen atmosphere (3 bar) in the presence of HDA (HDA/Ru molar ratio = 2), hexadecanoic acid (PA, PA/Ru molar ratio = 1) and 1.5 mL of the nanostar solution for 6 h. To synthesize the nanourchins, all compounds were added in a Fischer–Porter bottle at room temperature under Ar atmosphere. The solution was then pressurized under hydrogen, heated to 160°C at a rate of 10°C min^{-1}, and maintained at this temperature for 6 h under hydrogen atmosphere.

Once the reactions were complete, the solutions were cooled down to room temperature. Next, excess ethanol was added to the solution, producing a cloudy

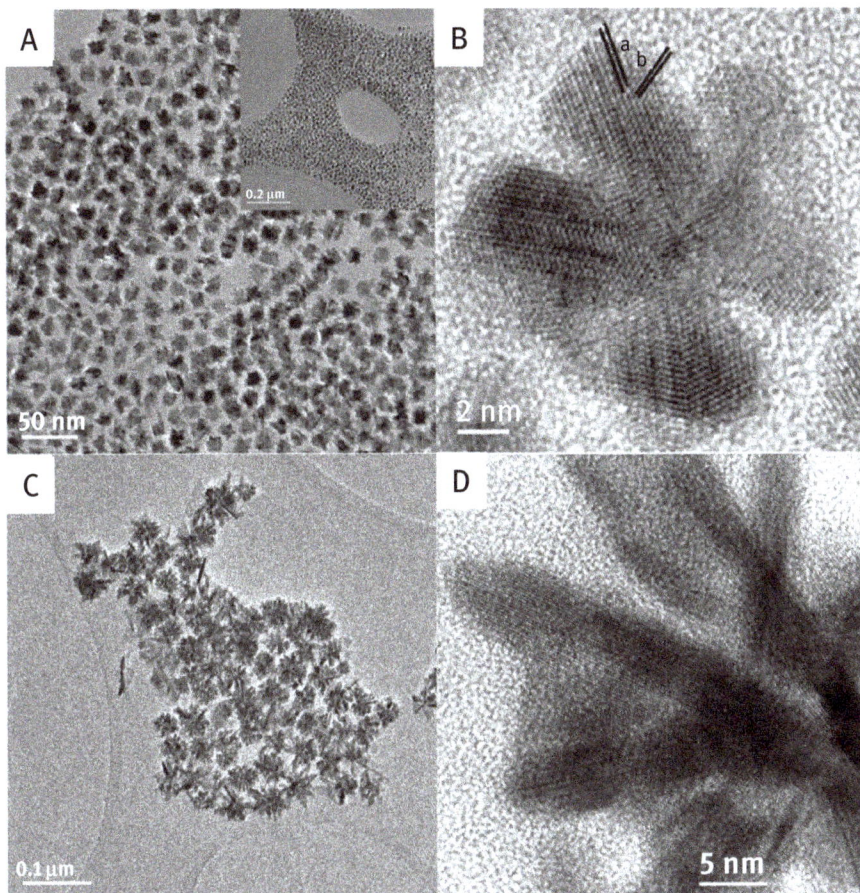

Figure 4.28: (a,c) TEM and (b,d) HRTEM images of the as-prepared (a,b) Ru nanostars and (c,d) nanourchins (Adapted with permission from ref. 227. Copyright 2012, American Chemical Society).

dark brown solution. A black product was obtained after centrifugation. Then, the Ru NCs were dispersed in hexane and precipitated out by addition of ethanol and centrifugation. The washing procedure was repeated two times, and the Ru NCs were finally re-dispersed in hexane.

4.2.5.4 Synthetic approach for Ru-capped columns

In a typical synthesis of Ru-capped columns (Figure 4.29), 0.12 mmol of $RuCl_3 \cdot xH_2O$, 80 mg of $Na_2C_2O_4$ and 100 mg of PVP were dissolved in 10 mL of water. Then, 0.4 mL of HCHO (40 wt%) was added, and the total volume of the solution was adjusted to 15 mL with water. The homogeneous black solution was transferred to a 25 mL Teflon-lined stainless steel autoclave and sealed. The autoclave was then heated at 160°C for 8 h or at 150°C for 24 h before it was cooled down to room temperature. The

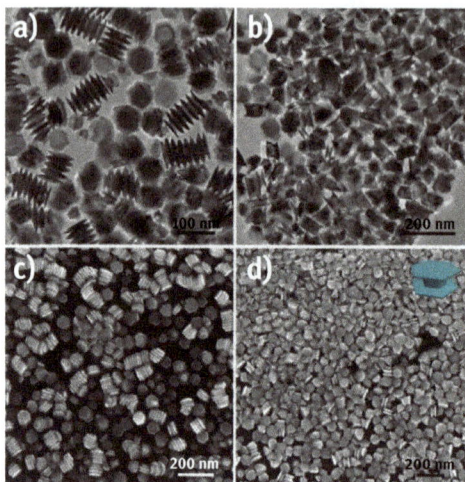

Figure 4.29: TEM (a,b) and SEM (c,d) images of Ru-capped columns prepared under 160 °C (a,c) and 150 °C (b,d). Inset in panel d is the geometric model of the capped columns (Adapted with permission from ref. 202. Copyright 2012, American Chemical Society).

black nanoparticles were centrifuged at 7500 rpm for 10 min with importing 45 mL of acetone and washed with water/acetone three times [202].

4.2.5.5 Synthetic approach for hourglass Ru NCs

To synthesize Ru NCs with hourglass morphologies (Figure 4.30), reaction solutions were made up containing 0.1 mmol ruthenium (II) acetylacetonate and 5 molar equivalents dodecylamine dissolved in 2 mL mesitylene in a small 10 mL glass vial. The vial was then placed in a Fischer–Porter bottle which was sealed, evacuated and filled with 1 bar hydrogen. The Fischer–Porter bottle was further evacuated and refilled with hydrogen twice more before finally sealing at a pressure of 3 bar hydrogen. The Fischer–Porter bottle was then placed in an oven to react for up to 70 hours at 140 °C. Following the reaction the solutions were allowed to cool down to room temperature before the Fischer–Porter bottles were opened to air. An equivalent volume of 50:50 dichloromethane and methanol was then added to the reaction mixture to flocculate and precipitate the Ru NCs. The mixture was centrifuged at 14,000 rpm for 5 min to isolate a brown to black solid. The solid was re-dispersed in dichloromethane and the precipitation and centrifugation steps repeated. The final Ru product was collected and stored as a powder [228].

4.2.5.6 Synthetic approach for rambutan-like Ru NCs

In a seed-mediated growth [113], the reduction of Ru metal precursors in the presence of Au seeds in oleylamine resulted in a rambutan-like structure, in which the worm-like Ru branches with diameter of ca. 1.6 nm are distributed evenly on the surface of

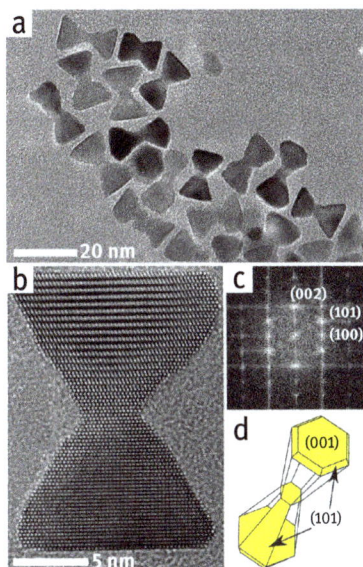

Figure 4.30: Hourglass Ru NCs synthesized after 70 h. (a) TEM image of hourglass Ru NCs. (b) HRTEM image of a single, highly crystalline, hourglass Ru NC. (c) FFT of the hourglass Ru NC shown in (b). (d) Schematic of a single hourglass Ru NC showing termination by {001} and {101} facets (Adapted with permission from ref. 228. Copyright 2013, American Chemical Society).

Au seeds, as shown in Figure 4.31(a) and (b) for the microscopic graphs. In detail, to prepare Au seeds, 152 mg of $AuCl_3$ (0.5 mmol) was dissolved in 20 mL of oleylamine placed in a three-necked flask equipped with a condenser and stir bar. The solution was heated to 150°C under flowing N_2 and kept at this condition for 1 h for the

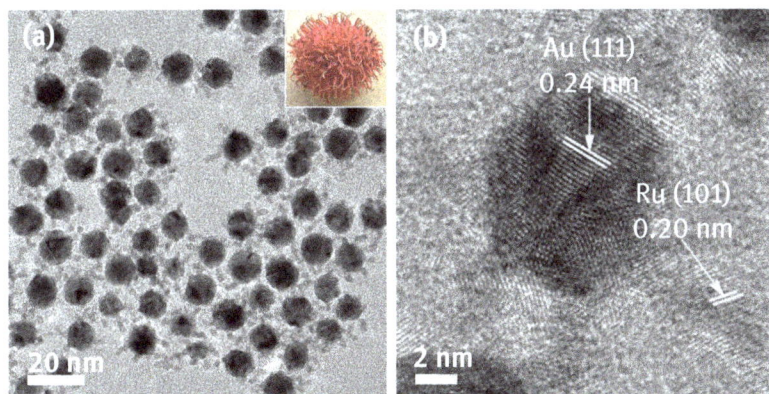

Figure 4.31: TEM images (a) and HRTEM images (b) of as-prepared rambutan-like Ru NCs via a seed-mediated growth. Inlet shows the digital photo of a rambutan (Adapted with permission from ref. 113. Copyright 2013, Royal Society of Chemistry).

reduction of Au^{3+} ions. Subsequently, 0.5 mmol of $RuCl_3$ was added swiftly, followed by heating and keeping at 320°C for 1 h under flowing N_2 to complete the reduction of Ru^{3+} ions. After reaction, the Au-Ru products were purified by precipitation with methanol, centrifugation, washing with methanol and re-dispersed in 20 mL of chloroform.

4.2.5.7 Synthetic approach for dendritic Ru NCs

The synthesis of dendritic Ru NCs with an overall size of 16.7 nm (Figure 4.32) was conducted in oleylamine using ruthenium (III) acetylacetonate ($Ru(acac)_3$) as the precursor [229]. In a typical synthesis, 0.5 mmol of $Ru(acac)_3$ was added to 20 mL of oleylamine in a three-necked flask equipped with a condenser and a stir bar. The solution was brought to and kept at 160°C for 2 h under flowing N_2 for the reduction of metal precursor by oleylamine. After reaction, the dendritic Ru NCs were purified by precipitation with methanol, centrifugation, washing with methanol and re-dispersed in 20 mL of toluene.

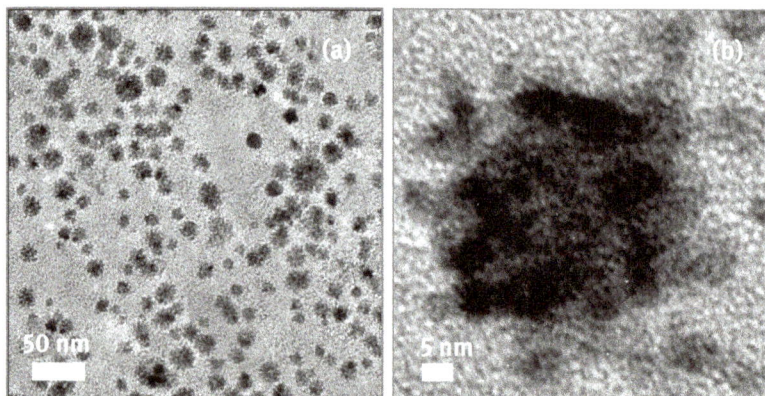

Figure 4.32: TEM image (a) and HRTEM image (b) of the dendritic Ru NCs synthesized in oleylamine at temperature of 160 °C (Adapted with permission from ref. 229. Copyright 2014, Royal Society of Chemistry).

4.2.5.8 Synthetic approach for Ru nanowires

The Ru nanowires shown by TEM and HRTEM images in Figure 4.33 were synthesized through a one-pot method [50]. Typically, 8 mL of EG containing 100 mg of PVP was hosted in a 20-mL glass vial and preheated to 200°C in an oil bath under magnetic stirring for 10 min. Then, 3 mL of an EG solution containing 45 mg of $RuCl_3 \cdot xH_2O$ was quickly added to the reaction solution using a pipette. The reaction was allowed to continue at 200°C for 3 h. After being washed with acetone once and ethanol twice via centrifugation, the final product was re-dispersed in 5 mL of deionized water.

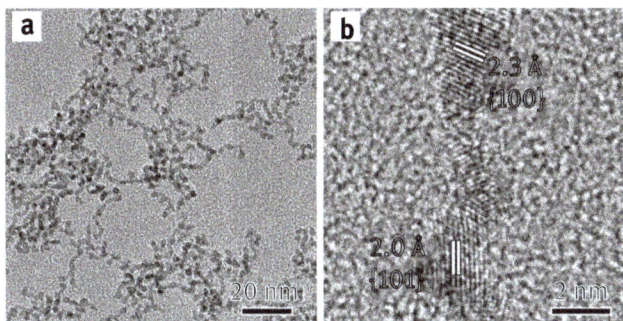

Figure 4.33: (a) Typical TEM image of Ru nanowires that were prepared in ethylene glycol by a one-pot method. (b) HRTEM image recorded from part of an individual Ru nanowire, showing the hexagonal close-packed (hcp) structure (Adapted with permission from ref. 50. Copyright 2016, American Chemical Society).

4.3 Characterization Methodologies and Instrumentation Techniques

The progress of nanoscience is largely owing to the rapid development of material characterization techniques, which allow researchers to directly observe the structure of the materials at the nanometer scale as well as to measure and manipulate the physical and chemical properties of the nanomaterials.

4.3.1 Spectral analysis

An important optical property of metal NCs is that they have strong local electromagnetic field enhancement effect. Many metal NCs have their own characteristic plasma absorption peaks under UV irradiation. For example, Au and Ag NCs with regular spherical morphologies usually have characteristic absorption at ca. 520 and 400 nm, respectively [221–223].

The process of metal reduction could be monitored by UV-visible spectrophotometry based on the corresponding UV-visible absorption spectra of the species presented in the liquid medium. Using a UV-visible spectrophotometer Cary 5E having a spectral resolution of 0.01 nm between 300 and 800 nm and EG as the reference, Figure 4.34 shows the typical UV-visible spectra of an EG solution of $RuCl_3$ in the presence of PVP before and after microwave-solvothermal reaction [224]. Before microwave irradiation, the broad absorption peak centered at 370 nm could be assigned to the Ru^{3+} ions (Figure 4.34(a)). After microwave irradiation for 10 or 30 min (Figure 4.34(b) and (c)), this band disappears and the spectra exhibit unstructured, continuous absorptions without any peaks in the visible range, indicating that the Ru^{3+} ions were entirely reduced to Ru metal.

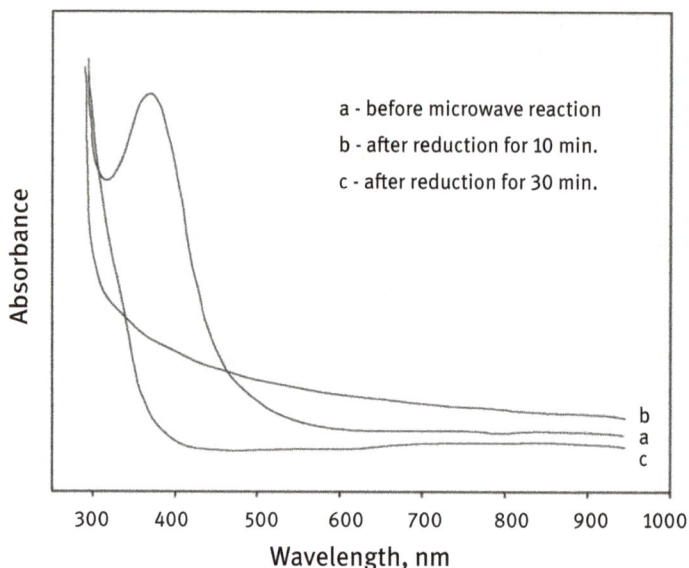

Figure 4.34: UV–visible absorption spectra of an ethylene glycol solution of RuCl₃ in the presence of PVP before and after microwave irradiation (Adapted with permission from ref. 233. Copyright 2008, Elsevier B. V.).

Similar UV-visible spectra (Figure 4.35) were observed during reduction of RuCl₃ in EG, but by using an oil bath [48]. The UV-visible spectrum of the freshly prepared EG solution of RuCl₃ displays a broad absorption maximum for Ru³⁺ ions at 398 nm (Figure 4.35(a)). This peak shifts to a shorter wavelength (328 nm) after 5–8 min of reduction (Figure 4.35(b)). It is noteworthy that a new absorption maximum is formed at 275 nm after 8 min of reduction (Figure 4.35(c)). The intensity of the new peak at 275 nm does not change by further heating, whereas the intensity of the peak at 328 nm decreases with time (Figure 4.35(d)). This peak at 275 nm may be due to the formation of some intermediate species, such as $[Ru(H_2O)_6]^{3+}$, $[Ru(H_2O)_5Cl]^{2+}$, or cis- and trans-$[Ru(H_2O)_4Cl_2]^+$, during the process of RuCl₃ reduction. The UV-visible spectrum of the final gray colloid shows broad adsorption continua that extend throughout the visible-near-UV region (Figure 4.35(e)). By calculating using the Mie theory with the optical constants of the metal, this spectrum agrees well with the spectrum of the Ru colloid having a diameter of 10 nm.

Typical transmittance Fourier transform infrared spectroscopy (FT-IR) spectra are often used to examine the surfactants, ligands and functional groups on the surface of metal NCs. As a typical example, Figure 4.36 shows the FT-IR spectra of pure PVP, PVP-capped Ru NCs as well as them on the γ-Al₂O₃ substrates [224]. The IR fingerprints of the PVP remain evident, indicating the presence of the polymer on the surface of Ru NCs. Comparing with the spectrum of PVP, the resonance peak of C=O (at 1652 cm⁻¹) shifts to slightly higher wavenumbers, while the peaks of C–N

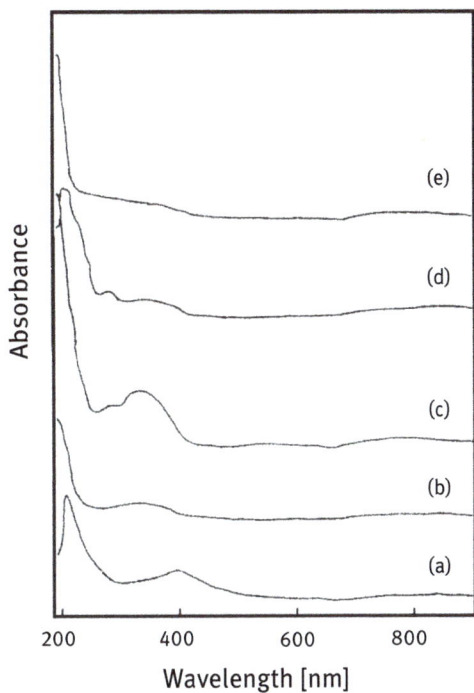

Figure 4.35: UV-vis spectra of the ethylene glycol solution of RuCl$_3$ heated by an oil bath for (a) 0 min, (b) 5 min, (c) 8 min, (d) 11 min, and (e) 15 min (Adapted with permission from ref. 48. Copyright 2001, Elsevier B. V.).

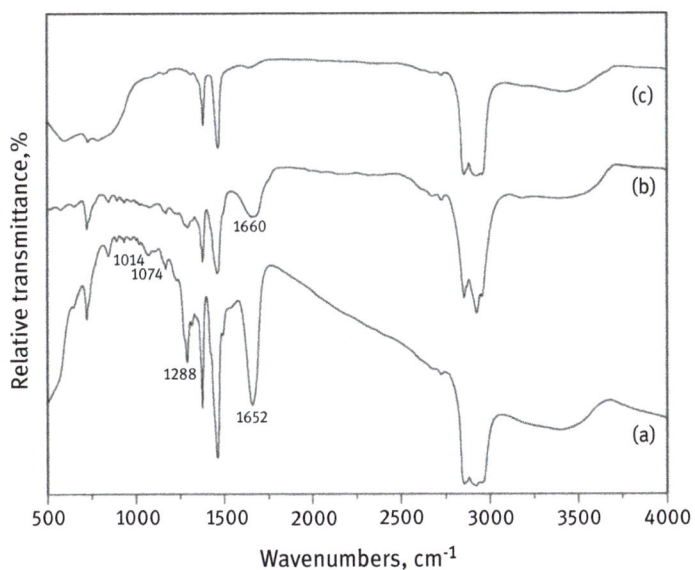

Figure 4.36: FT-IR spectra of (a) PVP, (b) PVP-capped Ru NCs, and (c) PVP-Ru/γ-Al$_2$O$_3$ (Adapted with permission from ref. 233. Copyright 2008, Elsevier B. V.).

(at 1014 and 1074 cm^{-1}) do not change, and the peak of N–OH complex (at 1288 cm^{-1}) has little red shifts. These changes suggest that nitrogen-containing hetero-cyclic ring of the PVP may coordinate more weakly than the carbonyl group with the surface of Ru NCs.

In another by Chen et al. [80] the FT-IR spectrum of the PVP-Ru^{3+}-EG solution both before and after refluxing shows a peak at 1652 cm^{-1} (curve b and c in Figure 4.37), which is assigned to a carbonyl stretching band. This new peak replaces at 1660 cm^{-1} for the pure PVP (curve a in Figure 4.37), indicating interaction between the carbonyl groups and Ru^{3+} ions or zerovalent Ru atoms. This result is comparable to that reported by Teranishi et al. [225]

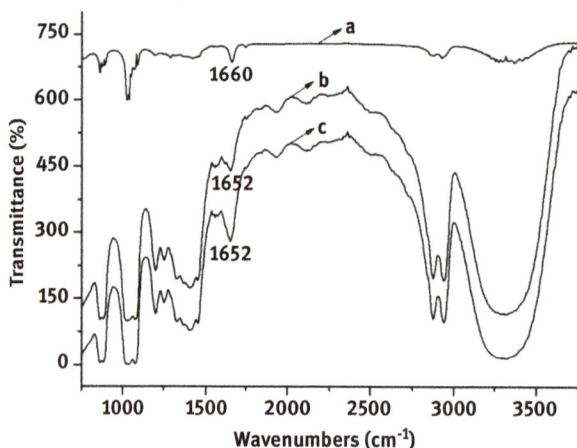

Figure 4.37: FT-IR spectra of (a) PVP, (b) PVP–Ru(NO)(NO$_3$)$_3$–ethylene glycol, and (c) PVP stabilized Ru colloid (Adapted with permission from ref. 80. Copyright 2008, Elsevier B. V.).

4.3.2 X-ray diffraction (XRD)

XRD is a commonly used tool to characterize the crystal phase of metal nanomaterials due to their unique XRD patterns. For the metal NCs dispersed in nonpolar organic solvents such as toluene, dichloromethane and hexane, sample preparation for XRD analysis usually begins with concentrating the organic solution of the metal NCs to 0.5 ml using flowing Ar or N$_2$. Methanol or ethanol is then added to precipitate the metal NCs, which are then recovered by centrifugation and washed with methanol or ethanol several times to remove nonspecifically bound capping agents e.g. oleylamine, alkylamine and alkanethiol [110, 218, 226]. The metal NCs were finally dried at room temperature in vacuum. For the metal NCs dispersed in polar solvents such as water and EG, the metal NCs could be collected by direct centrifugation or firstly transferred into toluene, followed by recovering using the strategy mentioned above.

As a typical example, Figure 4.38 displays XRD patterns of Ru NCs with different sizes, which were recorded on a Bruker D8 GADDS diffractometer using Co K radiation (λ = 1.79 Å)[54]. As observed, the line widths of the diffraction peaks become sharper with the increase in Ru particle size. The crystalline sizes of Ru NCs calculated from the XRD line width by Debye–Scherrer equation are smaller than those from TEM images. This is the usual case because many other factors, e.g. stacking faults and twining defects, can affect the peak width besides the particle size. Interestingly, it is observed that the small Ru NCs exhibit diffraction lines corresponding to the hcp phase structure, whereas bigger Ru particles show the mixed phase of fcc and hcp phase structures.

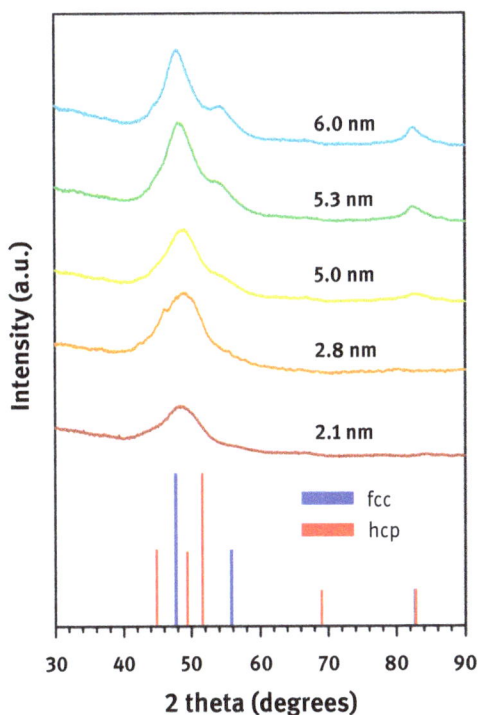

Figure 4.38: XRD patterns for Ru NCs of different sizes with assignment of hexagonal close-packed (hcp) and face-centered cubic (fcc) faces of Ru (Adapted with permission from ref. 54. Copyright 2010, American Chemical Society).

Using a Bruker D8 Advance diffractometer with Cu Kα radiation (λ = 1.54056 Å), Kusada et al. investigated the crystal structure of the Ru NCs with powder XRD analysis [227]. As shown in Figure 4.39, the XRD patterns suggest the diffraction peaks became sharper as the mean particle size of Ru NCs increases. One of the most noteworthy outcomes in this study is that the structure of the Ru NCs could be controlled simply by

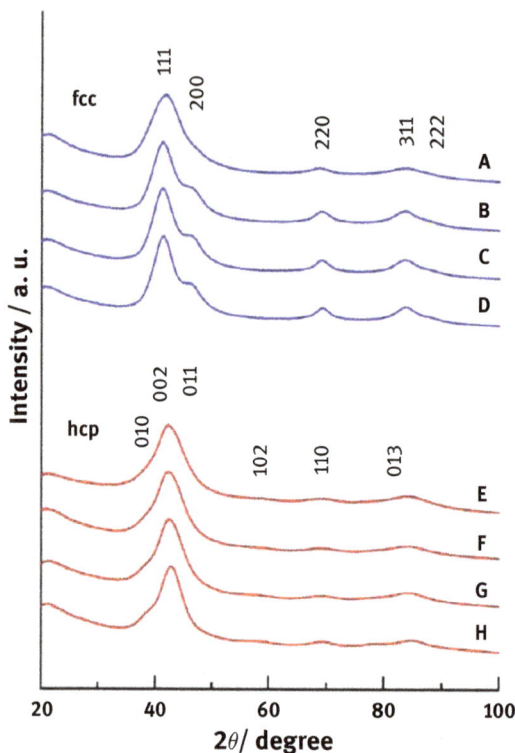

Figure 4.39: Powder XRD patterns for (A – D) fcc and (E – H) hcp Ru nanoparticles at room temperature (Adapted with permission from ref. 236. Copyright 2013, American Chemical Society).

choosing adequate combinations of the metal precursor and the reducing agent. All of the Ru NCs synthesized with Ru(acac)$_3$ and TEG adopt an fcc structure, whereas those synthesized with RuCl$_3$·nH$_2$O and EG have an hcp structure.

4.3.3 X-ray photoelectron spectroscopy (XPS)

XPS is an electron spectrum based on photoelectric effect. It uses X-ray photons to excite the inner electrons of the surface atoms of the substance, and obtain the energy spectrum by energy analysis of these electrons. In the field of nanoscience and technology, the most common application of XPS is the analysis of surface composition and chemical valence of the NCs. Sample preparation for XPS is analogous to that for XRD. After collection, the metal NCs are pasted on a sample holder for XPS measurements.

Figure 4.40 shows a typical XPS spectrum, which was obtained using a XR6 monochromated Alkα X-ray source (hν = 1486.6 eV) with a 650 μm spot size and a mass energy of 20 eV, for the metastable Ru NCs. Forty-eight hours after reduction

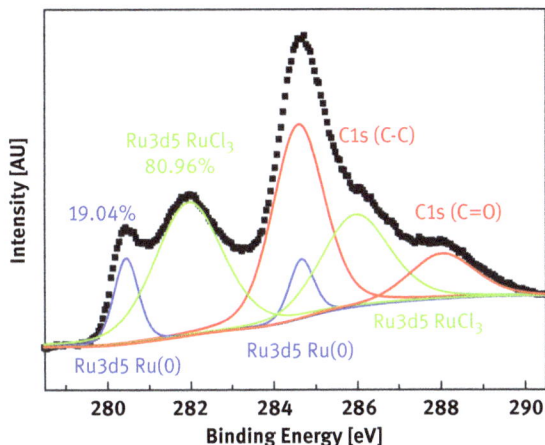

Figure 4.40: XPS spectra collected on a metastable Ru sample with 0.5 mM Ru in solution (Adapted with permission from ref. 58. Copyright 2014, Wiley-VCH Verlag).

[58] it shows the C1s-Ru3d XPS spectrum of a sample with 0.5 mM Ru in solution. The spectrum consists of two C1s peaks and Ru3d doublets; the low energy Ru3d doublet (Ru3d5 peak at binding energy of 280.4 eV) is assigned to zerovalent Ru metal. Only 19.04% of the Ru signal originates from zerovalent Ru, while the remaining Ru originates from a high-energy Ru3d doublet (Ru3d5 peak at binding energy of 281.9 eV), which is assigned to $RuCl_3$. Additionally, C1s peaks at 285.6 eV (C−C) and 288.0 eV (C = O) are observed in the XPS spectrum.

Yang et al. used XPS to analyze the surface composition of the positively charged Ru NCs precipitated from their hydrosol with NaOH [101]. To avoid the interference induced by the overlap of the $Ru3d_{3/2}$ peak with the C1s peak, they analyzed the $Ru3p_{3/2}$ signal. As indicated by Figure 4.41, the $Ru3p_{3/2}$ XPS signal could be deconvoluted into two peaks of different intensities at 461.8 and 465.1 eV, which correspond to Ru at zerovalent state and higher oxidation states such as Ru^{6+} in RuO_3, respectively.

Figure 4.42 shows 3d XPS spectrum of Ru NCs immobilized on MMT by ILs [147]. Although the Ru3d signal is obscured by the C1s signal of a carbon contaminant at 284.6 eV, the deconvoluted spectrum shows a doublet for two chemically different Ru entities with peak binding energies at 280.3 eV (Ru $3d_{5/2}$) and 284.5 eV (Ru $3d_{3/2}$), respectively, confirming the presence of zerovalent Ru metal in Ru/MMT samples. In addition, the peak at 281.1 eV suggests the presence of a Ru-O component, which probably results from oxidation of the Ru NCs upon exposure to air. The XPS results manifest that the $RuCl_3$ loaded on MMT is reduced to zerovalent Ru by hydrogen at 220°C within 2 h. The reduction of Ru^{3+} ions at relatively low temperature might be induced by the interaction of $RuCl_3$ with the IL and/or the MMT substrate.

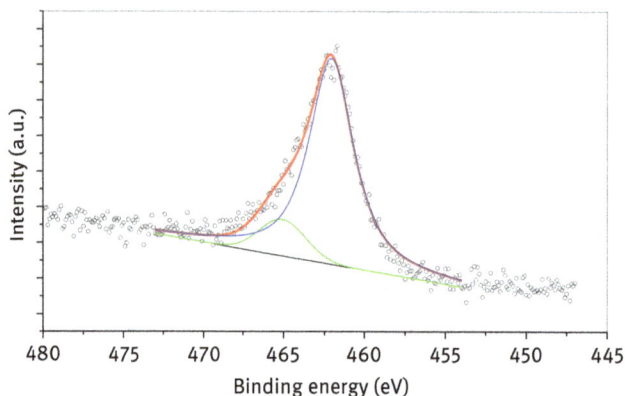

Figure 4.41: The Ru 3p$_{3/2}$ XPS spectrum for the positively Ru NCs recovered from their hydrosol by NaOH (Adapted with permission from ref. 101. Copyright 2004, Elsevier B. V.)

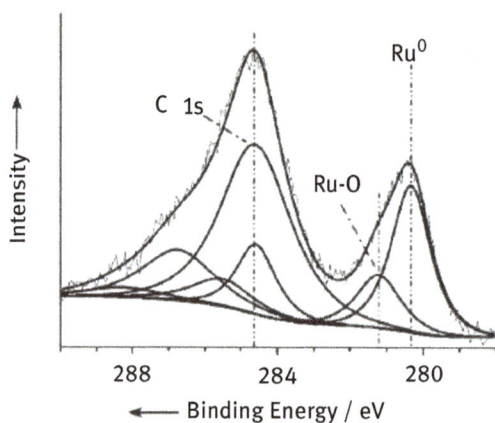

Figure 4.42: 3d XPS spectra of Ru NCs immobilized on montmorillonite by ionic liquids. The three vertical lines indicate the peak positions of the binding energies of C 1S, Ru–O, and Ru, respectively (Adapted with permission from ref. 149. Copyright 2006, Wiley-VCH Verlag).

XPS is also a powerful means for characterizing the lattice strain effect and electronic coupling effect in metal-based nanomaterials with complicated or composite nanostructures [228–232]. Through strong interaction between different domains in these complicated or composite nanostructures, the lattice strain effect and electronic coupling effect often induce the shifts in binding energies of the corresponding metal components, which can be sensitively detected by XPS.

4.3.4 Transmission electron microscopy

TEM is an indispensable tool for the characterization of nanostructured materials. It can offer the most intuitive description of the scale, scale distribution, morphology and structure of nanomaterials. For preparing samples for TEM measurements, a drop of the colloidal solution is usually dispensed onto a 3 mm carbon-coated copper grid, and excessive solution is removed by an absorbent paper. Then the sample is then dried under vacuum at room temperature.

A typical feature of nanomaterials is their small particle size. Although the use of optical spectroscopy, XRD and X-ray photoelectron spectroscopy can show some structural features of the nanomaterials, but only the use of TEM is possible to obtain the visual image of particles at nanoscale ranges. The TEM is unique and indispensable because it provides a true spatial image of the nanomaterials and their surface atoms distribution. Up to now, TEM has evolved into a multi-function instrument, which not only provides atomic resolution of the lattice image but also gives the material structure and chemical information in the 1 nm or even higher spatial resolution, so that directly identifying the chemical composition of a single NC is possible. TEM can not only give an intuitive description of the morphology for a single particle, but also can provide its whole mappings, e.g. inner structure, composition, defects and lattice features, by combining with energy dispersive X-ray spectroscopy (EDX), selected area electron diffraction, high resolution and high-angle annular dark-field scanning modes. A number of typical examples of applying TEM for characterization of Ru NCs were presented in this section.

The most common application of TEM is an intuitive description of the size, morphology and size distribution of NCs. The observation of the same nanometer system at different times or stages through TEM can reveal the evolution of the physical properties of NCs [226, 233].

Ru NCs were prepared by the reduction of $RuCl_3 \cdot 3H_2O$ with hydrogen in the presence of poly(ethylene glycol). As depicted in Figure 4.43 [234], the TEM image (obtained on a Tecnai $G^2 20$ Spirit microscope at an accelerating voltage of 120 kV) and particle size histogram show that the Ru NCs thus obtained display a homogeneous distribution in poly(ethylene glycol) with an average particle diameter of 1.6 ± 0.4 nm, evidencing a good stabilization by poly(ethylene glycol).

With the increase in particle size, the changes in particle morphologies could be discerned easily by TEM observation. Figure 4.44 shows the TEM images of uniform Ru NCs ranging from 2 to 6 nm with narrow particle size distributions (relative standard deviation below 15%)[54]. As exhibited, the smaller Ru NCs are mostly composed of spherical particles, while the larger ones contain a portion of well-faceted particles. In general, the concentration of the Ru precursor and the reduction temperatures are critical factors in controlling the size of Ru NCs. The higher

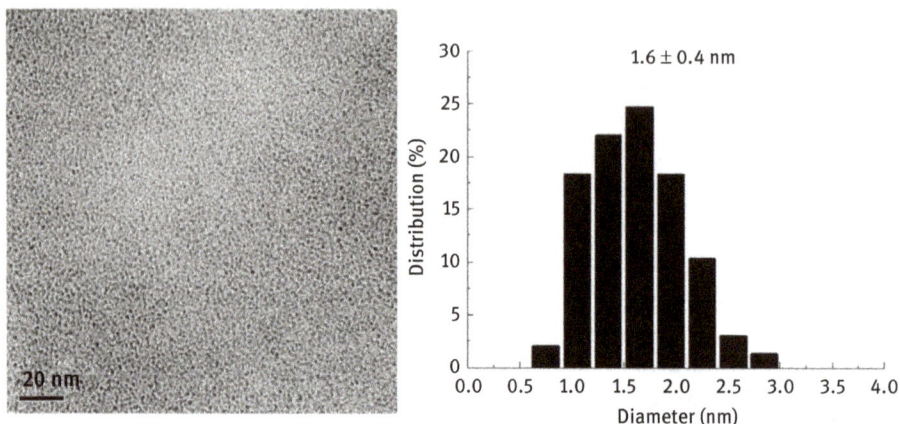

Figure 4.43: TEM image and particle size histogram of poly(ethylene glycol)-stabilized Ru NCs (Adapted with permission from ref. 243. Copyright 2013, Elsevier B. V.).

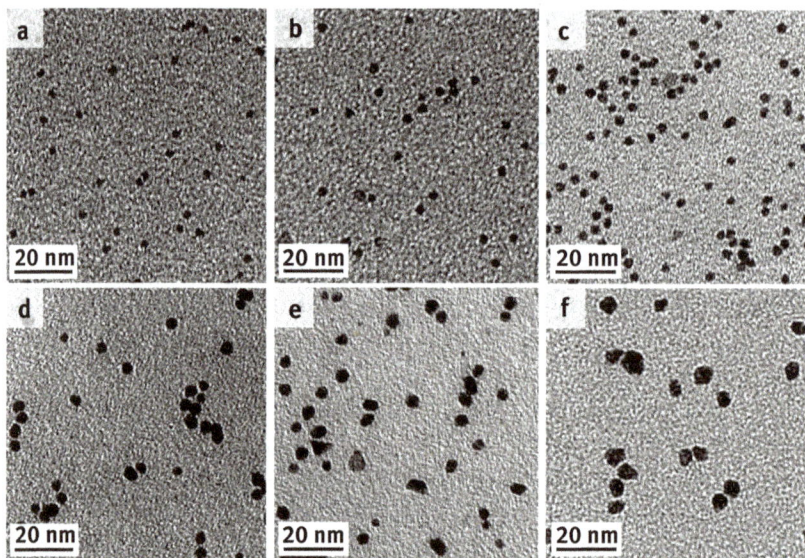

Figure 4.44: TEM images of Ru NCs: (a) 2.1, (b) 2.8, (c) 3.1, (d) 3.8, (e) 5.0, and (f) 6.0 nm. TEM images were taken using a Philips FEI Tecnai 12 machine, operated at 100 kV (Adapted with permission from ref. 54. Copyright 2010, American Chemical Society).

precursor concentration and lower reduction temperature usually yield larger Ru NCs, and this is consistent with the results from previous syntheses of noble metal NCs using polyol methods [79, 94, 235].

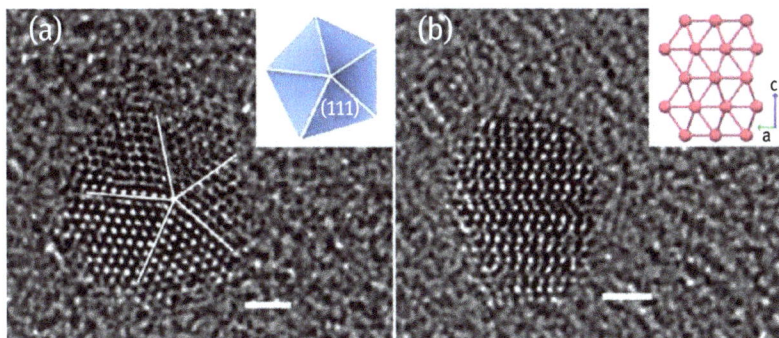

Figure 4.45: (a) HRTEM image of an fcc Ru nanocrystal. The inset is an illustration of the decahedral structure. (b) HRTEM image of an hcp Ru nanocrystal. The inset is an illustration of the hcp lattice viewed along the [100] direction. The scale bars in each image are 1.0 nm (Adapted with permission from ref. 236. Copyright 2013, American Chemical Society).

The details in particle structure can be revealed through analyzing the TEM images in high-resolution mode. As a typical example, Figure 4.45 shows the HRTEM images (obtained with a Hitachi HT7700 TEM instrument operated at 100 kV) of fcc and hcp structure of two different single Ru particles, which were prepared by adjusting the synthetic precursors in polyols [227]. The image of a hcp Ru particle recorded along the [100] direction (Figure 4.45(b)) clearly shows the hcp ABABAB … stacking sequence. On the other hand, Figure 4.45(a) shows a Ru particle representing the fcc planes. This is a single fivefold-symmetry twinned particle having a decahedral structure consisting of five tetrahedrons.

For the Ru NCs synthesized via seed-mediated growth, by combing the TEM observation at scanning mode with the EDX analysis, the structural information could be clarified. As a typical example, Figure 4.46 shows the element profiles of an individual Ag@Ru core@shell particle, which were obtained by EDX analyses of an arbitrarily chosen single particle in the high-angle annular dark-field scanning TEM mode (Inset of Figure 4.46). As indicated, the Ru signal is present throughout the particle whereas the Ag signal is only detected only in the core region. Then, after removing the Ag core from core@shell Ag@Ru NCs using BSPP or saturated NaCl aqueous solution, the resulted Ru NCs with hollow interiors could also be discerned by the strong brightness contrast between the core and shell regions in their TEM images, as shown by Figure 4.24 in previous section.

It is noteworthy that characterization of the nanomaterials using TEM requires the particle image must be obtained with right focus or only a slight deviation from the focus (with respect to the proficiency of the operators). If large deviation from the right focuses (i.e. large under focus or over focus) occurs, the image information obtained for the samples will be completely wrong.

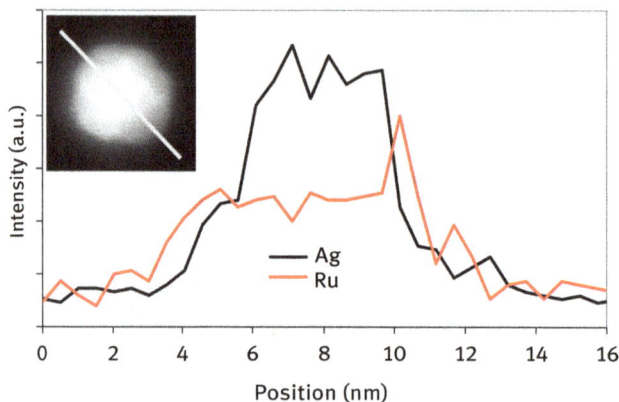

Figure 4.46: Element profiles of core@shell Ag@Ru NCs prepared in oleylamine using seed-mediated growth method. Inset is the STEM image of an individual Ag@Ru core@shell particle (Adapted with permission from ref. 223. Copyright 2012, American Chemical Society).

4.4 General growth mechanisms of size- and shape-controlled Ru nanocrystals

Over the past decades, various metal NCs with controllable particle sizes, compositions and shapes have been synthesized by a wide range of chemical synthetic procedures [5, 9, 12–16, 18, 21, 67, 87, 236]. The understanding of the formation mechanism accounting for monodispersed NCs is necessary because it is helpful to develop improved synthetic methods that can be universally applicable to various kinds of metal materials.

The study for preparing uniform colloidal particles could be dated back to the 1940s. LaMer and Dinegar proposed the concept of "burst nucleation" through investigating the preparation of a variety of oil aerosols and sulfur hydrosols [14, 237, 238]. In this process, they assumed that many nuclei are generated at the same time, and then these nuclei start to grow without the need for additional nucleation. Because all of the particles are nucleated almost simultaneously, their growth histories are nearly the same. This is the essence of the "burst-nucleation" process, which makes it possible to control the size distribution of the ensemble of particles as a whole during the process of growth. Otherwise, if nucleation process also occurs during the formation of particles, the growth histories of the particles would differ largely from one another, and would consequently make a great difference to the size distribution.

Upon "burst nucleation" theory, it is necessary to induce a single nucleation event and prevent additional nucleation during the subsequent growth process for the preparation of highly uniform colloidal solution. This synthetic strategy, often referred to as "separation of nucleation and growth" has been extensively used to synthesize monodispersed NCs [237, 239, 240]. The seed-mediated growth method is the most apparent case for the separation of nucleation and growth, wherein nucleation is physically separated from growth by using preformed NCs as seed nuclei. This method utilizes heterogeneous nucleation to suppress the formation of additional nuclei by homogeneous nucleation [241–244]. In this method, the preformed nuclei, which are requested to uniform in size, are introduced into the reaction solution and then the monomers are supplied to precipitate on the surface of the existing nuclei. The monomer concentration is kept low during growth to suppress homogeneous nucleation. Seed-mediated growth is further divided into two categories: the synthesis of homogeneous particles [241, 244] and the production of heterogeneous structures, such as core@shell structures [242, 243].

In the growth stage of NCs, the agglomeration or aggregation of small particles is inevitable in the absence of stabilizers due to the thermodynamics favor the maximization of the surface/volume ratio. Alternatively, the surface energy can be well controlled by absorbing some capping agents [14, 245, 246]. In this sense, knowledges and studies in surface science would be helpful to design NCs with desired sizes/shapes, which might be featured with specific facets. In addition, both thermodynamics and kinetics can affect the growth of metal NCs. Compared with the growth tendency, which is determined by thermodynamics, the crystal growth paths are mainly dependent on the kinetics. Therefore, metal NCs at various transition states that are metastable thermodynamically could be generated by carefully controlling the growth kinetics.

Although these general concepts have made significant successes for preparing noble metal NCs with controlled sizes/shapes, gaining a comprehensive understanding of how Ru NCs form is still very challenging. With respect to a large number of wet-chemistry methods for synthesizing Ru NCs, in this section, various growth mechanisms accounting for Ru NCs with different sizes and shapes are introduced.

4.4.1 Stabilization of Ru NCs in polyol synthesis

As depicted in previous sections, "polyol synthesis" has been established as an effective method to prepare well-dispersed Ru NCs. Wang *et al.* reported the preparation of "unprotected" Ru NCs in EG [81]. In their synthesis, an EG solution of NaOH was added to a glycol or aqueous RuCl$_3$ solution under stirring to obtain a transparent hydroxide or oxide colloidal solution, which was subsequently heated at 160°C for 3 h to obtain a transparent homogeneous colloidal solution. The stabilization of Ru NCs was attributed to the adsorption of EG and simple anions such as OH$^-$ on their surfaces. Komarneni *et al.* ever proposed a mechanism shown below for the reduction of metal salts in EG [247]:

$$HOCH_2CH_2OH \rightarrow CH_3CHO + H_2O \tag{4.1}$$

$$2CH_3CHO + M(OH)_2 \rightarrow CH_3-CO-CO-CH_3 + 2H_2O + M \tag{4.2}$$

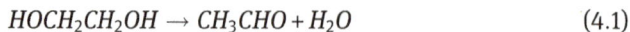

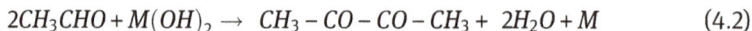

However, this mechanism did not explain the stability of Ru NCs in EG. Actually, Komaneni *et al.* used PVP to stabilize the metal NCs they prepared, although it is not absolutely necessary. As reported by Yang *et al.*, the reduction of RuCl$_3$ is equally complete in the absence of OH$^-$ anions [248]. However, the Ru organosols thus obtained are not stable, and the Ru NCs would precipitate in several minutes. It is therefore apparent that OH$^-$ contributed towards the stabilization of Ru NCs in EG. However, they also experimentally found that once the Ru organosol was formed, the neutralization with acid would not bring about precipitation of the Ru NCs.

Based on their observation that sodium acetate could stabilize Ru NCs in the aqueous environment, and that acetate is a common product from aldehyde oxidation, Yang *et al.* surmise that acetate may be produced as an intermediate and a stabilizer in the formation of Ru NCs in EG. To confirm their hypothesis, they performed ion chromatographic analysis to examine the chemical compositions of the Ru organosols prepared in the presence of NaOH. The ion chromatograms clearly show the presence of acetate in Ru organosol, which could not be explained by the mechanism proposed by Komarneni *et al.*

Further, Yang *et al.* verified that by replacing NaOH with sodium acetate in EG process, stable Ru organosols could also be obtained. The TEM image in Figure 4.47 shows the acetate-stabilized Ru NCs prepared in EG. The average diameter for the Ru NCs is 1.1 nm, which is same as the value for the Ru NCs obtained by Wang *et al.* [81] where OH$^-$ was explicitly used in the preparation of stabilized Ru NCs. Therefore,

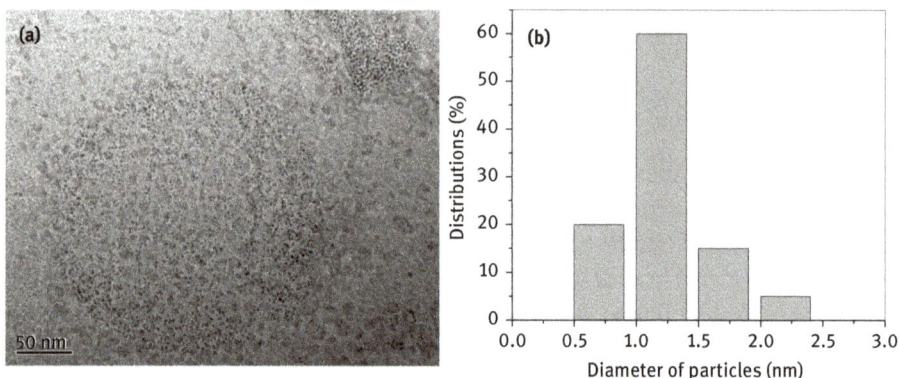

Figure 4.47: TEM image and histogram to show the size distribution of acetate-stabilized Ru NCs prepared in ethylene glycol (Adapted with permission from ref. 257. Copyright 2004, American Chemical Society).

Yang et al. put forward the following synthetic mechanism to interpret the stabilization of Ru NCs in EG [248]:

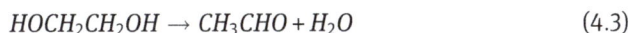

$$HOCH_2CH_2OH \rightarrow CH_3CHO + H_2O \tag{4.3}$$

$$3CH_3CHO + 2Ru^{3+} + 9OH^- \rightarrow 3CH_3COO^- + 2Ru + 6H_2O \tag{4.4}$$

In their modified mechanism, the OH$^-$ ions do not have a direct stabilization effect, and they are necessary to generate acetate ions, which are responsible for keeping the Ru NCs stable in EG. Their studies demonstrate that the "unprotected" Ru NCs in EG are actually metal particles stabilized by acetates.

4.4.2 Phase transfer-mediated synthesis of Ru NCs

In the ethanol-mediated phase transfer process of preformed Ru NCs [109], the citrate stabilized Ru NCs cannot be transferred directly to toluene by mixing the Ru hydrosol with a toluene solution of dodecylamine. Prolonged stirring only produces a milky mixture of Ru hydrosol and toluene, but no particle transfer takes place. The failure in transferring Ru NCs was assumed to be led by the poor contact between the two phases, which is not favorable for sufficient exchange between dodecylamine and citrate ions. To enhance the ligand exchange, ethanol, which is water-miscible and a good solvent for dodecylamine, was used to replace toluene to increase the interfacial contact between citrate stabilized Ru NCs and dodecylamine. The process of displacing citrate ions by dodecylamine from the surface of Ru NCs can be schematically depicted by Figure 4.48. In brief, the transfer process involves four steps. Step A is representative of adsorbed citrate ions in equilibrium with their surroundings. The progressive displacement of citrate by dodecylamine is depicted in steps B through D where it is inherently assumed that the binding of alkylamine to the surface of Ru NCs is more irreversible than the adsorptive interaction between the citrate ions and the Ru surface.

After stirring the mixture of ethanolic solution of dodecylamine and Ru hydrosol for 2 min, the dodecylamine displaces citrate from the surface of Ru NCs, resulting in quick extraction by nonpolar organic solvents, e.g. toluene and hexane. This transfer mechanism can also be used to interpret the preparation of thiolated oligonucleotide functionalized Au NCs from citrate stabilized Au NCs [249–251].

For the ethanol-mediated transfer of Ru^{3+} ions from aqueous phase to toluene, followed by reduction reaction for the synthesis of Ru NCs, A metal complex between Ru^{3+} ions and dodecylamine was speculated to be formed in the process, which could be easily extracted by toluene [110]. The formation of Ru^{3+}-dodecylamine was verified by FTIR spectra of the compounds recovered from toluene after phase transfer (Figure 4.49). In comparison with pure DDA, differences are observed in the N–C and N–H stretching regions, demonstrating that deodecylamine is bound to the Ru^{3+}

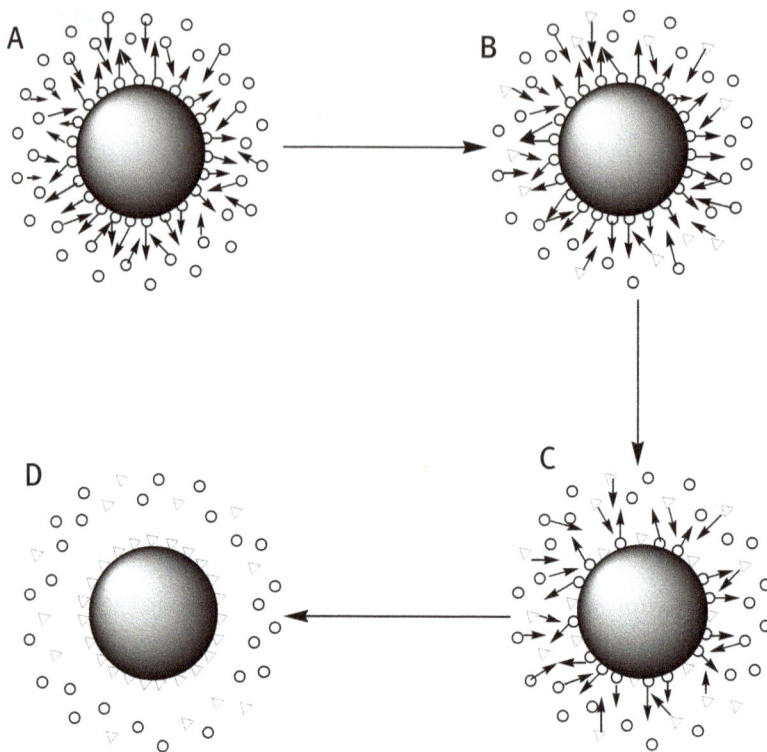

Figure 4.48: Schematic illustration to show the replacement of citrate by dodecylamine from the surface of Ru NCs. o: citrate ions; Δ: dodecylamine (Adapted with permission from ref. 109. Copyright 2013, Elsevier B. V.).

ions by its amino (NH_2–) group. After coordinating with Ru^{3+} ions, the nonpolar tail of dodecylamine enables the compounds to dissolve easily in nonpolar organic solvents, e.g. toluene or hexane.

4.4.3 Hollow Ru NCs from core-shell Ag@Ru nanostructures

The synthesis of hollow Ru NCs using core@shell Ag@Ru nanostructures as starting materials is based on an inside-out diffusion of Ag in core@shell nanostructures with Ag residing in the core regions. The inside-out diffusion of Ag in core@shell nanostructures was firstly observed by Liu *et al.*, who noticed that after storage in toluene for seven months at room temperature, Ag diffused out from the interior of the core@shell Ag-Ru nanostructures, resulting in a product composed of isolated Ag particles and hollow Ru NCs [218].

The inside-out diffusion of Ag is promoted by the structure of the Ag seeds, which are nm multiply twinned decahedral particles with an average size of ca. 9 nm. The twinned structures are inherently not stable and could slowly be etched by dissolved

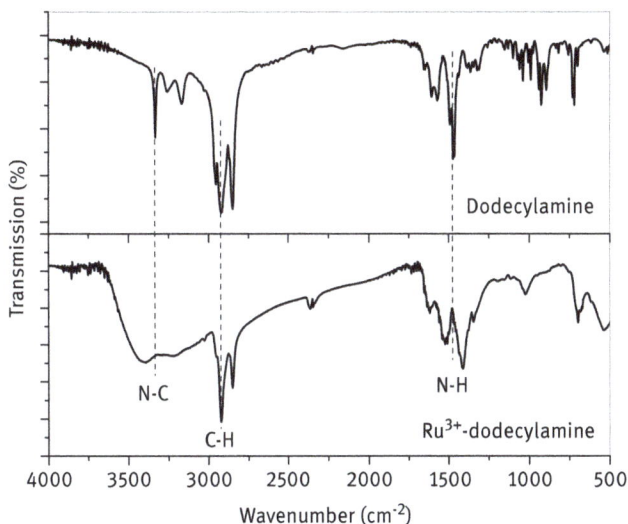

Figure 4.49: FTIR spectra of pure dodecylamine and Ru^{3+} ion-dodecylamine complexes (Adapted with permission from ref. 111. Copyright 2009, Nature Publishing Group).

O_2 and Cl^- dissociated from the Ru precursors, i.e. $RuCl_3$. The Ag^+ ions generated by the O_2/Cl^- etching of twinned Ag particles could be re-reduced by excess oleylamine in the solution to single crystalline Ag particles. Further, as confirmed by Xia *et al.*, [252] single crystalline Ag particles are stable and would continue to grow even in the presence of etching agents at the expense of their twinned seeds. Thus, Yang et al. proposed a mechanism to rationalize the inside-out diffusion of Ag in core@shell nanostructures [218]. As shown in Figure 4.50, because of the concentration gradient

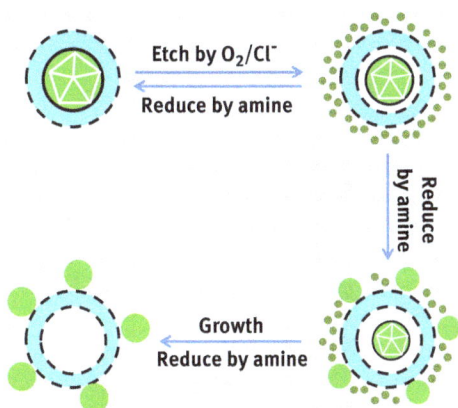

Figure 4.50: Schematic illustration to show mechanism for the inside-out diffusion of Ag in core@-shell Ag-Ru nanostructures (Adapted with permission from ref. 223. Copyright 2012, American Chemical Society).

between the interior of the core@shell structures and the surrounding solution, the Ag^+ ions generated from the O_2/Cl^- etching of twinned Ag seeds diffuse out through the discontinuous Ru shell, and are reduced by oleylamine to form isolated single crystalline Ag particles in the colloidal solution. Eventually, the core@shell Ag-Ru nanostructures disappear due to the etching and outward diffusion of Ag, leaving behind a physical mixture of hollow Ru NCs and single crystalline Ag particles.

The scheme in Figure 4.50 is then used to generate hollow Ru NCs. The inside-out diffusion process of Ag in core@shell Ag@Ru nanostructures is accelerated by introducing a chemical reagent BSPP, which binds strongly with Ag^+ ions to drive the outward diffusion of Ag in the core@shell nanostructures [218, 253], enabling the completion of the inside-out diffusion of Ag in ca. 24 h. More importantly, the BSPP-Ag^+ coordination compounds are water soluble and their continuous formation leaves behind an organosol containing sole hollow Ru NCs.

4.4.4 Rambutan-like Ru NCs from Au seed-mediated growth

By tracking the growth of other noble metals on the surface of multi-twinned Au seeds using TEM, the mechanism for the formation of rambutan-like Ru NCs from Au seed-mediated growth method was rationally proposed [226]. As schematically illustrated by Figure 4.51, the crystallographic defects in the multi-twinned Au seeds, which account for many interesting electronic, optical and catalytic properties [254–256], are essential for the formation of rambutan-like Ru NCs. The anisotropic growth of Ru on the surface of Au twinned seeds is natural due to the selective growth of Ru at the high-energy twin boundary sites. With the increase of the Ru atoms through the successive oleylamine reduction of Ru metal precursors, a number of extended Ru branches are finally formed on surface of Au seeds, which constitute the rambutan-like morphology of the final Ru products.

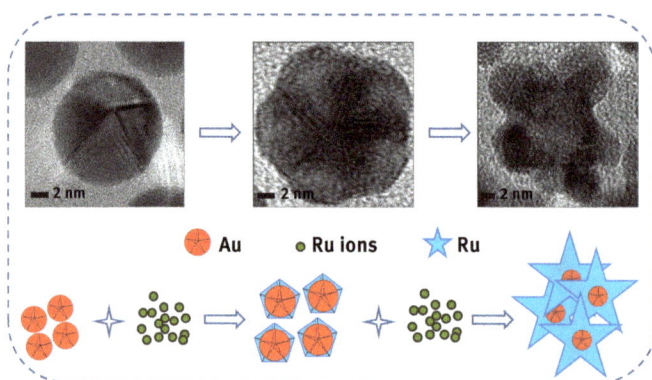

Figure 4.51: Schematic illustration to demonstrate the mechanism for the formation of rambutan-like Ru NCs using multi-twinned Au seed-mediated growth (Adapted with permission from ref. 235. Copyright 2014, Royal Society of Chemistry).

4.4.5 Ru nanodendrites formed by oleylamine reduction

Oleylamine is a common reducing agent for the synthesis of metal nanoparticles [112, 257–260]. Through FI-IR and ^{1}H NMR (nuclear magnetic resonance spectroscopy) analyses, Liu *et al.* elucidated the mechanism for the metal NCs formed by oleylamine reduction [261]. In brief, amine ligands are firstly oxidized to amides during the reduction of metal precursors, and then form a protecting layer of hydrogen bond network on the surface of metal NCs. Therefore, the mechanism for producing Ru nanodendrites in oleylamine could be depicted by the scheme in Figure 4.52 [220]. At initial stage, the Ru(acac)$_3$ are reduced into Ru atoms by oleylamine, which grow into Ru particles, while oleylamides are simultaneously formed from oleylamine as the capping agent to stabilize the Ru particles. Then the particle aggregation would compete with the oleylamide passivation, resulting in the formation of a large number of Ru particle aggregates in the solution (step (a) in Figure 4.52). Finally, the Ru particle aggregates continuously grow with the expense of small particles in solution *via* a ripening process to form larger and more stable dendritic Ru NCs (step (b) and (c) in Figure 4.52).

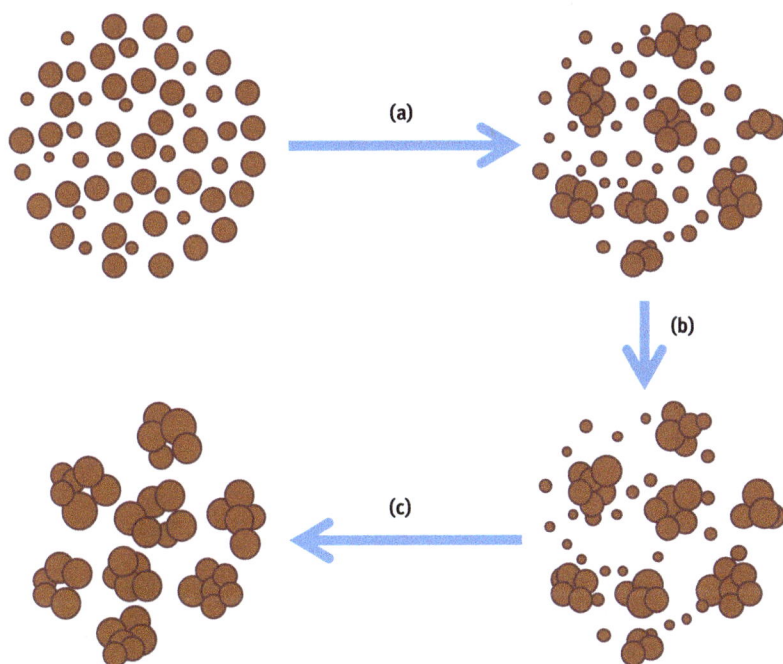

Figure 4.52: Schematic illustration to show the mechanism for forming dendritic Ru NCs *via* the reduction of Ru(acac)$_3$ in oleylamine at elevated temperature: (a) Competition between particle aggregation and oleylamide passivation results in the formation of particle aggregates; (b,c) particle aggregates grow into nanodendrites *via* a ripening process (Adapted with permission from ref. 229. Copyright 2014, Royal Society of Chemistry).

The competition between particle aggregation and oleylamide passivation at initial stage of the synthesis is essential for the formation of Ru nanodendrites, which might be affected by the reaction temperature since the temperature would have significant influence on the reduction kinetics of the Ru metal ions. For examples, when the reduction reaction of Ru(acac)$_3$ in oleylamine were conducted at 320°C and 350°C, worm-like and quasi-spherical Ru NCs are formed as dominant products, respectively, instead of Ru nanodendrites [113].

4.4.6 Organometallic approaches for synthesizing Ru NCs

As we have introduced in previous sections, organometallic approaches, which employ organometallic or metal–organic complexes as the metal sources in combination with an appropriate choice of ligands, have been extensively used to synthesize Ru NCs at room or elevated temperatures. To accomplish the synthesis of controlled NCs and further to try to explore the synthetic mechanisms, Amiens *et al.* developed modular approaches using organic ligands able to coordinate to the growing nanoparticles surfaces as stabilizing agents (Figure 4.53) [262]. Since the capping ligands are introduced from the beginning of the reaction with the metal precursors (most often at low temperature to avoid the formation of any intermediate complex difficult to decompose), they can interact with the metallic surface as soon as the nuclei are formed in the reaction medium and thus can influence their growth into NCs. They can also self-assemble in solution and afford soft, sometimes anisotropic, molecular templates inside which the nanostructures grow. Until now,

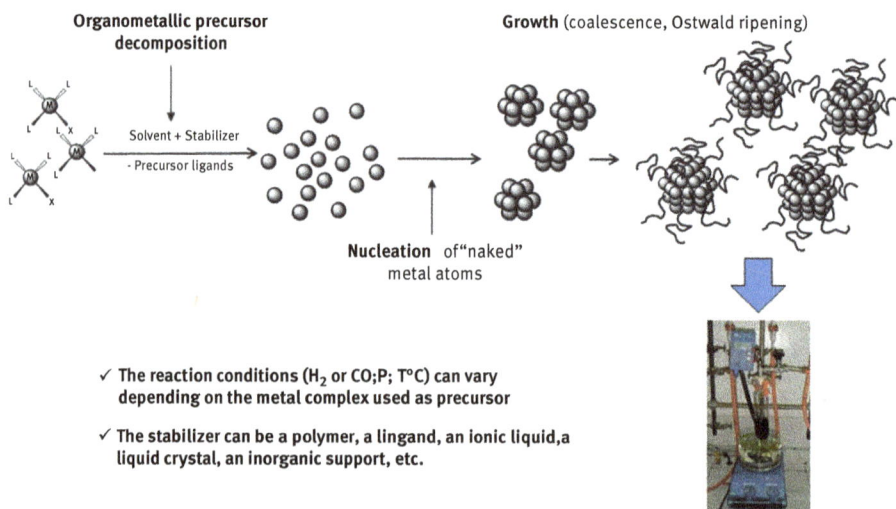

Figure 4.53: Schematic representation of the organometallic approaches for the synthesis of metal nanostructures (Adapted with permission from ref. 271. Copyright 2013, Royal Society of Chemistry).

various types of ligands have been employed for the synthesis of Ru NCs, in which the choice of the ligand is crucial for the control of the size/shape of the NCs.

4.4.7 Stabilization of Ru NCs in ionic liquids

Through *in situ* labeling and spectroscopic experiments, Campbell et al. explained the key points in the stabilization of Ru NCs generated in imidazolium-based ILs by decomposition of an organometallic precursors Ru(COD)(COT), under dihydrogen [263]. As shown in Figure 4.54, the main findings could be summarized to be (1) the presence of hydrides at the surface of Ru NCs; (2) the confinement of Ru NCs in the non-polar domains of the structured ILs, induced by the rigid three dimensional organization. Their results suggest a rational stabilization model for Ru NCs synthesized in "IL-only" solvent (i. e. without additional stabilizing agents).

4.4.8 Growth mechanism for two-dimensional (2D) Ru NCs

With the assistance of DFT (density functional theory) calculations, Yin et al. demonstrated the growing mechanism for the formation of 2D Ru NCs [200]. They calculated the surface energies of Ru(0001) (basal facets) and Ru(10 − 10) (prism facets). Two kinds of Ru(10 − 10) surfaces (denoted as (10 − 10)$_a$ and (10 − 10)$_b$) were cleaved. As shown in Figure 4.55, the surface energy of Ru(0001) was 1.06 eV/ surface atom (2.65 J m^{-2}), while the surface energy was 2.64 or 2.18 eV/surface atom (3.61 or 2.98 J m^{-2}) for Ru(10 − 10)$_a$ or Ru(10 − 10)$_b$, respectively. The surface energy of Ru(0001) was 0.96 or 0.33 J m^{-2} lower than that of Ru(10 − 10)$_a$ or Ru(10 − 10)$_b$, respectively. These calculation results are in good consistency with previous results reported by Skriver et al. (0.93 or 0.30 J m^{-2} lower for Ru(0001) than Ru(10 − 10)$_a$ or Ru(10 − 10)$_b$, respectively)[264]. The much lower surface energy of Ru(0001) than those of Ru(10 − 10) suggests that the intrinsic growth of Ru NCs under thermodynamic control would attempt to expose the most stable (0001) facets to minimize their total surface free energy. Therefore, these Ru NCs would adopt the radial growth manner (growth in the *ab* plane of hcp crystal) and grow into 2D nanoplates with the absence of any further morphology regulations.

4.5 Critical Safety Considerations

In this Chapter, most of the chemical agents for the synthesis of Ru NCs are nontoxic, and can ensure compliance with experimental safety regulations. However, some chemicals and experiment conditions should be paid special attention.
1. For all the synthetic procedures in the lab, safety glasses, lab coat, long pants, gloves, and fume hoods are required.
2. Most of the experiments should be conducted in fume hoods.

Figure 4.54: Schematic illustration to show the stabilization model for Ru NCs in ionic liquids (Adapted with permission from ref. 272. Copyright 2010, Royal Society of Chemistry).

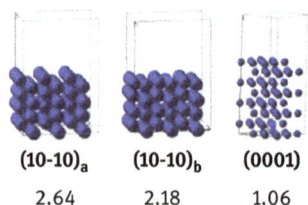

(10-10)$_a$	(10-10)$_b$	(0001)
2.64	2.18	1.06

Figure 4.55: Surface energies (eV/surface atom) of Ru(10 − 10)$_a$, Ru(10 − 10)$_b$, and Ru(0001), respectively (Adapted with permission from ref. 202. Copyright 2012, American Chemical Society).

3. If using high reaction temperature rather than the room temperature, please make sure the instruments work under proper condition. Additional attention should be paid when operating hot vessels.

4. Special attention should be paid when the organic solvents (e.g. oleylamine, dodecylamine, dodecanethiol and toluene) are used. These organic solvents may cause serious discomfort and pain if inhalation or in contact with skin, or swallowed.

5. Special attention should be paid when CO and H$_2$ gas is used. They are the gases easy to explode, while the CO is toxic, may cause death if inhalation.
6. The alkanethiols, e. g. dodecanethiol is very smelly, and please make sure they are not used in open spaces.

4.6 Conclusions and Future Perspective

Over the past few decades, the synthesis, characterization and application of Ru NCs have undergone tremendous developments. It has been well documented that the physicochemical properties of Ru NCs can be finely tuned by tailoring their size and/ or shapes. This chapter highlights the wet-chemistry-based state-of-the-art research progresses in size- and shape-controlled synthesis of Ru NCs, including their synthetic approaches and growth mechanisms. The key synthetic parameters and detailed procedures were depicted based on precise classification of Ru nanostructures. The systematic summarization on the growth mechanisms of Ru NCs provide a powerful guideline to rationally design and produce Ru NCs with desired properties.

Although considerable successes have been achieved in synthesizing Ru NCs for the scientific communities, there is still much more space to search effective methodologies to make highly efficient Ru NCs for various catalytic applications. Future research challenges in heterogeneous nanomaterials may include: (1) to develop effective strategies to prepare Ru NCs with more control on the sizes and shapes. After decades of intense effort, Ru NCs can now be produced with fairly good control. A number of particle geometries such as spheres, wires, prisms, and hollow structures can be routinely synthesized by wet-chemistry methods. However, compared with other noble metals, e.g. Ag, Au, Pt and Pd, the size- and shape-controlled synthesis of Ru NCs is still very limited. Future research challenges lie in the effective routes for more Ru NCs with good control in sizes and shapes. This challenge might be met by a combination of existing control strategies with new synthetic approaches. The successful control in Ru NCs would provide for significant improvements in their catalytic applications; (2) to understand the chemistry behind the controlled synthesis. The development of *in situ* surface characterization techniques and corresponding theoretical simulations would help improve our understanding of the underlying chemistry for the controlled synthesis, which might be valuable for the development of more well-defined Ru NCs with interesting architectures and tailored properties; (3) to explore the combination with other chemical components for enhancing performances. The integration of a noble metal and another chemically different material have been attracted great research interest due to their unique properties induced by synergistic effect between their different components [68, 70, 71, 265–271]. Besides alloying with other transition or noble metals, the synthesis of

Ru-based nanomaterials with heterogeneous nanostructures is very limited. Research challenges in this area are the effective synthetic approaches, powerful characterization tools, deep understanding of the synthetic mechanisms, and evaluation in various catalytic applications; (4) to identify other scientific-related issues. Many interesting scientific findings might be derived from the synthesis and characterization of the Ru NCs. For example, the inside-out diffusion of Ag in core@shell Ag@Ru nanostructures was noticed during their TEM characterization. This finding not only enables Ru NCs with hollow interiors to be made in a facile way, but also satisfy everlasting curiosity of human being.

Funding: We gratefully acknowledge the financial supports from the National Natural Science Foundation of China (Grant No.: 21376247, 21573240), and the Center for Mesoscience, Institute of Process Engineering, Chinese Academy of Sciences (Grant No: COM2015A001).

References

[1]   Narayanan R, El-Sayed MA. Effect of catalytic activity on the metallic nanoparticle size distribution: electron-transfer reaction between $Fe(CN)_6$ and thiosulfate ions catalyzed by PVP-platinum nanoparticles. J Phys Chem B. 2003;107(45):12416–24.

[2]   Narayanan R, El-Sayed MA. Shape-dependent catalytic activity of platinum nanoparticles in colloidal solution. Nano Lett. 2004;4(7):1343–48.

[3]   Narayanan R, El-Sayed MA. Changing catalytic activity during colloidal platinum nanocatalysis due to shape changes: electron-transfer reaction. J Am Chem Soc. 2004;126 (23):7194–95.

[4]   Daniel MC, Astruc D. Gold nanoparticles: assembly, supramolecular chemistry, quantum-size-related properties, and applications toward biology, catalysis, and nanotechnology. Chem Rev. 2004;104(1):293–346.

[5]   Burda C, Chen X, Narayanan R, El-Sayed MA. Chemistry and properties of nanocrystals of different shapes. Chem Rev. 2005;105(4):1025–102.

[6]   Wang C, Daimon H, Lee Y, Kim J, Sun S. Synthesis of monodisperse Pt nanocubes and their enhanced catalysis for oxygen reduction. J Am Chem Soc. 2007;129(22):6974–75.

[7]   Tian N, Zhou ZY, Sun S, Ding Y, Wang ZL. Synthesis of tetrahexahedral platinum nanocrystals with high-index facets and high electro-oxidation activity. Science. 2007;316 (5825):732–35.

[8]   Zhang Q, Xie J, Yang J, Lee JY. Monodisperse icosahedral Ag, Au, and Pd nanoparticles: size control strategy and superlattice formation. ACS Nano. 2008;3(1):139–48.

[9]   Wang C, Daimon H, Onodera T, Koda T, Sun S. A general approach to the size-and shape-controlled synthesis of platinum nanoparticles and their catalytic reduction of oxygen. Angew Chem Int Ed. 2008;47(19):3588–91.

[10] Antolini E, Lopes T, Gonzalez ER. An overview of platinum-based catalysts as methanol-resistant oxygen reduction materials for direct methanol fuel cells. J Alloys Compd. 2008; 461(1):253–62.

[11] Tsung CK, Kuhn JN, Huang W, Aliaga C, Hung LI, Somorjai GA, et al. Sub-10 nm platinum nanocrystals with size and shape control: catalytic study for ethylene and pyrrole hydrogenation. J Am Chem Soc. 2009;131(16):5816–22.

[12] Chen J, Lim B, Lee EP, Xia Y. Shape-controlled synthesis of platinum nanocrystals for catalytic and electrocatalytic applications. Nano Today. 2009;4(1):81–95.

[13] Peng Z, Yang H. Designer platinum nanoparticles: control of shape, composition in alloy, nanostructure and electrocatalytic property. Nano Today. 2009;4(2):143–64.

[14] Xia Y, Xiong Y, Lim B, Skrabalak SE. Shape-Controlled Synthesis of metal nanocrystals: simple chemistry meets complex physics?. Angew Chem Int Ed. 2009;48(1):60–103.

[15] Chen A, Holt-Hindle P. Platinum-based nanostructured materials: synthesis, properties, and applications. Chem Rev. 2010;110(6):3767–804.

[16] An K, Somorjai GA. Size and shape control of metal nanoparticles for reaction selectivity in catalysis. Chem Cat Chem. 2012;4(10):1512–24.

[17] Calle-Vallejo F, Martínez JI, García-Lastra JM, Sautet P, Loffreda D. Fast prediction of Adsorption properties for platinum nanocatalysts with generalized coordination numbers. Angew Chem Int Ed. 2014;53(32):8316–19.

[18] Cao S, Tao FF, Tang Y, Li Y, Yu J. Size- and shape-dependent catalytic performances of oxidation and reduction reactions on nanocatalysts. Chem Soc Rev. 2016;45(17):4747–65.

[19] Fenger R, Fertitta E, Kirmse H, Thünemann AF, Rademann K. Size dependent catalysis with CTAB-stabilized gold nanoparticles. Phys Chem Chem Phys. 2012;14(26):9343–49.

[20] Narayanan R, El-Sayed MA. Catalysis with transition metal nanoparticles in colloidal solution: nanoparticle shape dependence and Stability. J Phys Chem B. 2005;109(26):12663–76.

[21] Zhang H, Jin M, Xiong Y, Lim B, Xia Y. Shape-controlled synthesis of Pd nanocrystals and their catalytic applications. Acc Chem Res. 2013;46(8):1783–94.

[22] Brown KR, Walter DG, Natan MJ. Seeding of colloidal Au nanoparticle solutions. 2. Improved control of particle size and shape. Chem Mater. 2000;12(2):306–13.

[23] Xiong Y, Chen J, Wiley B, Xia Y, Aloni S, Yin Y. Understanding the role of oxidative etching in the polyol synthesis of Pd nanoparticles with uniform shape and size. J Am Chem Soc. 2005;127 (20):7332–33.

[24] Xiong Y, Chen J, Wiley B, Xia Y. Size-dependence of surface plasmon resonance and oxidation for Pd nanotubes synthesized *via* a seed etching process. Nano Lett. 2005;5(7): 1237–42.

[25] Xiong Y, Cai H, Wiley BJ, Wang J, Kim MJ, Xia Y. Synthesis and mechanistic study of palladium nanobars and nanorods. J Am Chem Soc. 2007;129(12):3665–75.

[26] Lim B, Xiong Y, Xia Y. A water-based synthesis of octahedral, decahedral, and icosahedral Pd nanocrystals. Angew Chem Int Ed. 2007;46(48):9279–82.

[27] Viswanath B, Kundu P, Halder A, Ravishankar N. Mechanistic aspects of shape selection and symmetry breaking during nanostructure growth by wet chemical methods. J Phys Chem C. 2009;113(39):16866–83.

[28] Chen DH, Yeh JJ, Huang TC. Synthesis of platinum ultrafine particles in AOT reverse micelles. J Colloid Interface Sci. 1999;215(1):159–66.

[29] Devarajan S, Bera P, Sampath S. Bimetallic nanoparticles: a single step synthesis, stabilization, and characterization of Au-Ag, Au-Pd, and Au-Pt in sol-gel derived silicates. J Colloid Interface Sci. 2005;290(1):117–29.

[30] Xiong Y, McLellan JM, Chen J, Yin Y, Li Z-Y, Xia Y. Kinetically controlled synthesis of triangular and hexagonal nanoplates of palladium and their SPR/SERS properties. J Am Chem Soc. 2005;127(48):17118–27.

[31] Xiong Y, Washio I, Chen J, Cai H, Li Z-Y, Xia Y. Poly(vinyl pyrrolidone): a dual functional reductant and stabilizer for the facile synthesis of noble metal nanoplates in aqueous phase. Langmuir. 2006;22(20):8563–70.

[32] Wiley BJ, Xiong Y, Li Z-Y, Yin Y, Xia Y. Right bipyramids of silver: a new shape derived from single twinned seeds. Nano Lett. 2006;6(4):765–68.

[33] Washio I, Xiong Y, Yin Y, Xia Y. Reduction by the end groups of poly(vinyl pyrrolidone): a new and versatile route to the kinetically controlled synthesis of Ag triangular nanoplates. Advanced Mater. 2006;18(13):1745–49.

[34] Wiley BJ, Wang Z, Wei J, Yin Y, Cobden DH, Xia Y. Synthesis and electrical characterization of silver nanobeams. Nano Lett. 2006;6(10):2273–78.

[35] Xiong Y, Cai H, Yin Y, Xia Y. Synthesis and characterization of fivefold twinned nanorods and right bipyramids of palladium. Chem Phys Lett. 2007;440(4-6):273–78.

[36] Habas SE, Lee H, Radmilovic V, Somorjai GA, Yang P. Shaping binary metal nanocrystals through epitaxial seeded growth. Nat Mater. 2007;6(9):692–97.

[37] Chen Y-H, Huang H, Huang MH. Seed-mediated synthesis of palladium nanorods and branched nanocrystals and their use as recyclable Suzuki coupling reaction. J Am Chem Soc. 2009;131 (25):9114–21.

[38] Yu Y, Zhang Q, Lu X, Lee JY. Seed-mediated synthesis of monodisperse concave trisoctahedral gold nanocrystals with controllable sizes. J Phys Chem C. 2010;114(25):11119–26.

[39] Wang A, Peng Q, Li Y. Rod-shaped Au-Pd core-shell nanostructures. Chem Mater. 2011;23 (13):3217–22.

[40] Yang J, Lee JY, Too H-P, Valiyaveettil S. A bis(*p*-sulfonatophenyl)- phenylphosphine-based synthesis of hollow Pt nanospheres. J Phys Chem B. 2006;110(1):125–29.

[41] Peng Z, Wu J, Yang H. Synthesis and oxygen reduction electrocatalytic property of platinum hollow and platinum-on-silver nanoparticles. Chem Mater. 2009;22(3):1098–106.

[42] Su FB, Lv L, Lee FY, Liu T, Cooper AI, Zhao XS. Thermally reduced ruthenium nanoparticles as a highly active heterogeneous catalyst for hydrogenation of monoaromatics. J Am Chem Soc. 2007;129(46):14213–23.

[43] Lu F, Liu J, Xu J. Synthesis of chain-like Ru nanoparticle arrays and its catalytic activity for hydrogenation of phenol in aqueous media. Mater Chem Phys. 2008;108(2–3):369–74.

[44] Liu X, Meng C, Han Y. Substrate-mediated enhanced activity of Ru nanoparticles in catalytic hydrogenation of benzene. Nanoscale. 2012;4(7):2288–95.

[45] Shawkataly OB, Jothiramalingam R, Adam F, Radhika T, Tsao TM, Wang MK. Ru-nanoparticle deposition on naturally available clay and rice husk biomass materials-Benzene hydrogenation catalysis and synthetic strategies for green catalyst development. Catal Sci Technol. 2012;2 (3):538–46.

[46] Liu Q, Joshi UA, Uber K, Regalbuto JR The control of Pt and Ru nanoparticle size on high surface area supports. Phys Chem Chem Phys. 2014;16(48):26431–35.

[47] Easterday R, Sanchez-Felix O, Losovyj Y, Pink M, Stein BD, Morgan DG, et al. Design of ruthenium/iron oxide nanoparticle mixtures for hydrogenation of nitrobenzene. Catalysis Sci Technol. 2015;5(3):1902–10.

[48] Miyazaki A, Balint I, Aika K, Nakano Y. Preparation of Ru nanoparticles supported on γ-Al$_2$O$_3$ and its novel catalytic activity for ammonia synthesis. J Catal. 2001;204(2):364–71.

[49] Rosowski F, Hornung A, Hinrichsen O, Herein D, Muhler M, Ertl G. Ruthenium catalysts for ammonia synthesis at high pressures: preparation, characterization, and power-law kinetics. Appl Catalysis A: Gen. 1997;151(2):443–60.

[50] Ye H, Wang Q, Catalano M, Lu N, Vermeylen J, Kim MJ, et al. Ru nanoframes with an fcc structure and enhanced catalytic properties. Nano Lett. 2016;16(4):2812–17.

[51] Aßmann J, Crihan D, Knapp M, Lundgren E, Löffler E, Muhler M, et al. Understanding the structural deactivation of ruthenium catalysts on an atomic scale under both oxidizing and reducing conditions. Angew Chem Int Ed. 2005;44(6):917–20.

[52] Assmann J, Narkhede V, Breuer NA, Muhler M, Seitsonen AP, Knapp M, et al. Heterogeneous oxidation catalysis on ruthenium: bridging the pressure and materials gaps and beyond. J Phys Condens Matter. 2008;20(18):184017.

[53] Gao F, Wang YL, Cai Y, Goodman DW. CO oxidation over Ru(0001) at near-atmospheric pressures: from chemisorbed oxygen to RuO_2. Surf Sci. 2009;603(8):1126–34.

[54] Joo SH, Park JY, Renzas JR, Butcher DR, Huang W, Somorjai GA. Size effect of ruthenium nanoparticles in catalytic carbon monoxide oxidation. Nano Lett. 2010;10(7):2709–13.

[55] Qadir K, Joo SH, Mun BS, Butcher DR, Renzas JR, Aksoy F, et al. Intrinsic relation between catalytic activity of CO oxidation on Ru nanoparticles and Ru oxides uncovered with ambient pressure XPS. Nano Lett. 2012;12(11):5761–68.

[56] Abe T, Tanizawa M, Watanabe K, Taguchi A. CO_2 methanation property of Ru nanoparticle-loaded TiO_2 prepared by a polygonal barrel-sputtering method. Energy & Environmental Science. 2009;2(3):315–21.

[57] Gupta S, Giordano C, Gradzielski M, Mehta SK. Microwave-assisted synthesis of small Ru nanoparticles and their role in degradation of congo red. J Colloid Interface Sci. 2013;411: 173–81.

[58] Abo-Hamed EK, Pennycook T, Vaynzof Y, Toprakcioglu C, Koutsioubas A, Scherman OA. Highly active metastable ruthenium nanoparticles for hydrogen production through the catalytic hydrolysis of ammonia borane. Small. 2014;10(15):3145–52.

[59] Balint I, Miyazaki A, Aika K. The relevance of Ru nanoparticles morphology and oxidation state to the partial oxidation of methane. J Catal. 2003;220(1):74–83.

[60] Koh T, Koo HM, Yu T, Lim B, Bae JW. Roles of ruthenium-support interactions of size-controlled ruthenium nanoparticles for the product distribution of Fischer-Tropsch synthesis. ACS Catal. 2014;4(4):1054–60.

[61] Nurunnabi M, Murata K, Okabe K, Inaba M, Takahara I. Effect of Mn addition on activity and resistance to catalyst deactivation for Fischer-Tropsch synthesis over Ru/Al_2O_3 and Ru/SiO_2 catalysts. Catal Commun. 2007;8(10):1531–37.

[62] Johnson MM, Nowack GP. Cyclic olefins by selective hydrogenation of aromatics. J Catal. 1975;38(1-3):518–21.

[63] Schoenmakerstolk MC, Verwijs JW, Don JA, Scholten JJF. The catalytic hydrogenation of benzene over supported metal catalysts: I. Gas-phase hydrogenation of benzene over ruthenium-on-silica. Appl Catal. 1987;29(1):73–90.

[64] da-Silva JW, Cobo AJG. The role of the titania and silica supports in Ru-Fe catalysts to partial hydrogenation of benzene. Appl Catalysis A: Gen. 2003;252(1):9–16.

[65] Su F, Lee FY, Lv L, Liu JJ, Tian XN, Zhao XS. Sandwiched ruthenium/carbon nanostructures for highly active heterogeneous hydrogenation. Adv Funct Mater. 2007;17(12):1926–31.

[66] Lewis LN, Lewis N. Preparation and structure of platinum group metal colloids: without solvent. Chem Mater. 1989;1(1):106–14.

[67] Chen G, Zhang J, Gupta A, Rosei F, Ma D. Shape-controlled synthesis of ruthenium nanocrystals and their catalytic applications. New J Chem. 2014;38(5):1827–33.

[68] Cozzoli PD, Pellegrino T, Manna L. Synthesis, properties and perspectives of hybrid nanocrystal structures. Chem Soc Rev. 2006;35(11):1195–208.

[69] Zhang J, Asakura H, Van Rijn J, Yang J, Duchesne P, Zhang B, et al. Highly efficient, NiAu-catalyzed hydrogenolysis of lignin into phenolic chemicals. Green Chem. 2014; 16(5):2432–37.

[70] Liu H, Feng Y, Chen D, Li C, Cui P, Yang J. Noble metal-based composite nanomaterials fabricated *via* solution-based approaches. J Mater Chem. 2015;3(7):3182–223.

[71] Qu J, Ye F, Chen D, Feng Y, Yao Q, Liu H, et al. Platinum-based heterogeneous nanomaterials via wet-chemistry approaches toward electrocatalytic applications. Adv Colloid Interface Sci. 2016;230:29–53.

[72] Tang J, Chen D, Yao Q, Xie J, Yang J. Recent advances in noble metal-based nanocomposites for electrochemical reactions. Mater Today Energy. 2017;6:115–27.

[73] Yu W, Liu M, Liu H, Ma X, Liu Z. Preparation, Characterization, and catalytic properties of polymer-stabilized ruthenium colloids. J Colloid Interface Sci. 1998;208(2):439–44.

[74] Liu M, Yu W, Liu H. Selective hydrogenation of o-chloronitrobenzene over polymer-stabilized ruthenium colloidal catalysts. J Mol Catalysis A: Chem. 1999;138(2–3):295–303.

[75] Sun S, Murray CB, Weller D, Folks L, Moser A. Monodisperse FePt nanoparticles and ferromagnetic FePt nanocrystal superlattices. Science. 2000;287(5460):1989–92.

[76] Pan C, Pelzer K, Philippot K, Chaudret B, Dassenoy F, Lecante P, et al. Ligand-stabilized ruthenium nanoparticles: synthesis, organization, and dynamics. J Am Chem Soc. 2001;123 (31):7584–93.

[77] Viau G, Fiévet-Vincent F, Fiévet F. Monodisperse iron-based particles: precipitation in liquid polyols. J Mater Chem. 1996;6(6):1047–53.

[78] Toneguzzo P, Viau G, Acher O, Fiévet-Vincent F, Fiévet F. Monodisperse Ferromagnetic particles for microwave applications. Advanced Mater. 1998;10(13):1032–35.

[79] Yan X, Liu H, Liew KY. Size control of polymer-stabilized ruthenium nanoparticles by polyol reduction. J Mater Chem. 2001;11(12):3387–91.

[80] Chen Y, Liew KY, Li J. Size-controlled synthesis of Ru nanoparticles by ethylene glycol reduction. Mater Lett. 2008;62(6–7):1018–21.

[81] Wang Y, Ren J, Deng K, Cui L, Tang Y. Preparation of tractable platinum, rhodium, and ruthenium nanoclusters with small particle size in organic media. Chem Mater. 2000; 12(6):1622–27.

[82] Yu W, Tu W, Liu H. Synthesis of nanoscale platinum colloids by microwave dielectric heating. Langmuir. 1999;15(1):6–9.

[83] Tu W, Liu H. Rapid synthesis of nanoscale colloidal metal clusters by microwave irradiation. J Mater Chem. 2000;10(9):2207–11.

[84] Harpeness R, Gedanken A. Microwave synthesis of core-shell gold/palladium bimetallic nanoparticles. Langmuir. 2004;20(8):3431–34.

[85] Tsuji M, Hashimoto M, Nishizawa Y, Kubokawa M, Tsuji T. Microwave-assisted synthesis of metallic nanostructures in solution. Chem Eur J. 2005;11(2):440–52.

[86] Glaspell G, Fuoco L, El-Shall MS. Microwave synthesis of supported Au and Pd nanoparticle catalysts for CO oxidation. J Phys Chem B. 2005;109(37):17350–55.

[87] Mallikarjuna NN, Varma RS. Microwave-assisted shape-controlled bulk synthesis of noble nanocrystals and their catalytic properties. Cryst Growth Des. 2007;7(4):686–90.

[88] Grace AN, Pandian K. One pot synthesis of polymer protected Pt, Pd, Ag and Ru nanoparticles and nanoprisms under reflux and microwave mode of heating in glycerol — A comparative study. Mater Chem Phys. 2007;104(1):191–98.

[89] Abdelsayed V, Aljarash A, El-Shall MS. Microwave synthesis of bimetallic nanoalloys and CO oxidation on ceria-supported nanoalloys. Chem Mater. 2009;21(13):2825–34.

[90] Bilecka I, Niederberger M. Microwave chemistry for inorganic nanomaterials synthesis. Nanoscale. 2010;2(8):1358–74.

[91] Lebègue E, Baranton S, Coutanceau C. Polyol synthesis of nanosized Pt/C electrocatalysts assisted by pulse microwave activation. J Power Sources. 2011;196(3):920–27.

[92] Harish S, Baranton S, Coutanceau C, Joseph J. Microwave assisted polyol method for the preparation of Pt/C, Ru/C and PtRu/C nanoparticles and its application in electrooxidation of methanol. J Power Sources. 2012;214:33–39.

[93] Suryawanshi YR, Chakraborty M, Jauhari S, Mukhopadhyay S, Shenoy KT, Shridharkrishna R. Microwave irradiation solvothermal technique: an optimized protocol for size-control synthesis of Ru nanoparticles. Crystal Res Technol. 2013;48(2):69–74.

[94] Viau G, Brayner R, Poul L, Chakroune N, Lacaze E, Fiévet-Vincent F, et al. Ruthenium nanoparticles: size, shape, and self-assemblies. Chem Mater. 2003;15(2):486–94.

[95] Chakroune N, Viau G, Ammar S, Poul L, Veautier D, Chehimi MM, et al. Acetate- and thiol-capped monodisperse ruthenium nanoparticles: XPS, XAS, and HRTEM studies. Langmuir. 2005;21(15):6788–96.

[96] Tsukatani T, Fujihara H. New method for facile synthesis of amphiphilic thiol-stabilized ruthenium nanoparticles and their redox-active ruthenium nanocomposite. Langmuir. 2005; 21(26):12093–95.

[97] Lee JY, Yang J, Deivaraj TC, Too H-P. A novel synthesis route for ethylenediamine-protected ruthenium nanoparticles. J Colloid Interface Sci. 2003;268(1):77–80.

[98] Ataka K, Yotsuyanagi T, Osawa M. Potential-dependent reorientation of water molecules at an electrode/electrolyte interface studied by surface-enhanced infrared absorption spectroscopy. J Phys Chem. 1996;100(25):10664–72.

[99] Chen N, Blowers P, Masel RI. Formation of hydronium and water-hydronium complexes during coadsorption of hydrogen and water on (2×1)Pt(110). Surf Sci. 1999;419:150–7.

[100] Olivera PP, Ferral A, Patrito EM. Theoretical investigation of hydrated hydronium ions on Ag (111). J Phys Chem B. 2001;105(30):7227–38.

[101] Yang J, Lee JY, Deivaraj TC, Too H-P. Preparation and characterization of positively charged ruthenium nanoparticles. J Colloid Interface Sci. 2004;271(2):308–12.

[102] García-Antón J, Axet MR, Jansat S, Philippot K, Chaudret B, Pery T, et al. Reactions of olefins with ruthenium hydride nanoparticles: NMR characterization, hydride titration, and room-temperature C–C bond activation. Angew Chem Int Ed. 2008;47(11):2074–78.

[103] Guerrero M, Roucoux A, Denicourt-Nowicki A, Bricout H, Monflier E, Collière V, et al. Alkyl sulfonated diphosphines-stabilized ruthenium nanoparticles as efficient nanocatalysts in hydrogenation reactions in biphasic media. Catalysis Today. 2012;183(1):34–41.

[104] Guerrero M, Coppel Y, Chau NTT, Roucoux A, Denicourt-Nowicki A, Monflier E, et al. Efficient Ruthenium Nanocatalysts in Liquid–liquid Biphasic Hydrogenation Catalysis: towards a Supramolecular Control through a Sulfonated Diphosphine–Cyclodextrin Smart Combination. Chem Cat Chem. 2013;5(12):3802–11.

[105] Herbois R, Noël S, Léger B, Tilloy S, Menuel S, Addad A, et al. Ruthenium-containing β-cyclodextrin polymer globules for the catalytic hydrogenation of biomass-derived furanic compounds. Green Chem. 2015;17(4):2444–54.

[106] Yang J, Lee JY, Deivaraj TC, Too H-P. An improved procedure for preparing smaller and nearly monodispersed thiol-stabilized platinum nanoparticles. Langmuir. 2003;19(24):10361–65.

[107] Yang J, Deivaraj TC, Too H-P, Lee JY. An alternative phase-transfer method of preparing alkylamine-stabilized platinum nanoparticles. J Phys Chem B. 2004;108(7):2181–85.

[108] Yang J, Lee JY, Ying JY. Phase transfer and its applications in nanotechnology. Chem Soc Rev. 2011;40(3):1672–96.

[109] Yang J, Lee JY, Deivaraj TC, Too H-P. A highly efficient phase transfer method for preparing alkylamine-stabilized Ru, Pt, and Au nanoparticles. J Colloid Interface Sci. 2004; 277(1):95–99.

[110] Yang J, Lee JY, Deivaraj TC, Too H-P. An improved Brust's procedure for preparing alkylamine stabilized Pt, Ru nanoparticles. Colloids Surf A Physicochem Eng Asp. 2004;240(1-3):131–34.

[111] Yang J, Sargent EH, Kelley SO, Ying JY. A general phase-transfer protocol for metal ions and its application in nanocrystal synthesis. Nat Mater. 2009;8(8):683–89.

[112] Mourdikoudis S, Liz-Marzán LM. Oleylamine in nanoparticle synthesis. Chem Mater. 2013; 25(9):1465–76.

[113] Ye F, Liu H, Yang JH, Cao H, Yang J. Morphology and structure controlled synthesis of ruthenium nanoparticles in oleylamine. Dalton Trans. 2013;42(34):12309–16.

[114] Wang C, Ye F, Liu C, Cao H, Yang J. Ag facilitated shape control of transition-metal nanoparticles. Colloids Surf A Physicochem Eng Asp. 2011;385(1-3):85–90.

[115] Ma Z, Yu J, Dai S. Preparation of inorganic materials using ionic liquids. Advanced Mater. 2010;22(2):261–85.

[116] Dupont J, Scholten JD. On the structural and surface properties of transition-metal nanoparticles in ionic liquids. Chem Soc Rev. 2010;39(5):1780–804.

[117] Richter K, Birkner A, Mudring A-V. Stabilizer-free metal nanoparticles and metal–metal oxide nanocomposites with long-term stability prepared by physical vapor deposition into ionic liquids. Angew Chem Int Ed. 2010;49(13):2431 –35.

[118] Von Prondzinski N, Cybinska J, Mudring A-V. Easy access to ultra long-time stable, luminescent europium(II) fluoride nanoparticles in ionic liquids. Chem Commun. 2010;46(24):4393–95.

[119] Vollmer C, Janiak C. Naked metal nanoparticles from metal carbonyls in ionic liquids: easy synthesis and stabilization. Coord Chem Rev. 2011;255(17-18):2039–57.

[120] Krossing I, Slattery JM, Daguenet C, Dyson PJ, Oleinikova A, Weingärtner H. Why are ionic liquids liquid? A simple explanation based on lattice and solvation energies. J Am Chem Soc. 2006;128(41):13427–34.

[121] Dahl JA, Maddux BLS, Hutchison JE. Toward greener nanosynthesis. Chem Rev. 2007; 107(6):2228–69.

[122] Lara P, Philippot K, Chaudret B. Organometallic ruthenium nanoparticles: A comparative study of the influence of the stabilizer on their characteristics and reactivity. Chem Cat Chem. 2013; 5(1):25–48.

[123] Scholten JD, Leal BC, Dupont J. Transition metal nanoparticle catalysis in ionic liquids. ACS Catal. 2012;2(1):184–200.

[124] Campbell PS, Prechtl MHG, Santini CC, Haumesser P-H. Ruthenium nanoparticles in ionic liquids – A saga. Curr Org Chem. 2013;17(4):414–29.

[125] Prechtl MHG, Scariot M, Scholten JD, Machado G, Teixeira SR, Dupont J. Nanoscale Ru(0) particles: arene hydrogenation catalysts in imidazolium ionic liquid. Inorg Chem. 2008;47 (19):8995–9001.

[126] Prechtl MHG, Scholten JD, Dupont J. Tuning the selectivity of ruthenium nanoscale catalysts with functionalised ionic liquids: hydrogenation of nitriles. J Mol Catalysis A: Chem. 2009; 313(1-2):74–78.

[127] Julis J, Hölscher M, Leitner W. Selective hydrogenation of biomass derived substrates using ionic liquid-stabilized ruthenium nanoparticles. Green Chem. 2010;12(9):1634–39.

[128] Luska KL, Moores A. Ruthenium nanoparticle catalysts stabilized in phosphonium and imidazolium ionic liquids: dependence of catalyst stability and activity on the ionicity of the ionic liquid. Green Chem. 2012;14(6):1736–42.

[129] Gutel T, Garcia-Antōn J, Pelzer K, Philippot K, Santini CC, Chauvin Y, et al. Influence of the self-organization of ionic liquids on the size of ruthenium nanoparticles: effect of the temperature and stirring. J Mater Chem. 2007;17(31):3290–92.

[130] Gutel T, Santini CC, Philippot K, Padua A, Pelzer K, Chaudret B, et al. Organized 3D-alkyl imidazolium ionic liquids could be used to control the size of in situ generated ruthenium nanoparticles?. J Mater Chem. 2009;19(22):3624–31.

[131] Campbell PS, Santini CC, Bayard F, Chauvin Y, Collière V, Podgoršek A, et al. Olefin hydrogenation by ruthenium nanoparticles in ionic liquid media: does size matter?. J Catal. 2010; 275(1):99–107.

[132] Salas G, Santini CC, Philippot K, Collière V, Chaudret B, Fenet B, et al. Influence of amines on the size control of in situ synthesized ruthenium nanoparticles in imidazolium ionic liquids. Dalton Trans. 2011;40(17):4660–68.

[133] Salas G, Podgoršek A, Campbell PS, Santini CC, Pádua AAH, Gomes MFC, et al. Ruthenium nanoparticles in ionic liquids: structure and stability effects of polar solutes. Phys Chem Chem Phys. 2011;13(30):13527–36.

[134] Krämer J, Redel E, Thomann R, Janiak C. Use of ionic liquid for the synthesis of iron, ruthenium, and osmium nanoparticles from their metal carbonyl precursors. Organometallics. 2008; 27(9):1976–78.

[135] Vollmer C, Redel E, Abu-Shandi K, Thomann R, Manyar H, Hardacre C, et al. Microwave irradiation for the facile synthesis of transition-metal nanoparticles (NPs) in ionic liquid (ILs) from metal-carbonyl precursors and Ru-, Rh-, and Ir-NP/IL dispersions as biphasic liquid-liquid hydrogenation nanocatalysts for cyclohexene. Chem Eur J. 2010;16(12):3849–58.

[136] Yoo J. Propene hydrogenation over truncated octahedral Pt nanoparticles supported on alumina. J Catal. 2003;214(1):1–7.

[137] Bronstein LM, Goerigk G, Kostylev M, Pink M, Khotina IA, Valetsky PM, et al. Structure and catalytic properties of Pt-modified hyper-cross-linked polystyrene exhibiting hierarchical porosity. J Phys Chem B. 2004;108(47):18234–42.

[138] Denicourt-Nowicki A, Roucoux A, Wyrwalski F, Kania N, Monflier E, Ponchel A. Carbon-supported ruthenium nanoparticles stabilized by methylated cyclodextrins: a new family of heterogeneous catalysts for the gas-phase hydrogenation of arenes. Chemistry-A Eur J. 2008; 14(27):8090–93.

[139] Wyrwalski F, Léger B, Lancelot C, Roucoux A, Monflier E, Ponchel A. Chemically modified cyclodextrins as supramolecular tools to generate carbon-supported ruthenium nanoparticles: an application towards gas phase hydrogenation. Appl Catalysis A: Gen. 2011;391(1–2): 334–41.

[140] Biard P-F, Werghi B, Soutrel I, Organd R, Couvert A, Denicourt-Nowicki A, et al. Efficient catalytic ozonation by ruthenium nanoparticles supported on SiO_2 or TiO_2: towards the use of a non-woven fiber paper as original support. Chem Eng J. 2016;289:374–81.

[141] Wen B, Ma J, Chen C, Ma W, Zhu H, Zhao J. Supported noble metal nanoparticles as photo/sono-catalysts for synthesis of chemicals and degradation of pollutants. Sci China Chem. 2011;54(6):887–97.

[142] Holade Y, Sahin NE, Servat K, Napporn TW, Kokoh KB. Recent advances in carbon supported metal nanoparticles preparation for oxygen reduction reaction in low temperature fuel cells. Catalysts. 2015;5(1):310–48.

[143] Li J, Liu H, Deng Y, Liu G, Chen Y, Yang J. Emerging nanostructured materials for the catalytic removal of volatile organic compounds. Nanotechnology Rev. 2016;5(1): 147–81.

[144] Hulea V, Brunel D, Galarneau A, Philippot K, Chaudret B, Kooyman PJ, et al. Synthesis of well-dispersed ruthenium nanoparticles inside mesostructured porous silica under mild conditions. Microporous Mesoporous Mater. 2005;79(1-3):185–94.

[145] Huang J, Jiang T, Han B, Wu W, Liu Z, Xie Z, et al. A novel method to immobilize Ru nanoparticles on SBA-15 firmly by ionic liquid and hydrogenation of arene. Catal Lett. 2005;103(1–2):59–62.

[146] Song L, Li X, Wang H, Wu H, Wu P. Ru nanoparticles entrapped in mesopolymers for efficient liquid-phase hydrogenation of unsaturated compounds. Catal Lett. 2009; 133(1-2):63–69.

[147] Miao S, Liu Z, Han B, Huang J, Sun Z, Zhang J, et al. Ru nanoparticles immobilized on montmorillonite by ionic liquids: A highly efficient heterogeneous catalyst for the hydrogenation of benzene. Angew Chem Int Ed. 2006;45(2):266–69.

[148] Okal J, Zawadzki M, Kepiński L, Krajczyk L, Tylus W. The use of hydrogen chemisorption for the determination of Ru dispersion in Ru/γ-alumina catalysts. Appl Catalysis A: Gen. 2007;319:202–09.

[149] Li L, Wang Y, Vanka S, Mu X, Mi Z, Li C-J. Nitrogen photofixation over III-nitride nanowires assisted by ruthenium clusters of low atomicity. Angew Chem Int Ed. 2017;56(30):8701–05.

[150] Abe T, Akamaru S, Watanabe K. Surface modification of Al_2O_3 ceramic grains using a new RF sputtering system developed for powdery materials. J Alloys Compd. 2004; 377(1–2):194–201.

[151] Abe T, Akamaru S, Watanabe K, Honda Y. Surface modification of polymer microparticles using a hexagonal-barrel sputtering system. J Alloys Compd. 2005;402(1–2):227–32.

[152] He JH, Ichinose I, Kunitake T, Nakao A. In situ synthesis of noble metal nanoparticles in ultrathin TiO_2-gel films by a combination of ion-exchange and reduction processes. Langmuir. 2002; 18(25):10005–10.

[153] Koo IG, Lee MS, Shim JH, Ahn JH, Lee WM. Platinum nanoparticles prepared by a plasma-chemical reduction method. J Mater Chem. 2005;15(38):4125–28.

[154] Zou JJ, Zhang YP, Liu CJ. Reduction of supported noble-metal ions using glow discharge plasma. Langmuir. 2006;22(26):11388–94.

[155] Lee SW, Liang D, Gao XPA, Sankaran RM. Direct writing of metal nanoparticles by localized plasma electrochemical reduction of metal cations in polymer films. Adv Funct Mater. 2011; 21(11):2155–61.

[156] Xia Y, Yang P, Sun Y, Wu Y, Mayers B, Gates B, et al. One-dimensional nanostructures: synthesis, characterization, and applications. Advanced Mater. 2003;15(5):353–89.

[157] Koenigsmann C, Wong SS. One-dimensional noble metal electrocatalysts: a promising structural paradigm for direct methanol fuel cells. Energy & Environmental Science. 2011; 4(4):1161–76.

[158] Patete JM, Peng X, Koenigsmann C, Xu Y, Karn B, Wong SS. Viable methodologies for the synthesis of high-quality nanostructures. Green Chem. 2011;13(3):482–519.

[159] Koenigsmann C, Scofield ME, Liu H, Wong SS. Designing enhanced one-dimensional electrocatalysts for the oxygen reduction reaction: probing size- and composition-dependent electrocatalytic behavior in noble metal nanowires. J Phys Chem Lett. 2012; 3(22):3385–98.

[160] Zhou H, Zhou W, Adzic RR, Wong SS. Enhanced electrocatalytic performance of one-dimensional metal nanowires and arrays generated via an ambient, surfactantless synthesis. J Phys Chem C. 2009;113(14):5460–66.

[161] Koenigsmann C, Santulli AC, Sutter E, Wong SS. Growth mechanism, and size-dependent electrocatalytic behavior of high-quality, single crystalline palladium nanowires. ACS Nano. 2011;5(9):7471–87.

[162] Koenigsmann C, Sutter E, Chiesa TA, Adzic RR, Wong SS. Highly enhanced electrocatalytic oxygen reduction performance observed in bimetallic palladium-based nanowires prepared under ambient, surfactantless conditions. Nano Lett. 2012;12(4):2013–20.

[163] Wang P, Zhang X, Zhang J, Wan S, Guo S, Lu G, et al. Precise tuning in platinum-nickel/nickel sulfide interface nanowires for synergistic hydrogen evolution catalysis. Nat Commun. 2017;8:14580.

[164] Bu L, Guo S, Zhang X, Shen X, Su D, Lu G, et al. Surface engineering of hierarchical platinum-cobalt nanowires for efficient electrocatalysis. Nat Commun. 2016;11850:1–10.

[165] Lu X, Yavuz MS, Tuan H-Y, Korgel BA, Xia Y. Ultrathin gold nanowires can be obtained by reducing polymeric strands of oleylamine-AuCl complexes formed via aurophilic interaction. J Am Chem Soc. 2008;130(28):8900–01.

[166] Bernardi M, Raja SN, Lim SK. Nanotwinned gold nanowires obtained by chemical synthesis. Nanotechnology. 2010;21(28):285607.

[167] Halder A, Ravishankar N. Ultrathin single-crystalline gold nanowire arrays by oriented attachment. Advanced Mater. 2010;19(14):1854–58.

[168] Kang M, Lee H, Kang T, Kim B. Synthesis, properties, and biological application of perfect crystal gold nanowires: A review. J Mater Sci Technol. 2015;31(6):573–80.

[169] Sun Y, Yin Y, Mayers BT, Herricks T, Xia Y. Uniform silver nanowires synthesis by reducing AgNO₃ with ethylene glycol in the presence of seeds and poly(vinyl pyrrolidone). Chem Mater. 2002;14(11):4736–45.

[170] Sun Y, Gates B, Mayers B, Xia Y. Crystalline silver nanowires by soft solution processing. Nano Lett. 2002;2(2):165–68.

[171] Sun Y, Mayers B, Herricks T, Xia Y. Polyol synthesis of uniform silver nanowires: A plausible growth mechanism and the supporting evidence. Nano Lett. 2003;3(7):955–60.

[172] Caswell KK, Bender CM, Murphy CJ. Seedless, surfactantless wet chemical synthesis of silver nanowires. Nano Lett. 2003;3(5):667–69.

[173] Guo L, Chipara M, Zaleski JM. Convenient, rapid synthesis of Ag nanowires. Chem Mater. 2007;19(7):1755–60.

[174] Song Y, Garcia RM, Dorin RM, Wang H, Qiu Y, Coker EN, et al. Nano Lett. 2007;7(12):3650–55.

[175] Teng X, Han W-Q, Ku W, Hücker M. Synthesis of ultrathin palladium and platinum nanowires and a study of their magnetic properties. Angew Chem Int Ed. 2008;47(11):2055–58.

[176] Shui J, Li JCM. Platinum nanowires produced by electrospinning. Nano Lett. 2009;9(4):1307–14.

[177] Luo Z, Yuwen L, Bao B, Tian J, Zhu X, Weng L, et al. One-pot, low-temperature synthesis of branched platinum nanowires/reduced graphene oxide (BPtNW/RGO) hybrids for fuel cells. J Mater Chenmistry. 2012;22(16):7791–96.

[178] Wang R, Higgins DC, Hoque MA, Lee D, Hassan F, Chen Z. Controlled growth of platinum nanowire arrays on sulfur doped graphene as high performance electrocatalyst. Sci Rep. 2013;3:2431.

[179] Huang X, Zheng N. One-pot, high-yield synthesis of 5-fold twinned Pd nanowires and nanorods. J Am Chem Soc. 2009;131(13):4602–03.

[180] Lu N, Chen W, Fang G, Chen B, Yang K, Yang Y, et al. 5-Fold twinned nanowires and single twinned right bipyramids of Pd: utilizing small organic molecules to tune the etching degree of O₂/halides. Chem Mater. 2014;26(7):2453–59.

[181] Chen A, Ostrom C. Palladium-based nanomaterials: synthesis and electrochemical applications. Chem Rev. 2015;115(21):11999–2044.

[182] Ghosh S, Ghosh M, Rao CNR. Nanocrystals, nanorods and other nanostructures of nickel, ruthenium, rhodium and iridium prepared by a simple solvothermal procedure. J Cluster Sci. 2007;18(1):97–111.

[183] Ponrouch A, Bichat MP, Garbarino S, Maunders C, Botton G, Taberna PL, et al. Synthesis and characterization of well aligned Ru nanowires and nanotubes. ECS Trans. 2010;25(41):3–11.

[184] Ponrouch A, Garbarino S, Pronovost S, Taberna P-L, Simon P, Guay D. Electrodeposition of arrays of Ru, Pt, and PtRu alloy 1D metallic nanostructures. J Electrochem Soc. 2010;157(3): K59–K65.

[185] Koenigsmann C, Semple DB, Sutter E, Tobierre SE, Wong SS. Ambient synthesis of high-quality ruthenium nanowires and the morphology-dependent electrocatalytic performance of platinum-decorated ruthenium nanowires and nanoparticles in the methanol oxidation reaction. ACS Appl Mater Interfaces. 2013;5(12):5518–30.

[186] Huang X, Tang S, Mu X, Dai Y, Chen G, Zhou Z, et al. Freestanding palladium nanosheets with plasmonic and catalytic properties. Nat Nanotechnol. 2011;6(1):28–32.

[187] Huang X, Li S, Huang Y, Wu S, Zhou X, Li S, et al. Synthesis of hexagonal close-packed gold nanostructures. Nat Commun. 2011;2:292.

[188] Huang X, Li H, Li S, Wu S, Boey F, Ma J, et al. Synthesis of gold square-like plates from ultrathin gold square sheets: the evolution of structure phase and shape. Angew Chem Int Ed. 2011;50 (51):12245–48.

[189] Duan H, Yan N, Yu R, Chang C-R, Zhou G, Hu H-S, et al. Ultrathin rhodium nanosheets. Nat Commun. 2014;5:3093.

[190] Fan Z, Huang X, Han Y, Bosman M, Wang Q, Zhu Y, et al. Surface modification-induced phase transformation of hexagonal close-packed gold square sheets. Nat Commun. 2015;6:6571.
[191] Fan Z, Bosman M, Huang X, Huang D, Yu Y, Ong KP, et al. Stabilization of 4H hexagonal phase in gold nanoribbons. Nat Commun. 2015;6:7684.
[192] Fan Z, Huang X, Tan C, Zhang H. Thin metal nanostructures: synthesis, properties and applications. Chem Sci. 2015;6(1):95–111.
[193] Tan C, Zhang H. Wet-chemical synthesis and applications of non-layer structured two-dimensional nanomaterials. Nat Commun. 2015;6:7873.
[194] Fan Z, Luo Z, Huang X, Li B, Chen Y, Wang J, et al. Synthesis of 4H/*fcc* noble multimetallic nanoribbons for electrocatalytic hydrogen evolution reaction. J Am Chem Soc. 2016;138 (4):1414–19.
[195] Yang N, Zhang Z, Chen B, Huang Y, Chen J, Lai Z, et al. Synthesis of ultrathin PdCu alloy nanosheets used as a highly efficient electrocatalyst for formic acid oxidation. Advanced Mater. 2017;29(29):1700769.
[196] Saleem F, Zhang Z, Xu B, Xu X, He P, Wang X. Ultrathin Pt–cu nanosheets and nanocones. J Am Chem Soc. 2013;135(49):18304–07.
[197] Saleem F, Xu B, Ni B, Liu H, Nosheen F, Li H, et al. Atomically thick Pt-Cu nanosheets: self-assembled sandwich and nanoring-like structures. Advanced Mater. 2015;27(12):2013–18.
[198] Hong JW, Kim Y, Wi DH, Lee S, Lee S-U, Lee YW, et al. Ultrathin free-standing ternary-alloy nanosheets. Angew Chem Int Ed. 2016;55(8):2753–58.
[199] Zhang Y, Wang M, Zhu E, Zheng Y, Huang Y, Huang X. Seedless growth of palladium nanocrystals with tunable structures: from tetrahedra to nanosheets. Nano Lett. 2015;15(11): 7519–25.
[200] Yin A-X, Liu W-C, Ke J, Zhu W, Gu J, Zhang Y-W, et al. Ru Nanocrystals with shape-dependent surface-enhanced Raman spectra and catalytic properties: controlled synthesis and DFT calculations. J Am Chem Soc. 2012;134(50):20479–89.
[201] Zhang Z, Liu Y, Chen B, Gong Y, Gu L, Fan Z, et al. Submonolayered Ru deposited on ultrathin Pd nanosheets used for enhanced catalytic applications. Advanced Mater. 2016;28(46): 10282–86.
[202] Kim S-W, Kim M, Lee WY, Hyeon T. Fabrication of hollow palladium spheres and their successful application to recyclable heterogeneous catalyst for Suzuki coupling reactions. J Am Chem Soc. 2002;124(26):7642–43.
[203] Liang H-P, Zhang H-M, Hu J-S, Guo Y-G, Wan L-J, Bai C-L. Pt hollow nanospheres: facile synthesis and enhanced electrocatalysts. Angew Chem Int Ed. 2004;43(12):1540–43.
[204] Bai F, Sun Z, Wu H, Haddad RE, Xiao X, Fan H. Templated photocatalytic synthesis of well-defined platinum hollow nanostructures with enhanced catalytic performance for methanol oxidation. Nano Lett. 2011;11(9):3759–62.
[205] Chen D, Cui P, He H, Liu H, Yang J. Highly catalytic hollow palladium nanoparticles derived from silver@silver-palladium core-shell nanostructures for the oxidation of formic acid. J Power Sources. 2014;272:152–59.
[206] Chen D, Cui P, He H, Liu H, Ye F, Yang J. Carbon-supported hollow palladium nanoparticles with enhanced electrocatalytic performance. RSC Adv. 2015;5(15):10944–50.
[207] Chen J, Wiley B, McLellan J, Xiong Y, Li Z-Y, Xia Y. Optical properties of Pd-Ag and Pt-Ag nanoboxes synthesized *via* galvanic replacement reactions. Nano Lett. 2005;5(10): 2058–62.
[208] Kundu J, Le F, Nordlander P, Halas NJ. Surface enhanced infrared absorption (SEIRA) spectroscopy on nanoshell aggregate substrates. Chem Phys Lett. 2008;452(1-3):115–19.
[209] Bardhan R, Grady NK, Halas NJ. Nanoscale control of near-infrared fluorescence enhancement using Au nanoshells. Small. 2008;4(10):1716–22.

[210] Park J, Estrada A, Sharp K, Sang K, Schwartz JA, Smith DK, et al. Two-photon-induced photo-luminescence imaging of tumors using near-infrared excited gold nanoshells. Opt Express. 2008;16(3):1590–99.

[211] Acevedo R, Lombardini R, Halas NJ, Johnson BR. Plasmonic enhancement of Raman optical activity in molecules near metal nanoshells. J Phys Chem. 2009;113(47):13173–83.

[212] Zhang Q, Cobley CM, Zeng J, Wen L-P, Chen J, Xia Y. Dissolving Ag from Au-Ag alloy nanoboxes with H_2O_2: A method for both tailoring the optical properties and measuring the H_2O_2 concentration. J Phys Chem C. 2010;114(14):6396–400.

[213] Chen J, Wang D, Xi J, Au L, Siekkinen A, Warsen A, et al. Immuno gold nanocages with tailored optical properties for targeted photothermal destruction of cancer cells. Nano Lett. 2007; 7(5):1318–22.

[214] Gobin AM, Lee MH, Halas NJ, James WD, Drezek RA, West JL. Near-infrared resonant nanoshells for combined optical imaging and photothermal cancer therapy. Nano Lett. 2007;7(7):1929–34.

[215] Song KH, Kim C, Cobley CM, Xia Y, Wang LV. Near-infrared gold nanocages as a new class of tracers for photoacoustic sentinel lymph node mapping on a rat model. Nano Lett. 2009; 9(1):183–88.

[216] Zhang JZ. Biomedical applications of shape-controlled plasmonic nanostructures: A case study of hollow gold nanospheres for photothermal ablation therapy of cancer. J Phys Chem Lett. 2010;1(4):686–95.

[217] Xia Y, Li W, Cobley CM, Chen J, Xia X, Zhang Q, et al. Gold nanocages: from synthesis to theranostic applications. Acc Chem Res. 2011;44(10):914–24.

[218] Liu H, Qu J, Chen Y, Li J, Ye F, Lee JY, et al. Hollow and cage-bell structured nanomaterials of noble metals. J Am Chem Soc. 2012;134(28):11602–10.

[219] Liu H, Ye F, Yang J. A universal and cost-effective approach to the synthesis of carbon-supported noble metal nanoparticles with hollow interiors. Ind Eng Chem Res. 2014;53(14):5925–31.

[220] Feng Y, Ma X, Han L, Peng Z, Yang J. A universal approach to the synthesis of nanodendrites of noble metals. Nanoscale. 2014;6(11):6173–79.

[221] Link S, Wang ZL, El-Sayed MA. Alloy formation of gold-silver nanoparticles and the dependence of the plasmon absorption on their composition. J Phys Chem B. 1999;103(18):3529–33.

[222] Link S, El-Sayed MA. Size and temperature dependence of the plasmon absorption of colloidal gold nanoparticles. J Phys Chem B. 1999;103(21):4212–17.

[223] Kariuki NN, Luo J, Maye MM, Hassan SA, Menard T, Naslund HR, et al. Composition-controlled synthesis of bimetallic gold-silver nanoparticles. Langmuir. 2004;20(25):11240–46.

[224] Zawadzki M, Okal J. Synthesis and structure characterization of Ru nanoparticles stabilized by PVP or gamma-Al_2O_3. Mater Res Bull. 2008;43(11):3111–21.

[225] Teranishi T, Hosoe M, Tanaka T, Miyake M. Size control of monodispersed Pt nanoparticles and their 2D organization by electrophoretic deposition. J Phys Chem B. 1999;103:3818–27.

[226] Feng Y, Liu H, Yang J. Bimetallic nanodendrites via selective overgrowth of noble metals on multiply twinned Au seeds. J Mater Chem. 2014;2(17):6130–37.

[227] Kusada K, Kobayashi H, Yamamoto T, Matsumura S, Sumi N, Sato K, et al. Discovery of face-centered-cubic ruthenium nanoparticles: facile size-controlled synthesis using the chemical reduction method. J Am Chem Soc. 2013;135(15):5493–96.

[228] Yang J, Ying JY. Diffusion of gold from the inner core to the surface of Ag_2S nanocrystals. J Am Chem Soc. 2010;132(7):2114–15.

[229] Yang J, Ying JY. Nanocomposites of Ag_2S and noble metals. Angew Chem Int Ed. 2011; 50(20):4637–73.

[230] Ye F, Liu H, Hu W, Zhong J, Chen Y, Cao H, et al. Heterogeneous Au–pt nanostructures with enhanced catalytic activity toward oxygen reduction. Dalton Trans. 2012; 41(10):2898–903.

[231] Chen D, Cui P, Liu H, Yang J. Heterogeneous nanocomposites composed of silver sulfide and hollow structured Pd nanoparticles with enhanced catalytic activity toward formic acid oxidation. Electrochim Acta. 2015;153:461–67.

[232] Cui P, He H, Liu H, Zhang S, Yang J. Heterogeneous nanocomposites of silver selenide and hollow platinum nanoparticles toward methanol oxidation reaction. J Power Sources. 2016;327:432–37.

[233] Liu H, Ye F, Yao Q, Cao H, Xie J, Lee JY, et al. Stellated Ag-Pt bimetallic nanoparticles: an effective platform for catalytic activity tuning. Sci Rep. 2014;4:3969.

[234] Niu M, Wang Y, Li W, Jiang J, Jin Z. Highly efficient and recyclable ruthenium nanoparticle catalyst for semihydrogenation of alkynes. Catal Commun. 2013;38:77–81.

[235] Zhang YW, Grass ME, Habas SE, Tao F, Zhang TF, Yang P, et al. One-step polyol synthesis and langmuir-blodgett monolayer formation of size-tunable monodisperse rhodium nanocrystals with catalytically active (111) surface structures. J Phys Chem C. 2007;111(33):12243–53.

[236] Park J, Joo J, Kwon SG, Jang Y, Hyeon T. Synthesis of monodisperse spherical nanocrystals. Angew Chem Int Ed. 2007;46(25):4630–60.

[237] Lamer VK, Dinegar RH. Theory, production and mechanism of formation of monodispersed hydrosols. J Am Chem Soc. 1950;72(11):4847–54.

[238] Nasilowski M, Mahler B, Lhuillier E, Ithurria S, Dutret B. Two-dimensional colloidal nanocrystals. Chem Rev. 2016;116(18):10934–82.

[239] Peng X, Wickham J, Alivisatos AP. Kinetics of II-VI and III-V colloidal semiconductor nanocrystal growth: "Focusing" of size distributions. J Am Chem Soc. 1998;120(21):5343–44.

[240] Peng Z, Peng X. Nearly monodisperse and shape-controlled CdSe nanocrystals via alternative routes: nucleation and growth. J Am Chem Soc. 2002;124(13):3343–53.

[241] Jana NR, Gearheart L, Murphy CJ. Evidence for seed-mediated nucleation in the chemical reduction of gold salts to gold nanoparticles. Chem Mater. 2001;13(7):2313–22.

[242] Yu H, Gibbons PC, Kelton KF, Buhro WE. Heterogeneous seeded growth: A potentially general synthesis of monodisperse metallic nanoparticles. J Am Chem Soc. 2001;123(37):9198–99.

[243] Wilcoxon JP, Provencio PP. Heterogeneous growth of metal clusters from solutions of seed nanoparticles. J Am Chem Soc. 2004;126(20):6402–08.

[244] Park J, Lee E, Hwang NM, Kang M, Kim SC, Hwang Y, et al. One-nanometer-scale size-controlled synthesis of monodisperse magnetic iron oxide nanoparticles. Angew Chem Int Ed. 2005; 44(19):2872–77.

[245] Grzelczak M, Perez-Juste J, Mulvaney P, Liz-Marzan LM. Shape control in gold nanoparticle synthesis. Chem Soc Rev. 2008;37(9):1783–91.

[246] Tao AR, Habas S, Yang P. Shape control of colloidal metal nanocrystals. Small. 2008;4(3): 310–25.

[247] Arico AS, Creti P, Kim H, Mantegna R, Giordano N, Antonucci V. Analysis of the electrochemical characteristics of a direct methanol fuel cell based on a Pt-Ru/C anode catalyst. J Electrochem Soc. 1996;143(12):3950–59.

[248] Yang J, Deivaraj TC, Too H-P, Lee JY. Acetate stabilization of metal nanoparticles and its role in the preparation of metal nanoparticles in ethylene glycol. Langmuir. 2004;20(10):4241–45.

[249] Mucic RC, Storhoff JJ, Mirkin CA, Letsinger RL. DNA-directed synthesis of binary nanoparticle network materials. J Am Chem Soc. 1998;120(48):12674–75.

[250] Storhoff JJ, Lazarides AA, Mucic RC, Mirkin CA, Letsinger RL, Schatz GC. What controls the optical properties of DNA-linked gold nanoparticle assemblies?. J Am Chem Soc. 2000; 122(19):4640–50.

[251] Storhoff JJ, Elghanian R, Mirkin CA, Letsinger RL. Sequence-dependent stability of DNA-modified gold nanoparticles. Langmuir. 2002;18(17):6666–70.

[252] Wiley B, Herricks T, Sun Y, Xia Y. Polyol Synthesis of silver nanoparticles: use of chloride and oxygen to promote the formation of single-crystal, truncated cubes and tetrahedrons. Nano Lett. 2004;4(9):1733–39.

[253] Tan Y-N, Yang J, Lee JY, Wang DIC. Mechanistic study on the bis(p-sulfonatophenyl)phenylphosphine synthesis of monometallic Pt hollow nanoboxes using Ag*–Pt core–shell nanocubes as sacrificial templates. J Phys Chem C. 2007;111(38):14084–90.

[254] Wang ZL. Transmission electron microscopy of shape-controlled nanocrystals and their assemblies. J Phys Chem B. 2000;104(6):1153–75.

[255] Tang Y, Ouyang M. Tailoring properties and functionalities of metal nanoparticles through crystallinity engineering. Nat Mater. 2007;6(10):754–59.

[256] Brodersen SH, Grønbjerg U, Hvolbæk B, Schiøtz J. Understanding the catalytic activity of gold nanoparticles through multi-scale simulations. J Catal. 2011;284(1):34–41.

[257] Hiramatsu H, Osterloh FE. A simple large-scale synthesis of nearly monodisperse gold and silver nanoparticles with adjustable sizes and with exchangeable surfactants. Chem Mater. 2004;16(13):2509–11.

[258] Yu H, Chen M, Rice PM, Wang SX, White RL, Sun S. Dumbbell-like bifunctional Au-Fe$_3$O$_4$ nanoparticles. Nano Lett. 2005;5(2):379–82.

[259] Li Z, Tao J, Lu X, Zhu Y, Xia Y. Facile synthesis of ultrathin Au nanorods by aging the AuCl (oleylamine) complex with amorphous Fe nanoparticles in chloroform. Nano Lett. 2008;8 (9):3052–55.

[260] Xu Z, Shen C, Hou Y, Cao H, Sun S. Oleylamine as both reducing agent and stabilizer in a facile synthesis of magnetite nanoparticles. Chem Mater. 2009;21(9):1778–80.

[261] Liu X, Atwater M, Wang J, Dai Q, Zou J, Brennan JP, et al. J Nanosci Nanotechnol. 2007;7 (9):3126–33.

[262] Amiens C, Chaudret B, Ciuculescu-Pradines D, Collière V, Fajerwerg K, Fau P, et al. Organometallic approach for the synthesis of nanostructures. New J Chem. 2013;37(11): 3374–401.

[263] Campbell PS, Santini CC, Bouchu D, Fenet B, Philippot K, Chaudret B, et al. A novel stabilisation model for ruthenium nanoparticles in imidazolium ionic liquids: in situ spectroscopic and labelling evidence. Phys Chem Chem Phys. 2010;12(16):4217–23.

[264] Skriver HL, Rosengaard NM. Surface energy and work function of elemental metals. Phys Rev B. 1992;46(11):7157–68.

[265] Carbone L, Cozzoli PD. Colloidal heterostructured nanocrystals: synthesis and growth mechanisms. Nano Today. 2010;5(5):449–93.

[266] Buck MR, Bondi JF, Schaak RE. A total-synthesis framework for the construction of high-order colloidal hybrid nanoparticles. Nat Chem. 2012;4(1):37–44.

[267] Buck MR, Schaak RE. Emerging strategies for the total synthesis of inorganic nanostructures. Angew Chem Int Ed. 2013;52(24):6154–78.

[268] Sitt A, Hadar I, Banin U. Band-gap engineering, optoelectronic properties and applications of colloidal heterostructured semiconductor nanorods. Nano Today. 2013;8(5):494–513.

[269] Banin U, Ben-Shahar Y, Vinokurov K. Hybrid semiconductor–metal nanoparticles: from architecture to function. Chem Mater. 2014;26(1):97–110.

[270] Ben-Shahar Y, Banin U. Hybrid semiconductor–metal nanorods as photocatalysts. Top Curr Chem. 2016;374(4):54.

[271] Hodges JM, Morse JR, Fenton JL, Ackerman JD, Alameda LT, Schaak RE. Insights into the seeded-growth synthesis of colloidal hybrid nanoparticles. Chem Mater. 2017;29(1):106–19.

Bionotes

Linlin Xu received her BEng in Chemical Engineering and Technology at Beijing University of Chemical Technology in 2016. She is currently a Ph. D candidate at Institute of Process Engineering, Chinese Academy of Sciences (IPE-CAS). Her research focuses on the study of Palladium-based bimetallic nanoalloys and their catalytic properties toward environmental and electrochemical reactions.

Jun Yang received his Ph. D in Chemical and Biomolecular Engineering in 2006 from National University of Singapore (with Professor Jim Yang LEE). After postdoctoral research at Boston College and University of Toronto, he joined the Institute of Bioengineering and Nanotechnology, Singapore in 2007. In 2010, he moved to Institute of Process Engineering, Chinese Academy of Sciences as the leader of Group of Materials for Energy Conversion and Environmental Remediation (MECER). His main research interests include (i) applied catalysis, (ii) composite nanomaterials for energy conversion and environmental remediation, (iii) electrocatalysis and (iv) separation techniques.

Nirmal Kumar Das and Saptarshi Mukherjee

5 Size-controlled atomically precise copper nanoclusters: Synthetic protocols, spectroscopic properties and applications

Abstract: Noble metal nanoclusters (NCs) are a new class of nanomaterials which are considered being a missing link between isolated metal atoms and metal nanoparticles (NPs). The sizes of the NCs are comparable to the Fermi wavelength of the conduction electrons, and this renders them to be luminescent in nature. They exhibit size-dependent fluorescence properties spanning almost the entire breath of the visible spectrum. Among all the noble metal NCs being explored, copper NCs (CuNCs) are the most rarely investigated primarily because of their propensity of getting oxidised. In this chapter, we have given a comprehensive understanding as to why these NCs are luminescent in nature. We have also given a detailed overview regarding the various templates used for the synthesis of these CuNCs along with the respective protocols being followed. The various instrumental techniques used to characterize these CuNCs are discussed which provides an in-depth understanding as to how these CuNCs can be properly examined. Finally, we have highlighted some of the most recent applications of these CuNCs which make them unique to serve as the next-generation fluorophores.

This article has previously been published in the journal *Physical Sciences Reviews*. Please cite as: Das, N. K., Mukherjee, S. Size-controlled atomically precise copper nanoclusters: Synthetic protocols, spectroscopic properties and applications. *Physical Sciences Reviews* [Online] **2018**, 3. DOI: 10.1515/psr-2017-0081

https://doi.org/10.1515/9783110636666-005

Graphical Abstract:

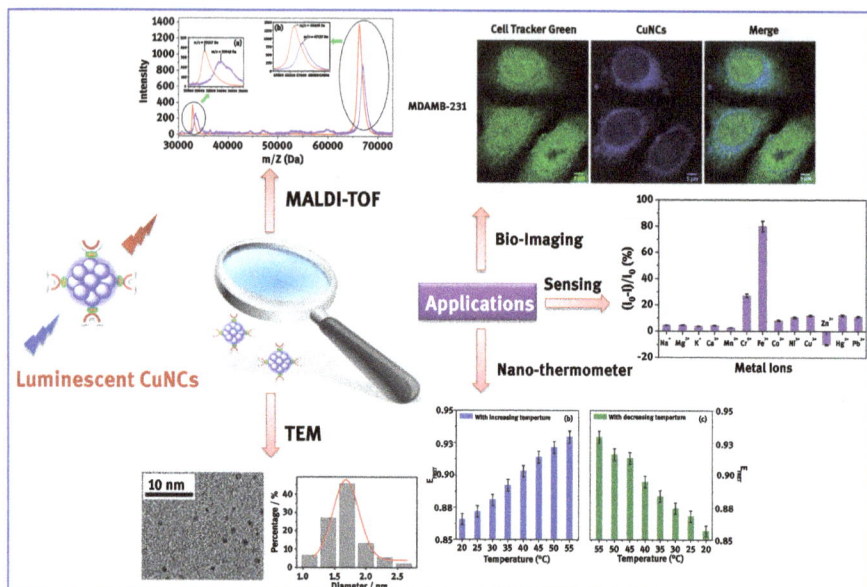

The Graphical Abstract highlights some of the key spectroscopic signatures of the CuNCs and their applications.

Keywords: luminescent nanoclusters, spectroscopic properties, synthetic protocols, templates, instrumentations, sensing, bio-imaging, nano-thermometry

5.1 Introduction

The last few decades really saw an upsurge in the research related to nanomaterials. Broadly, these materials can be classified into three categories, namely large and small nanoparticles (NPs) and nanoclusters (NCs) mainly depending on their size domains [1, 2]. The coinage metals (namely gold, silver and copper) have always been widely explored since ages owing to their metallic surface cluster, usage in ornamental purposes and high monetary values. A large number (~10^6) of atoms/molecules bind together and results in the formation of NPs having a wide size distribution of 1–100 nm [3]. Thus, these NPs have sizes greater than individual atoms/molecules and smaller than bulk materials which thereby results in having properties in between these two classes [3, 4]. When the size (R) of the NPs is in comparison to the wavelength of light used (λ) for excitation, i. e., $R \approx \lambda$, then the optical properties exhibited by these NPs are almost similar to those observed for bulk metals [5]. However, upon decreasing the size of the NPs, i. e., $R < \lambda$, their optical features become a function of their size [5]. Further reduction in size, whereby, the size becomes comparable to the Fermi wavelength of the conduction electrons, leads to the formation of a new class of nanomaterials, conventionally termed as NCs [6, 7]. The optical and chemical properties of these metal NCs are vastly different from those of the NPs as they are composed of a fewer (yet definite) number of atoms in their zero-valent states [8]. These sub-nanometer-sized NCs (typically ~2 nm in diameter) serve as "molecular species" with discrete electronic states that result in strong luminescent properties as compared to the larger sized NPs [9–13]. The metal NCs serve to bridge the gap between the single noble metal atoms which show distinct optical properties and metal NPs exhibiting a typical surface plasmon absorbance [14] as depicted in Figure 5.1. The luminescence properties of the NCs can be attributed to (i) the "quantum size effects" through which the NCs behave like molecules having different and tunable absorbance and emission characteristics depending on the number of metal atoms comprising the NC [15] and (ii) the "surface-ligand effects" through which the emissive properties can be controlled via the ligands surrounding the NCs [16]. The quantum confinement of the free electrons in the case of NCs can be best understood in terms of the free electron model. In 1900, Paul Drude proposed the "Drude Model" in order to explain the electrical and thermal conductivities of bulk metals [17–19]. Drude made the following assumptions: (i) the electronic attractions and repulsions were considered to be negligible between collisions, (ii) the electron–electron scattering was neglected and (iii) the collision frequency of the electrons was

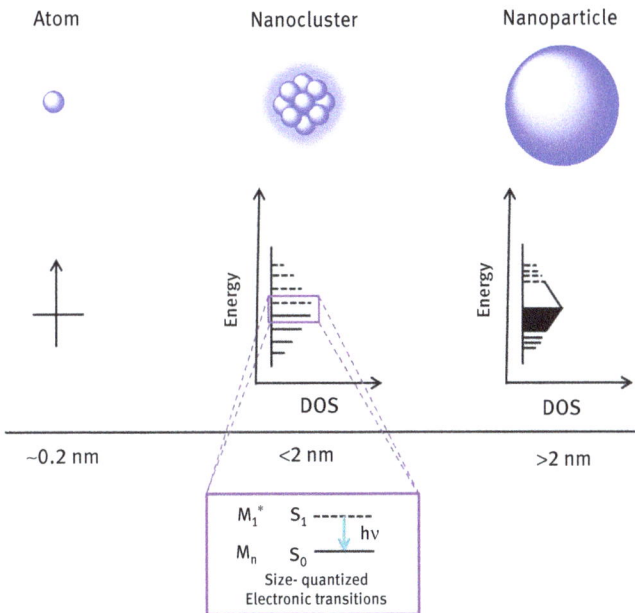

Figure 5.1: Metal NCs are the missing link between isolated metal atoms and plasmonic metal NPs.

considered to be $1/t$; where t is the interval between two successive collisions [18]. Due to strong electron screening effects, the valance electrons of the metal were considered to be free electrons, delocalized within the bulk metal and are not specific to any metal atoms. Based on the "one-particle approximation" and the abovementioned assumptions of the Drude model, the total Hamiltonian of the metal NC can be expressed as the sum of non-interacting Hamiltonians of the free electrons. The one-particle approximation is related to the particle in a box model which is used to explain the behaviour of free electrons in metals. Hence, the energy levels of the metals can be expressed by the following equation:

$$E = \frac{n^2 h^2}{8ml^2}$$
(5.1)

where n is the principle quantum number, E is the energy of a quantized level, l is the size of a metal, m is the electron mass and h is the Plank's constant. This equation is limited to the boundary condition that the l is of the nanoscale dimension as in the bulk state, metals are characterized by periodic crystal structures and thus are devoid of any definite length scales.

As stated earlier, NCs are an aggregate of metal atoms in their zero-valent states. The number of metal atoms comprising a NC varies which consequently results in different spectroscopic features. The atomistic composition of a NC is theoretically estimated by the Jellium model which is a quantum mechanical model comprising of interacting electrons and a positive background charge [19–26]. According to this model, the valance electrons of the metal atoms are considered to be "free electrons" because the electronic attractions and repulsions are neglected, an assumption which is similar to the Drude model. Here, the metal cluster is conceived to behave as a uniform positively charged sphere surrounded by delocalized electronic shells which are filled by the free electrons. These free electrons are contributed from the valance electrons of the individual metal atoms. Unlike the electronic structure of the single atoms, the electron density of the NCs does not depend on the number of free electrons; rather these free electrons are delocalized in different electronic shells in analogy to the single atoms thereby following the Pauli's exclusion principle [21]. Wood and Ashcroft used this concept of the Jellium model in correlating the quantum confinement and optical properties of the NCs and proposed that the atomic composition of the NCs can be estimated by using the following relation [27–30]:

$$h v = E_f / N^{1/3} = E_f r_s / R$$
(5.2)

where v denotes the emission frequency of the NC, E_f denotes the Fermi energy of the concerned metal, R is the radius of the metal NC, N is the number of atoms comprising a NC and r_s is the Wigner–Seitz radius of the metal.

In most of the cases, the formation of NCs with a specific composition is not exclusive, i. e. NCs with slightly different compositions may also be formed but with a very low occurrence. Hence, the Jellium model which is often used to have an estimate of the chemical composition of the NCs is surely an approximation and it only provides us with an estimate. A more accurate technique is perhaps the matrix-assisted laser desorption ionization-time of flight (MALDI-TOF) spectroscopy (which has been discussed in details in Section 5.3.4). If the composition has to be accurately established taking into consideration the HOMO/LUMO band gap energies, then a more detailed calculation of their optimized structural geometries would be required along with a full knowledge of their molecular orbital parameters.

Contrary to the NPs, most of the metal NCs are luminescent as the size of these nanomaterials is comparable to the Femi wavelength of the conduction electrons. Besides being luminescent in nature, these NCs are water soluble, generally non-toxic and highly photo-stable and exhibit large Stokes shift [31–36]. These unique properties make the NCs behave as the next-generation fluorophores having wide and versatile applications in almost all branches of science and technology. Among the several NCs investigated, gold NCs (AuNCs) and silver NCs (AgNCs) are the most commonly explored [37–50]. On the contrary, copper NCs (CuNCs) are far less studied owing to their propensity to undergo oxidation and the limitations in controlling their ultra-fine size which make the CuNCs less stable [51]. Additionally, the quantum yields of CuNCs are far less as compared to Au/Ag counterparts, which limits their usage as conventional fluorophores [34, 35, 52]. However, metallic Cu has more significance in biological studies and it is much cheaper as compared to Au/Ag. As stated before, the CuNCs are very rarely explored in comparison to AgNCs or AuNCs, primarily owing to their small size and propensity to get oxidized (leading to the formation of larger sized nanoparticles as by-products which are mostly non-luminescence in nature). Hence, researchers are always challenged to generate luminescent CuNCs which are stable and can safely be used as a fluorophore starting from cheaper copper salts as compared to the more expensive Ag/Au counterparts [34, 35, 53]. Besides, copper is one of the primary abundant essential trace metals in the human body, which plays frontier roles in various metabolic processes of living organisms [34]. Additionally, owing to the very high conductivity value of metallic copper, CuNCs have potential applications in industries as well. This thus necessitates the optimization of synthetic protocols for stable CuNCs having significant quantum yields besides being non-toxic in nature so that they can be used for studying systems having biological relevance. The pioneering works of Pileni and co-workers [54–56] on copper nanocrystals also provide us with valuable information regarding the structural and spectroscopic properties of copper nanomaterials. Although a few reviews are available on CuNCs [57–59] which highlight the rationale of synthesizing CuNCs and deal with specific synthetic protocols and applications, this book chapter delineates the various methodologies used for the preparation of

CuNCs using different templates highlighting the nuances of the protocols used therein. Additionally, we have provided details regarding the origin of the luminescent properties of CuNCs and how these materials are characterized with special emphasis on their spectroscopic and morphological properties. Finally, we have also highlighted the recent applications of CuNCs, taking into consideration that these CuNCs can serve as the next-generation fluorophores.

5.2 Methods of preparation

In this section, we will highlight the various synthetic approaches used for CuNCs. Broadly, the synthesis of NCs can be divided into two main categories: (i) top-down (NPs to NCs) approach and (ii) bottom-up (atoms to NCs) approach.

5.2.1 Top-down (NPs to NCs) approach

In this methodology, the metal NPs serve as the precursors. The process involves breaking down of a solid substance, and this can be achieved by either dry or wet grinding. The desired metal NCs can be obtained from the larger sized NPs by the etching method. Although this approach is more difficult and cumbersome to be carried out, the fact that there is very less chance of larger sized NPs being formed as side product(s) seems to be the only advantage [60–62]. Hence, the more practiced approach is the bottom-up approach.

5.2.2 Bottom-up (atoms to NCs) approach

This methodology involves the aggregation of smaller sized atoms to a cluster of these atoms to form the stable NCs. Normally, the metal ions when subjected to a chemical reduction results in the formation of zero-valent metal atoms. This process of reduction is followed by an aggregation of the atoms into a cluster having a definite composition [60, 63, 64]. Besides this reduction process, the procedure further calls for a protecting or stabilizing agent which serves as a template in controlling the synthetic aspects which eventually leads to the formation of metal NCs with a definite size and composition.

The various templates used for synthesizing CuNCs are enumerated below.

5.2.2.1 Proteins
The synthesis of NCs involves the reduction of the metal ions to atoms. This is generally done by a reducing agent or sometimes the template used play the dual role; it not only stabilizes the NCs through proper scaffolding but also helps in reducing the ions to their zero-valent states. Of all the templates used, proteins occupy a seminal position. In this section, we describe some of the proteins used as

templates and provide an account of the experimental protocols used therein for the synthesis of CuNCs.

5.2.2.1.1 Bovine serum albumin (BSA)

Wang and co-workers synthesized red-emitting CuNCs (emission peaked at 640 nm when excited at 320 nm) using the circulatory protein BSA as a template for offering stability and hydrazine monohydrate as the reducing agent [65]. BSA has several functional groups like $-NH_2$, $-SH$, $-OH$, etc. which initiates the complexation through electrostatic forces of attraction. Hydrazine monohydrate reduces the Cu^{2+} ions to Cu atoms and the protein matrix initiates the nucleation process and thereby stabilizes the CuNCs so formed [65]. For the synthesis of the CuNCs, 0.5 mL of 0.1 M $CuCl_2$ aqueous solution and 40 mL of BSA (5 mg mL^{-1}) were initially mixed. The reaction mixture was stirred for 10 min at 25 °C. Subsequently, 1 mL of 50 wt% hydrazine monohydrate ($N_2H_4.H_2O$) was added and the reaction mixture was stirred for another 5 h. This resulted in the formation of a deep yellow-coloured solution, indicating the formation of CuNCs. Finally, the mixture was precipitated with acetone and separated through centrifugation, washed with ethanol at 60 °C and redispersed in double-distilled water [65]. In a different report, Zhang and co-workers used the same protocol and synthesized red-emitting (emission peaked at 620 nm) CuNCs [66].

Pal and co-workers reported the synthesis of blue-emitting CuNCs using BSA as a template (Figure 5.2) [67]. These CuNCs had an emission maximum at 410 nm when excited at 325 nm characterized by a quantum yield of ~15 %. Aqueous $CuSO_4$ solution (1 mL, 20 mM) was mixed with BSA solution (5 mL, 15 mg mL^{-1}) at 25 °C under vigorous stirring for 2–3 min.

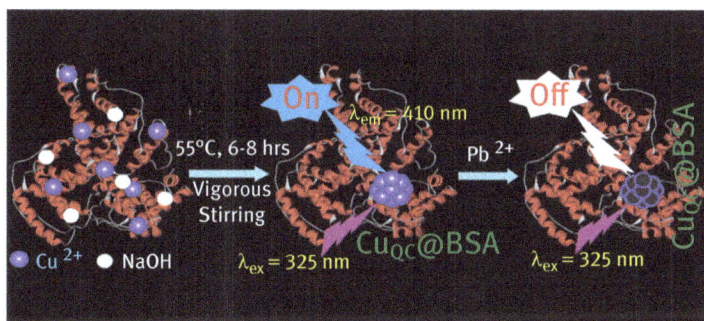

Figure 5.2: Schematic representation shows the formation of BSA-templated CuNCs and subsequent Pb^{2+} ion detection (Reference 67, copyright ACS).

Thereafter, NaOH was added to the reaction mixture so that the pH of the resulting solution reaches ~12. The colour of the solution so formed changes from blue to violet within 5 min. Finally, the mixture was allowed to incubate under mild stirring for another 6–8 h at 55 °C when the colour changes to light brown [67]. Gao et al. also did not use any additional reducing agents to obtain luminescent CuNCs as the BSA-templated NCs were generated *in situ* through reduction by the protein [68]. They also used the same technique as mentioned above. Aqueous $CuSO_4$ solution (1 mL, 20 mM) was mixed with BSA solution (5 mL, 15 mg mL^{-1}) accompanied by vigorous stirring at 25 °C. After 2 min, NaOH solution was added to the reaction mixture so as to attain a final pH of 12. The resulting mixture was incubated at 55 °C for another 8 h which resulted in the formation of CuNCs in solution. Finally, the solution was dialyzed against distilled water using a 3 kDa cut-off dialysis bag for 36 h to separate out the unreacted small molecules [68]. The CuNCs were characterized with an emission peak centred at 407 nm when excited at 330 nm. Pina-Luis and co-workers also used a similar protocol to establish that the red-emitting (emission peaked at 640 nm) CuNCs so formed can detect trace amounts of magniferin [69]. Using a slightly modified protocol, Zhao et al. developed BSA-templated CuNCs which exhibited an emission maximum at 408 nm [70]. In their methodology, 1 mL of 10 mM $CuSO_4$ was mixed initially with 1 mL of 2 mg mL^{-1} of BSA solution. 0.3 mL of 1 M NaOH was added under stirring when the colour gradually changes from blue to violet and finally it became light brown when stirred at 65 °C for 8 h. They further centrifuged the mixture at 16,000 rpm for 15 min and the supernatant was finally dialyzed (by dialysis membrane molecular weight cutoff (MWCO): 3,500 Da) to obtain the CuNCs [70].

Through the usage of hydrogen peroxide (H_2O_2) along with BSA as a template, CuNCs were synthesized which served as Hg^{2+} ion sensors [71]. The synthetic protocol was almost similar to what was used by Pal and co-workers, except for the addition of H_2O_2 which was done after NaOH was added to the reaction mixture. It has been opined that the addition of H_2O_2 accelerates the formation of CuNCs [71]. D-Penicillamine (DPA)–BSA-stabilized CuNCs have been reported by Zaijun and co-workers [72]. 24 mg of DPA was first dissolved in 16 mL of pure water, to which 1 mL of 4 mg mL^{-1} $Cu(NO_3)_2$ was added drop-wise under vigorous stirring. After 90 min, this mixture was filtered using 0.22 nm filter membrane to remove larger particles. The filtrate was dialyzed against deionized water for 24 h. Subsequently, 0.3 mL of 1 mg mL^{-1} BSA solution was added into the DPA-capped CuNCs under agitation. This was then stirred for another 10 min and centrifuged at 4,000 rpm [72]. The unique feature of these CuNCs so formed is that they exhibited dual fluorescence characteristics showing emission peaks centred at 420 nm (corresponding to BSA–CuNCs) and 642 nm (corresponding to DPA–CuNCs) [72]. Gao and co-workers have developed a technique for targeting orthotopic lung tumours using positron emission tomography imaging where they have used CuNCs as the fluorophore [73]. In their synthetic protocol, 1 mL of 20 mM $CuCl_2$ was added to 5 mL of 40 mg mL^{-1} BSA solution which was stirred for 2–3 min. Then, 0.4 mL of 0.5 M NaOH was added drop-wise to adjust the pH to ~12. This resulting mixture was kept at 55 °C for 5–6 h which eventually resulted in the formation of BSA-templated CuNCs [73]. In continuation to this strategy, they developed another CuNCs, where the template was

BSA conjugated with a peptide, luteinizing hormone-releasing hormone (LHRH). For the conjugation, 19.5 mg of LHRH peptide was mixed with 1.5 mg N-hydroxy-succinimide and 10.5 mg of coupling agent, 1-ethyl-3-(3-dimethylaminopropyl)car-bodiimide and dissolved in water. To this mixture, 100 mg of BSA was added, and it was left to incubate overnight at ambient conditions. This resulted in the formation of BSA–LHRH conjugate which was dialyzed against deionized water for 12 h using MWCO: 12,000 dialysis tube. Then, using this purified conjugate, CuNCs were synthesized following the same protocol as mentioned above except the concentration of BSA–LHRH was 42 mg mL^{-1}. However, both the synthesized CuNCs exhibited emission maxima centred at 410 nm [73].

5.2.2.1.2 Human serum albumin (HSA)

Unlike BSA, its analogue HSA is rarely used to serve as a template for CuNCs. Mukherjee and co-workers reported the synthesis of CuNCs using HSA as a template [34]. Initially, 5 mL of 10 mM Cu(NO$_3$)$_2$ aqueous solution was added to 0.75 mM HSA under vigorous stirring. On doing so, the mixture formed a white turbid solution due to the coordination between Cu^{2+} ions and the various electronegative functional groups of HSA. This turbidity gets over upon the drop-wise addition of 1 M NaOH (in order to maintain the pH of the solution ~11) solution whence the colour becomes light violet (Figure 5.3). The reaction mixture was stirred at 40 °C for another 12 h which resulted in the formation of CuNCs having an emission maximum at 414 nm [34].

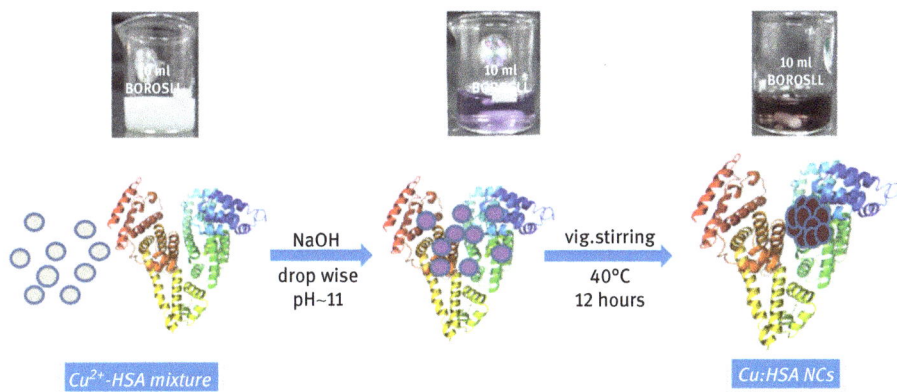

The formation of CuNCs with HSA as a template is triggered by the reduction of the Cu^{2+} ions to Cu atoms which aggregate and eventually leads to the formation of stable Cu/HSA NCs. This reduction is initiated by keeping the pH of the solution ~11, which thereby renders the 18 tyrosine amino acid residues (present inherently within HSA having pK$_a$ of 10.46) to behave as the reductant [34]. The Cu^{2+} ions get stabilized owing to the electrostatic interactions between the several ionic groups of the protein like –OH, –NH$_2$, –COOH, –SH and the bulky scaffolds of HSA. Thus, in this case, the template used serves both as a stabilizing and as a reducing agent, leading to the formation of CuNCs [34].

5.2.2.1.3 Lysozyme

Lysozyme is a small protein (14.3 kDa) having 129 amino acid residues including 8 cysteine residues. Chattopadhay and co-workers have synthesized blue-emitting CuNCs of varied compositions using hydrazine as reducing agent and lysozyme as a template [74]. Initially, CuSO$_4$ and lysozyme were mixed in water in the ratio 2:1 (w/w). The solution mixture was stirred for 10 min at 45 °C followed by the addition of 0.04 mL of 1 M NaOH in order to adjust the pH to ~11. To this, 0.01 mM of 80 % hydrazine was added, and the resultant mixture was again stirred for 6–12 h. They repeated the same procedure using different metal:protein ratios (4:1, 6:1 and 8:1) [74]. A little modified version of this protocol was used by Song and co-workers who also synthesized red-emitting CuNCs for the estimation of glucose [75]. 12.5 mg of lysozyme was mixed with 6.4 mg of CuSO$_4$.5H$_2$O in 5 mL of deionized water under vigorous stirring. Further, 1 mL of hydrazine monohydrate (N$_2$H$_4$.H$_2$O) was added to this mixture which was subsequently heated to 40 °C under stirring for 2 h. 1 M NaOH was added drop-wise until the pH reached a value of ~12. The solution was eventually cooled and the product was collected by precipitating with alcohol and centrifuged at 6,000 rpm [75]. Lysozyme, a protein with 129 amino acid residues, is characterized by the presence of large number of free carboxylic and amino groups besides four disulfide bonds. These groups have enhanced affinity to bind to the metal ions and thereby facilitate the formation of stable CuNCs [75].

5.2.2.1.4 Other proteins and peptides

Miao et al. used the protein Papain as a template in synthesizing red-emitting CuNCs [76]. In a typical synthetic procedure, 0.2 mL of 100 mM CuSO$_4$ solution was added to 4.8 mL of papain containing 20.8 mg mL^{-1} under vigorous stirring. After 10 min, 0.1 mL of 3 M NaOH solution and 0.2 mL diamidhydrate solution were added. The mixture was vortexed at 37 °C for 2 h in a water bath. The resulting aqueous solution was filtered using a 0.22 μm filter membrane to remove the larger products, and finally the CuNCs were collected by dialysis against deionized water using a dialysis membrane (MWCO: 1,000) for 24 h.

Zhao et al. used the protein transferrin as a template for synthesizing red-emitting CuNCs which were used for targeting imaging of transferrin receptor overexpressed cancer cells (Figure 5.4) [77]. Herein, they have mixed 1 mL of 10 mM CuSO$_4$ solution and 1 mL of transferrin solution containing 40 mg mL^{-1} of the protein under the condition of stirring at 25 °C for 5 min. In order to maintain the pH to ~12, 0.1 mL of 1 M NaOH solution was added followed by the injection of 0.5 mL of freshly prepared

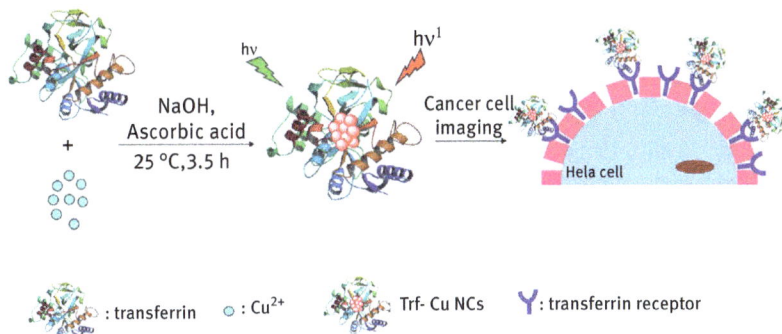

Figure 5.4: Schematic representation of transferrin-protected red-emitting CuNCs as bio-imaging probe for transferrin receptor overexpressed cancer cells (Reference 77, copyright RSC).

ascorbic acid (120 mM). The reaction was carried out under vigorous stirring at 25 °C for almost 4 h when CuNCs were formed [77].

Huang and co-workers used trypsin to prepare CuNCs which served as reversible pH sensors [78]. 8 mL of trypsin solution containing 100 mg of the protein was mixed with an aqueous solution of $CuCl_2$ (2mL, 100 mM) under stirring for 3 min at 25 °C. The mixture was then refluxed for 12 h at 100 °C when the colour of the solution gradually changed from light blue to brownish yellow, indicating the formation of CuNCs [78].

Instead of a protein, Huang et al. used the artificially synthesized peptide (CLEDNN) to prepare blue-emitting CuNCs [79]. 50 µL of 50 mM $CuSO_4$ solution was added to a 5 mL peptide solution (1.5 mg mL^{-1}) under vigorous stirring at 65 °C. After 5 min, 75 µL of 100 mM ascorbic acid solution was slowly added to the reaction mixture and kept for incubation for 6 h at 65 °C. Upon the completion of the reaction, the solution was cooled at room temperature and was further purified by centrifugation at 15,000 rpm and eventually dialyzed against water using a dialyzing membrane (with a cut-off of 1,000 Da) for 2 days [79].

5.2.2.2 Deoxyribonucleic acid (DNA)

Like proteins, DNA has also been widely used as a template for synthesizing metal NCs owing to the architectural features of the DNA that enables to fine tune the emission features of the synthesized NCs (varying from visible to near infrared (NIR)) [80]. Wang and co-workers have synthesized DNA-templated CuNCs for the identification of single-nucleotide polymorphism [81]. The CuNCs were prepared by reducing 100 µM $Cu(NO_3)_2$ with 1 mM ascorbic acid in the presence of DNA duplex. The stock solutions of the probe DNA and the target DNA were diluted to a specific concentration with 20 mM 3-(N-morpholino)propanesulfonic acid (MOPS) buffer (the buffer was prepared by mixing 20 mM MOPS, 300 mM NaCl at pH = 7.5), and these were mixed in the molar ratio 1:1. The subsequent mixture was heated at 90 °C for 10 min, and then it was slowly cooled down to room temperature (20 °C) in 3 h. Then, 5 µL of 100 mM ascorbic acid was added into 250 µL of the DNA duplex solutions. To this, 250 µL of 200 µM Cu(NO$_3$)$_2$ solution was added and kept for incubation for 15 min at room temperature that resulted in the formation of the CuNCs [81].

Dual-metal NCs comprising of Cu and Ag atoms using DNA template has been reported [82]. Reduction of AgNO$_3$ and Cu(NO$_3$)$_2$ by sodium borohydride (NaBH$_4$) and using the DNA template (5 -CCCTTAATCCC-3) resulted in the formation of bimetallic Cu/Ag NCs composed of 1 Cu and 2 Ag atoms. For the preparation of DNA-Cu/Ag NCs, initially AgNO$_3$ was added to the DNA solution in the molar ratio of 4.5:1. After that, NaBH$_4$ solution was added to bring in the reduction process. The mixture was incubated for 30 min under ice bath and then Cu(NO$_3$)$_2$ solution was added in the molar ratio of 1:5:1 (Cu^{2+}:DNA). This leads to the formation of bimetallic Cu/Ag NCs [82]. DNA–RNA heteroduplexes have also been used for the synthesis of CuNCs for estimating miRNA (Figure 5.5) [83]. For the preparation of the DNA–miRNA heteroduplexes-templated CuNCs, initially the DNA and the target miRNA were mixed in equivalent amounts (2 µL of 100 µM each) in 196 µL hybridization buffer (10 mM PBS buffer and 0.1 M NaCl at pH = 7.4). To ensure the formation of the heteroduplexes, the mixture was

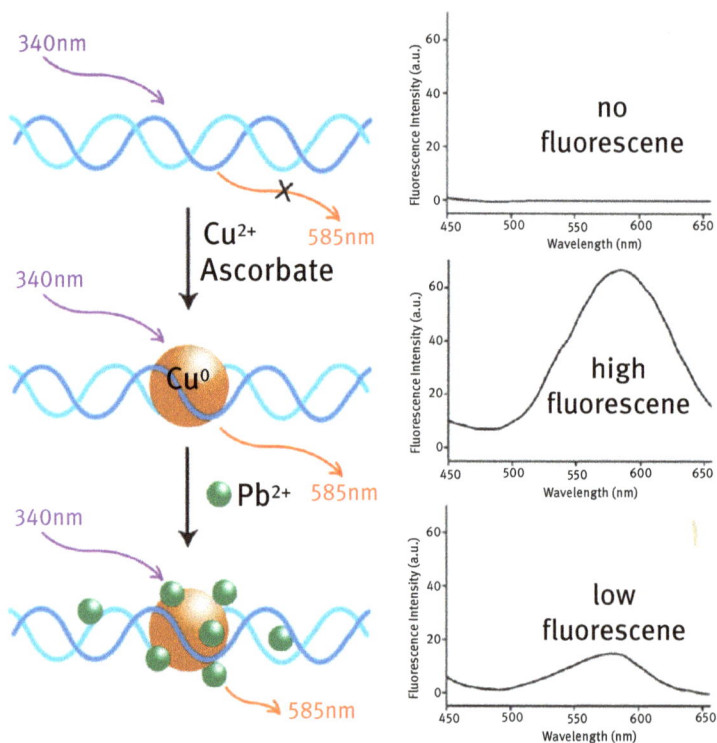

Figure 5.5: Schematic representation exhibits label-free Pb^{2+} ion detection by using heteroduplex-templated luminescent CuNCs (Reference 83, copyright RSC).

heated at 90 °C for 5 min and then slowly cooled down to room temperature. Then, 4 μL of 100 mM ascorbic acid was slowly added and kept for incubation at room temperature for 15 min. Finally, 200 μM Cu^{2+} ions in 200 μL formation buffer (20 mM MOPS, 300 mM NaCl and 2 mM $MgCl_2$ at pH = 7.5) was added and allowed to react in the dark for 20 min. This resulted in the formation of the desired CuNCs [83].

5.2.2.3 Thiols

Among several templates used for the synthesis of stable NCs, thiols are considered to be important owing to their small size and tailorable surface properties which provides the flexibility of fine tuning the spectral properties of the NCs so synthesized [84–92]. In one of the early works on CuNCs using thiols as a template, Chen and co-workers synthesized sub-nanometer-sized CuNCs using 2-mercapto-5-*n*-propylpyrimidine (MPP) as the stabilizing agent. For the synthesis, 0.025 g of $Cu(NO_3)_2$ were co-dissolved with 0.136 g of tetra-*n*-octylammonium bromide in 25 mL absolute alcohol under stirring at 80 °C for 30 min [85]. This solution was then subjected to cooling using ice-cold water. Under inert atmosphere (using a stream of Ar gas), 0.0766 g of MPP was added to this solution with vigorous stirring for another 6 h. Further, 0.047 g of $NaBH_4$ in 5 mL ethanol was added to this mixture, and the resulting solution was kept undisturbed for 7 h. Finally, it was centrifuged and washed repeatedly with ethanol and was vacuum dried, and the desired CuNCs were obtained by redispersing in toluene [85]. In a very systematic investigation of the effect of isomeric substitutions of thiol groups of mercaptobenzoic acid (MBA), Chang and co-workers studied the fluorescent properties of three CuNCs [86]. For synthesizing each CuNC, 300 μL of 0.1 M of thiosalicylic acid (TA) or 3-MBA or 4-MBA in Dimethylformamide (DMF) was mixed with 600 μL of Tetrahydrofuran (THF) under vigorous stirring. 100 μL aliquots of 0.1 M $Cu(NO_3)_2$ solution in 0.1 M HNO_3 were added to each of the mixtures. The mixtures were shaken vigorously for 5 s when precipitates were formed in all the mixtures and the reaction was continued for 30 min at 70 °C in the absence of light [86]. The resulting mixtures were all cooled to 25 °C when the TA–CuNCs became transparent, while the 3-MBA and 4-MBA-CuNCs were turbid in nature. Interestingly, these three CuNCs (templated by isomers of MBA) were characterized by different emission properties (the emission maxima for the TA–CuNCs, 3-MBA-CuNCs and 4-MBA-CuNCs were 420 nm, 668 nm and 646 nm, respectively) [86]. Recently, Mukherjee and co-workers have synthesized tripeptide glutathione (GSH)-capped CuNCs and demonstrated them as nuclear membrane markers for cancerous cells (HeLa, MDAMB-231 and A549) [35]. Additionally, these CuNCs can selectively and quantitatively detect Fe^{3+} ions in solution and in human haemoglobin [35]. For the synthesis of the CuNCs, 2 mL of 5 mM $Cu(NO_3)_2$ was mixed with 2 mL of 5 mM GSH solution in water under vigorous stirring followed by the addition of 30 μL of 1 M NaOH solution for reaching a pH ~10. This mixture was incubated at 40 °C for almost 14 h under stirring when light green solution of CuNCs was obtained (Figure 5.6) [35]. The formation of the CuNCs has been ascribed to the

Figure 5.6: Schematic representation exhibits the formation of GSH-templated CuNCs along with their corresponding photographic images (Reference 35, copyright ACS).

metal–ligand chelation process where the Cu^{2+} ions coordinate and subsequently get stabilized by the thiol and carboxylate groups of GSH (here the template again plays the dual role of a reductant and stabilizing agent). Additionally, since the pK_a of –SH group of GSH is ~9.35, the reduction process gets augmented as the pH of the solution is kept at 10 [35].

5.2.2.4 Polymers

Luminescent polymeric based films find applications in sensors, solar concentrators and light-emitting diodes [93–97]. Polymers have also been used as a conventional template in developing NCs having varied applications. Very recently, CuNCs have been fabricated using a polymeric mixture of polyvinylpyrrolidone (PVP) and poly-vinyl alcohol (PVA) and GSH in aqueous medium [98]. For the *in situ* fabrication of the CuNCs thin films, 6 mL aqueous solution of PVA (which was prepared *a priori* by dissolving 60 mg mL^{-1} of PVA in distilled water at 80 °C) was mixed with 1 mL of PVP solution (having 60 mg mL^{-1} of PVP) under stirring. To this, 2.5 mL of 0.1 M GSH was added followed by 0.5 mL of 0.1 M $Cu(NO_3)_2$ solution. For obtaining thin films, this resulting mixture was dropped on a clean glass surface which was vacuumed at 40 °C for 10 h [98]. Pellegrino and co-workers have developed a method for synthesizing highly fluorescent CuNCs using a living polymer chain which served a dual purpose of a stabilizing agent and a reductant [99]. For achieving the synthesis of the CuNCs, initially the living amphiphilic polymer, poly(ethylene glycol)-block-poly(propylene sulfide) (MPEG-b-PPS) was synthesized following the ring-opening polymerization mechanism. 0.2 mmol of this polymer in 20 mL THF was transferred into a flask containing CuBr (0.45 mol, 64 mg) which served as a metal precursor. Under nitrogen atmosphere, this mixture was stirred at room temperature for 30 min whereby the colourless solution turned pale yellow within 5 min which resulted in the formation of CuNCs [99]. Highly photo-stable red-emitting CuNCs have been prepared using the polymer PVP and dihydrolipoic acid as templates [100]. Luo and co-workers prepared polyethyleneimine (PEI)-capped CuNCs which displayed solvent effect [101]. 50 μL of

PEI (containing 0.094 g mL^{-1}) was added to 95 µL of water and stirred for 2 min. To this polymeric solution, 25 µL of 0.1 M CuSO$_4$ solution was added and stirred for another 2 min followed by the addition of 30 µL of hydrazine hydrate which serves as the reducing agent. The final resulting mixture was heated in a water bath at 95 °C for 12 h which resulted in the formation of CuNCs [101].

5.3 Characterization methodologies and instrumentations involved

The NCs are ultra-small nano-entities and mostly their luminescent properties are utilized from the applications point of view. The following techniques are mainly used for characterizing NCs.

5.3.1 Steady-state spectroscopy

Unlike the bigger nanomaterials, these NCs having sizes typically less than 2 nm (approaching the corresponding Fermi wavelength) do not exhibit a characteristic surface plasmon resonance band in the UV-visible region [102, 103]. For CuNCs, most of the time, the absorption spectrum is featureless, and hence drawing any conclusions from the absorption studies may not be that scientific and meaningful. Thus, most often, steady-state excitation and emission spectroscopy are used instead. The excitation spectroscopy is an alternate methodology of ascertaining the absorption profiles of a molecule. Hence, the solution of the NC is scanned in the spectrofluorimeter, varying the excitation wavelengths to obtain the specific wavelength at which the emission is a maximum. By doing so, the corresponding wavelength for maximum absorption of the NC is thereby ascertained. For emission studies, the NC solution is subsequently excited at this wavelength in order to get a maximum Quantum Yield (QY). Figure 5.7 displays the spectroscopic properties of a HSA-templated luminescence CuNCs [34].

The emission profiles/characteristics of the CuNCs depend on their atomistic composition which in turn controls the size [34, 35]. Hence, CuNCs having different compositions will display different emission wavelengths. The difference in composition is primarily due to the templates and also depends on the other experimental conditions used for the synthesis [34, 35].

Besides QY, the emission maxima of the CuNCs are also used for a very good estimation of their atomistic composition using the Jellium model as described in the section "Introduction" and eq. (2) [34, 35]. In general, an emission maximum in the red region corresponds to a larger sized NC as compared to a NC which emits at lower wavelengths. In analogy to the statement above, Mukherjee and co-workers have explored the effect of the enzyme trypsin on the blue-emitting AgNCs protected by has [104]. The HSA-protected AgNCs having a core composition of 9 Ag atoms when subjected to tryptic digestion resulted in the formation of a new meta-NC (AgTp NCs)

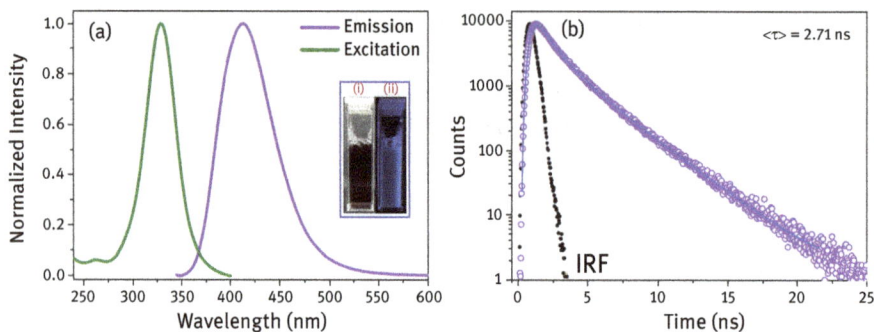

Figure 5.7: (a) The normalized PL excitation and emission spectra of the Cu:HSA NCs as marked in the figure. Inset shows the photographic images of the Cu:HSA NCs under (i) visible light and (ii) UV light. (b) Representative PL lifetime decay profile (in logarithmic scale) of the Cu:HSA NCs (Reference 34, copyright ACS).

composed of 28 Ag atoms as estimated using the Jellium model [104]. The AgNCs were characterized by an emission maximum of ~460 nm which undergoes a drastic red shift (centred at ~690 nm) once the AgTp NCs were formed [104]. Hence, emission spectroscopy can be used to unravel these phenomenon of change in composition of the NCs.

The stability of the CuNCs can also be ascertained by emission spectroscopy [34, 35, 74]. Synthesizing stable CuNCs is also one of the targets as otherwise these NCs cannot be used as fluorescent probes. The emission signature of these CuNCs should be appreciable for a considerable time in order to use them for imaging purposes. The photoluminescent intensities are monitored as a function of time, and interestingly, many CuNCs have been reported to be more photo-robust than some of the conventional fluorophores used in imaging/spectroscopy (Figure 5.8) [34, 35, 74].

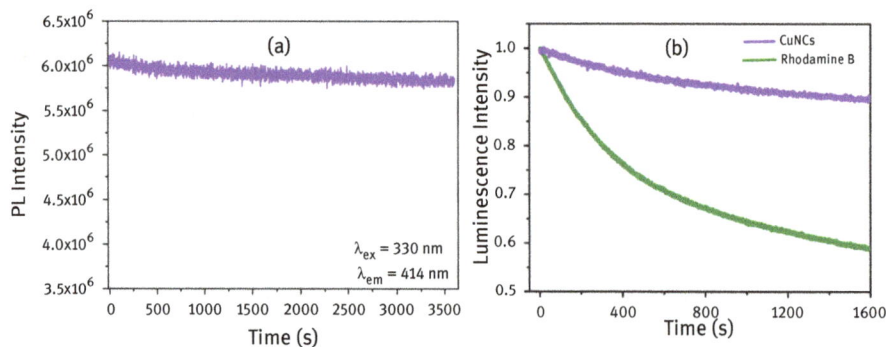

Figure 5.8: (a) Luminescence spectra of the HSA-templated CuNCs as a function of time (Reference 34, copyright ACS). (b) Simultaneous luminescence spectra of the GSH-protected CuNCs and rhodamine B as a function of time as marked in the figure (Reference 35, copyright ACS).

5.3.2 Excited-state lifetime measurements

The excited-state lifetime measurements are usually carried out using the technique of time-correlated single-photon counting [105, 106]. The lifetime decay transients are fitted using the following equation [74, 105, 106]:

$$I(t) = \sum a_i \exp(-t/\tau_i) \tag{5.3}$$

where a_i are the corresponding amplitudes of the decay constants having a lifetime, τ_i. The average lifetime of the metal NCs is estimated by the following equation [105, 106]:

$$\langle \tau \rangle = \frac{\sum_i a_i \tau_i^2}{\sum_i a_i \tau_i} \tag{5.4}$$

Mukherjee and co-workers have reported a lifetime of ~2.72 ns for CuNCs using HSA and GSH as templates (Figure 5.9) [34, 35]. However, longer lifetimes of CuNCs have also been reported when the protein lysozyme is used as a template [74]. Pal and co-workers have obtained much shorter lifetimes (~0.38 ns) for BSA-templated CuNCs [67]. Thus, depending on the nature of the template, the synthetic protocol used and the composition of the CuNCs, the lifetime varies.

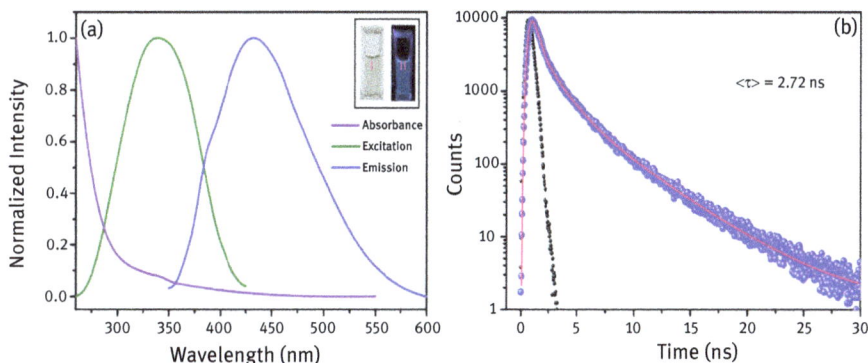

Figure 5.9: (a) The normalized absorption, excitation and emission spectra of the GSH-protected CuNCs as marked in the figure. Inset shows the photographic images of the CuNCs under (I) visible light and (II) UV light. (b) Representative lifetime decay profile (in logarithmic scale) of the CuNCs (Reference 35, copyright ACS).

5.3.3 Transmission electron microscopy

Transmission electron microscopy (TEM) is perhaps the most used technique to char-acterize the surface morphology of the metal NCs. From the TEM images, the existence of ultra-small metal NCs can be ascertained with specific information regarding their size and shape. Generally, for obtaining the TEM images, carbon-coated Cu grids (300 mesh) are used as a substrate for sample preparation. The CuNCs are first diluted to a nanomolar concentration and are drop-casted on the Cu grids. Subsequently, these are initially air dried and then kept under vacuum drying for almost 24 h prior to scanning. Although a characteristic size of ~2–3 nm is obtained from the TEM images along with metallic lattice fringes [34, 35] as seen from the Figure 5.10 (which comply very well with that of literature values), drawing a final conclusion from such a measurement may not always be that scientific. More often, the sample preparation of TEM results in some aggregation and that erroneously leads to an overestimation of the size of the CuNCs. Using high-resolution TEM (HR-TEM), Pal and co-workers reported the existence of CuNCs templated by BSA having a diameter of ~2.8 ± 0.5 nm [67]. HSA-templated CuNCs have also been reported to have a typical size ~3 nm [34]. Feng and co-workers [79] have reported the size of the peptide-templated CuNCs to be ~1.7 ± 0.4 nm which is in excellent agreement to those obtained from DNA-hosted CuNCs [81].

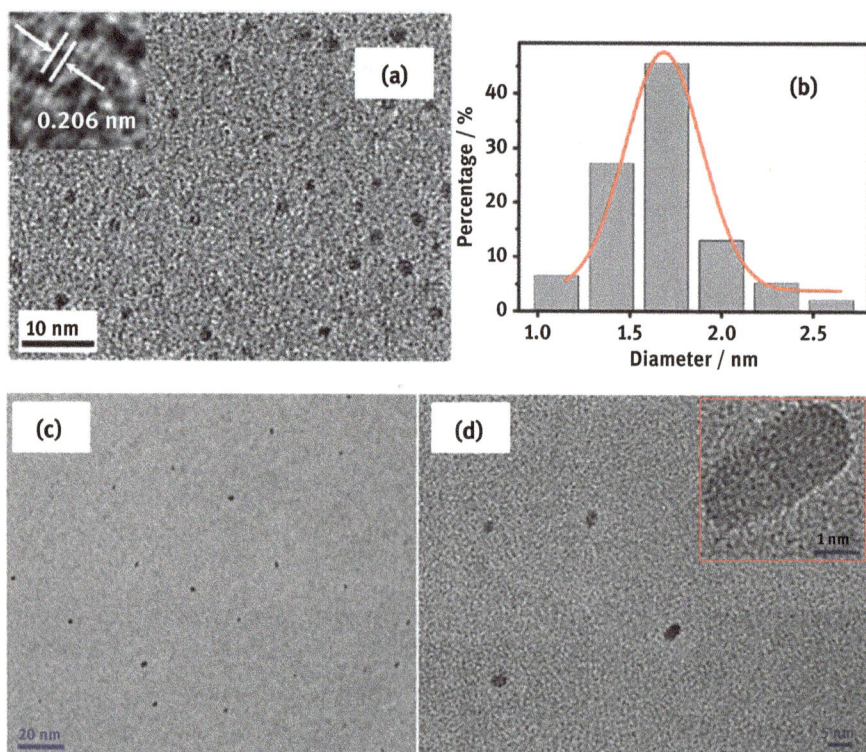

Figure 5.10: (a) TEM image of peptide-templated CuNCs. Inset represents associated HR-TEM image. (b) The corresponding size distribution histogram (Reference 79, copyright RSC). (c) TEM image of the HSA-templated CuNCs. (d) HR-TEM image of the CuNCs; the inset shows the image of a single NC exhibiting lattice fringes (Reference 34, copyright ACS).

5.3.4 Matrix-associated laser desorption ionization–time of flight

Another advanced technique for characterizing these small-sized NCs is mass spectrometry [8, 107]. In a seminal work, Schaaff et al. characterized thiol-protected AuNCs by laser desorption ionization (LDI) technique [108]. Later, Dass and co-workers successfully assigned the precise chemical composition of $Au_{25}SR_{18}$ clusters using trans-2-[3-(4-tert-butylphenyl)-2-methyl-2-propenyldidene] malononitrile as a matrix probed by mass spectrometry (MALDI-TOF-MS) technique [109]. They also found that, apart from matrix, laser pulse intensity and ion modes (positive or negative) also play major roles in MALDI-TOF-MS measurements [109]. MALDI is an ionization technique that uses a laser energy-absorbing matrix which creates ions from large molecules (generally biomolecules) with minimal fragmentation, and hence such a technique is also considered to be "soft" in nature [110]. Recently, MALDI-TOF mass spectrometry has become an indispensable tool in the characterization of metal (Au, Ag and Cu) NCs [111]. Generally, by the choice of a good matrix and by lowering the laser pulse intensity, prominent intact ions of NCs with different m/z peaks can be generated with minimal fragmentation [111]. Using MALDI-TOF analysis, Das et al. have ascertained that the most probable chemical composition of a GSH-protected CuNCs was $Cu_{15}GSH_4$. The molecular weight of GSH and atomic weight of Cu are 307.32 and 63.5, respectively. Hence, the most prominent base peak centred at m/z = 2,186 can be correlated to the chemical formula of $[Cu_{15}GSH_4 + 4\ H^+]$ [35]. Comparing the differences between the two base peaks of the MALDI-TOF data for HSA (m/z = 66,410 Da) and Cu/HSA NCs (m/z = 67,137 Da), Mukherjee and co-workers rationalized that the most stable atomic composition of the blue-emitting HSA-templated CuNCs comprised of 12 Cu atoms (Figure 5.11) [34].

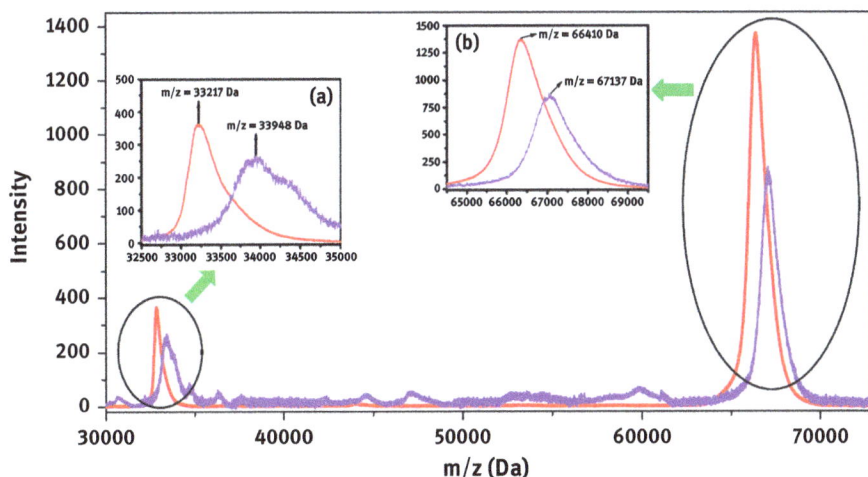

Figure 5.11: MALDI-TOF spectra of HSA (red) and HSA-templated CuNCs (violet) under same experimental conditions. Inset (a) represents the lower order mass peaks (m/z) due to doubly charged species and (b) shows the higher order mass peaks due to singly charged species (Reference 34, copyright ACS).

Using MALDI-TOF analysis, Chattopadhyay and co-workers have been characterized polydispersed lysozyme-coated CuNCs having different atomic compositions such as Cu_2, Cu_4 and Cu_9 NCs [74]. Pal and co-workers have reported the atomic compositions of BSA-templated CuNCs to be composed of 5 and 13 Cu atoms [67]. Thus, MALDI-TOF experiments are vastly used to ascertain the chemical composition of one NC with excellent resolution.

5.4 Critical safety considerations

Generally, the CuNCs are non-toxic in nature. However, as stated earlier, the synthetic protocols for generating CuNCs involved repeated iterations and fine tuning as the CuNCs so formed are often unstable. The challenge is to obtain stable CuNCs which do not undergo oxidation and also avoid generating larger sized nanoparticles as side products. Although the process of synthesis is not that rigorous, the following safety precautions should be taken:

i.   The purity of all chemicals used should be of the highest order, else the resultant CuNCs may not be having the best of stability and quantum yield.
ii.   It is advisable that all synthesis is carried out inside fume hoods.
iii. Proper safety measures such as laboratory coats, safety goggles, gloves etc. should be used.
iv.   The temperature and pH mentioned in the synthetic protocols should be strictly maintained in order to achieve the desired CuNCs having a definite composition.
v.   Special attention should be paid whenever organic solvents are used as a part of the synthesis. It is strongly advised that such solvents should not have any contact with skin in order to avoid chemical injuries.

5.5 Applications of the CuNCs

As stated earlier that the biocompatible metal NCs possess some superior optical properties in terms of high photo-stability, strong luminescence and large Stokes shift [28, 33]. Further, these NCs are characterized by having low toxicity which enables them to play frontier roles in biological and analytical applications [34, 35]. Recently, luminescent CuNCs expand their suitable applicability in various disciplines of science and technology, and some of them are enumerated below [28, 33–35, 59, 112–114].

5.5.1 Sensors

Detecting important chemical compounds have paramount significances in biomedical and analytical sciences [115, 116]. Hence, developing highly sensitive and biocompatible sensors is of utmost importance in chemistry, biology and material sciences [117, 118]. The non-toxic and biocompatible nature of the CuNCs as compared to the organic molecule/dyes enables them serve as potential metal ions sensors with high efficacy [35, 74, 119].

5.5.1.1 Metal ion sensors

Luminescent CuNCs have been used as potential probes for the detection of biologically relevant ions, including Hg^{2+}, Pb^{2+}, Cu^{2+}, Fe^{3+}, Cr^{6+} etc. Kalita and co-workers have developed CuNCs stabilized by citric acid and CTAB and further used them as a luminescence probe to detect Hg^{2+} based on the luminescence quenching of the CuNCs. They have reported the limit of detection (LOD) as 1 nM [119]. Recently, Xiaoqing et al. have developed BSA-templated CuNCs which exhibits strong luminescence [71]. The luminescence of CuNCs gets significantly quenched in the presence of Hg^{2+} ions through the generation of CuNC aggregations due to the formation of Hg-S covalent bond and a part destruction of Cu-S bonds in CuNCs by Hg^{2+} ions. The estimated LOD of Hg^{2+} ions was found to be 4.7 pM [71]. In 2011, Goswami et al. selectively detected Pb^{2+} ions with ppm level sensitivity [67]. They opined that the detection mechanism is based on the luminescence quenching of the CuNCs in the presence of Pb^{2+} ions due to the aggregation of CuNCs [67]. Using ds-DNA hosted CuNCs, Chen et al. also detected Pb^{2+} ions with LOD of 72 nM [83]. Li et al. developed DPA-protected red-emitting CuNCs [120]. These CuNCs were further used to detect Cu^{2+} ions with a LOD of 0.3 ppm [120]. Song and co-workers reported that the luminescence of the BSA-templated CuNCs can be quenched in the presence of Cu^{2+} ions through the formation of the BSA–CuNC aggregates due to the paramagnetic interactions of Cu^{2+} ions [121]. The estimated LOD for Cu^{2+} ions was found to be 1 nM [121]. Cao et al. synthesized luminescent CuNCs using tannic acid as a capping agent and applied them in selective Fe^{3+} ion sensing in blood serum and live cells using fluorescence imagining techniques [122]. They have reported that the estimated LOD of Fe^{3+} ions to be ~10 nM [122]. Chen and co-workers synthesized branched polyethyleneimine (BPEI)-capped CuNCs in the presence of ascorbic acid [123]. These synthesized luminescent CuNCs serve as selective Fe^{3+} ions sensor with a LOD of 340 nM. They have further ascribed this Fe^{3+} ion sensing behaviour to detect Fe^{3+} ions in human urine samples [123]. Recently, Mukherjee and co-workers synthesized GSH-stabilized CuNCs [35]. Besides high photo-stability and biocompatibility, these CuNCs serve as potential Fe^{3+} ion sensors in solution as well as human haemoglobin (Figure 5.12) [35]. The estimated LOD was found to be 25 nM.

An electron-transfer mechanism has been proposed from the surface of CuNCs to the Fe^{3+} ions, which further leads to the "turn-off" luminescence characteristics of CuNCs [35]. The outer electronic configuration of Fe^{3+} is $3d^5 4s^0$. Since the d-orbitals are half-filled, the Fe^{3+} ions gain a strong positive charge density which induces a strong electron-withdrawing character as compared to other metal ions. The addition of Fe^{3+} results in the electron transfer from the surface of the GSH-encapsulated CuNCs, thereby destabilizing the core of the NCs and consequently a "turn off"

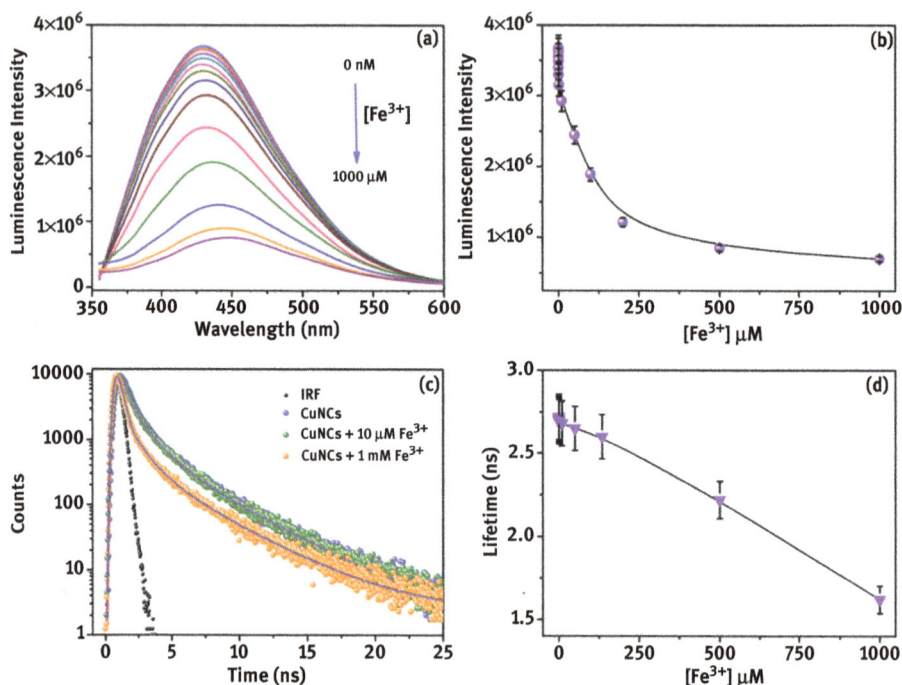

Figure 5.12: (a) Luminescence spectra of the GSH-stabilized CuNCs in the presence of increasing concentrations of Fe^{3+} ions. (b) The variation in luminescence intensity of CuNCs in the presence Fe^{3+} ions. (c) Representative luminescence lifetime decay profiles of CuNCs in the presence of increasing concentrations of Fe^{3+} ions as marked in the figure. (d) Variation in the average luminescence lifetimes of CuNCs in the presence of different concentrations of Fe^{3+} ions (Reference 35, copyright ACS).

behaviour in the luminescence is observed [35]. Cui et al. have developed cysteine-protected green-emitting CuNCs [124]. These CuNCs exhibit the potential applicability to detect Cr^{6+} ions with a LOD of 43 nM.

5.5.1.2 Anion sensors

Researchers have also focused the effect anions (CN^-, I^-, $(PO_4)^{3-}$, S^{2-} etc.) in the luminescent properties of CuNCs [59]. In a seminal contribution, Chang and co-workers have synthesized CuNCs using MBA and its isomers as reducing and capping agents [86]. The TA (one of the isomers of MBA)-protected CuNCs serve as selective CN^- ion sensors with a LOD od 5 nM (Figure 5.13). This estimated LOD is ~540 times lower than the maximum contaminant level of CN^- in drinking water permitted by the World Health Organization (WHO) [86]. Iodide (I^-) ion was detected by PEI-protected CuNCs with the LOD of 100 nM [125]. This method was further utilized to detect I^- ion in urine samples. Li et al. reported "turn-on" luminescence signal of cysteine-stabilized CuNCs in the presence of S^{2-} ions [126]. The estimated LOD was found to be 42 nM [126].

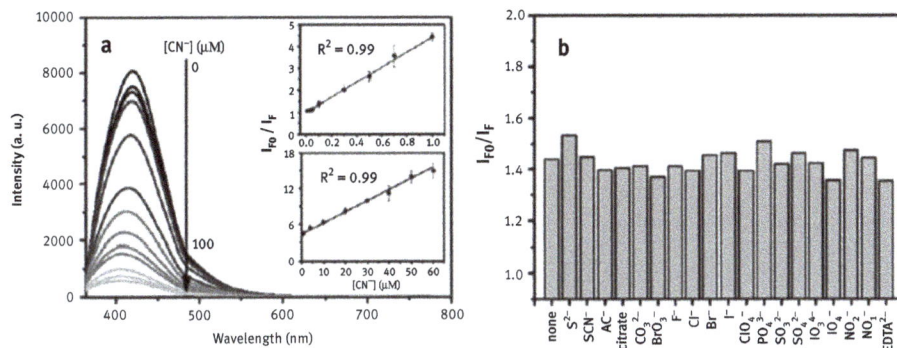

Figure 5.13: Detection of CN^- in standard solutions without (a) and with (b) containing interfering species using TA–CuNC aggregates). Insets to (a): plots of luminescence intensity ratio (I_F/I_{F0}) vs. CN^- concentration over the ranges of 0–1.0 mM (top) and 1.0–60.0 μM (bottom) (Reference 86, copyright RSC).

5.5.1.3 pH Sensors

The pH-sensing applicability using luminescent CuNCs was first shown by the Zhang and co-workers [66]. The BSA-templated CuNCs exhibit two times increment of the luminescence intensity when the pH was varied from 12 to 6. They also demonstrated this photo-switching property is reversible in nature [66]. Wang et al. also reported reversible pH responsive by GSH-stabilized CuNCs. These NCs show reversible pH responding behaviour in the pH range of 4 to 9 [127]. A reduction in pH value leads to an aggregation of the CuNCs due to the non-covalent molecular interactions, thereby leading to an enhanced emission [127]. Similarly, Liao et al. developed BSA-templated CuNCs for sensitive and pH sensing in the range of 2–14 [128].

5.5.1.4 Bio-sensors

Analogous to metal ion sensing, luminescent metal NCs are also used in the sensing of biological molecules as well [129]. The generated H_2O_2 may induce the aggregation of CuNCs, resulting in the quenching of luminescence and subsequently the levels of glucose can be detected. This method was well demonstrated by Jia et al. who synthesized DPA-mediated luminescent CuNCs exhibiting aggregation-induced emission in the NIR region [84]. These luminescent CuNCs showed excellent bio-sensing prospects by the detection of H_2O_2, an intermediate species in several biological processes. The detection was exemplified by the glucose oxidase-stimulated oxidation of glucose through the enzyme-catalysed generation of H_2O_2, which quenched the luminescence of the NCs [84]. Jia et al. developed DNA-templated CuNCs for identification of single-nucleotide polymorphisms [81]. The luminescence of the CuNCs was significantly enhanced by almost threefold when synthesized with mismatched DNA as compared to fully matched DNA [81]. Wang et al. designed a ds-DNA-hosted luminescent CuNCs for selective detection of miRNAs, which was based on the formation of new CuNCs

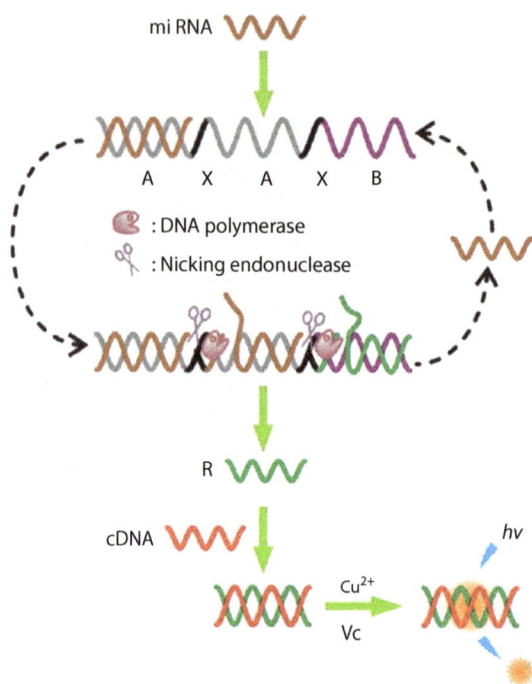

Figure 5.14: Schematic illustration of miRNA detection using a novel luminescence probe of ds-DNA-templated CuNCs as the signal output via the target triggered isothermal exponential amplification reaction (Reference 130, copyright RSC).

derived from the DNA duplex templates induced by hybridization between DNA and target miRNA (Figure 5.14) [130].

The resulting new CuNCs exhibited enhanced luminescence and the LOD was found to be 1 pM [130]. Zhang et al. developed a simple and sensitive label-free strategy to detect $3'-5'$ exonuclease activity of exonuclease III (Exo III) using ds-DNA-templated luminescent CuNCs [131]. In the presence of the Exo III, the luminescence of CuNCs decreased drastically due to the cleavage of DNA duplexes into fragments owing to the high exodeoxyribonuclease activity of Exo III. The LOD was estimated to be 0.02 U mL^{-1} for Exo III [131].

5.5.2 Bio-imaging

Highly stable metal NCs possess promising optical properties like enhanced photo-stability and significant quantum yield [60, 74]. In addition, biocompatible and non-toxic nature of these NCs make them excellent bio-imaging probes for both *in vitro* and *in vivo* systems [35, 60, 74]. Hence, these non-toxic and relatively less explored CuNCs are highly beneficial for cell imaging and can thus serve as bio-labels for exploring intercellular dynamics. Recently, Chattopadhyay and co-workers synthesized highly luminescent blue-emitting CuNCs using lysozyme as the template which act as a potential bio-imaging probe for imaging cervical cancer cells (HeLa) [74]. Mukherjee and co-workers have performed a cell-viability study and subsequent uptake assays of blue-emitting GSH-capped CuNCs on three different cancerous cell lines, like HeLa (malignant immortal cell line derived from cervical cancer), A549 (human lung carcinoma) and MDAMB-231 (human breast adenocarcinoma) [35]. Corresponding cell-viability study suggests that the synthesized CuNCs are not detrimental for the normal cell growth morphology of these cancerous cells. Further, cell-imaging study (Figure 5.15) reveals that these non-toxic CuNCs specifically localize in the nuclear membrane of the different cancerous cells [35]. The presence of GSH-recognizing motifs on the nuclear membrane may lead to the selective accumulation of the CuNCs [35].

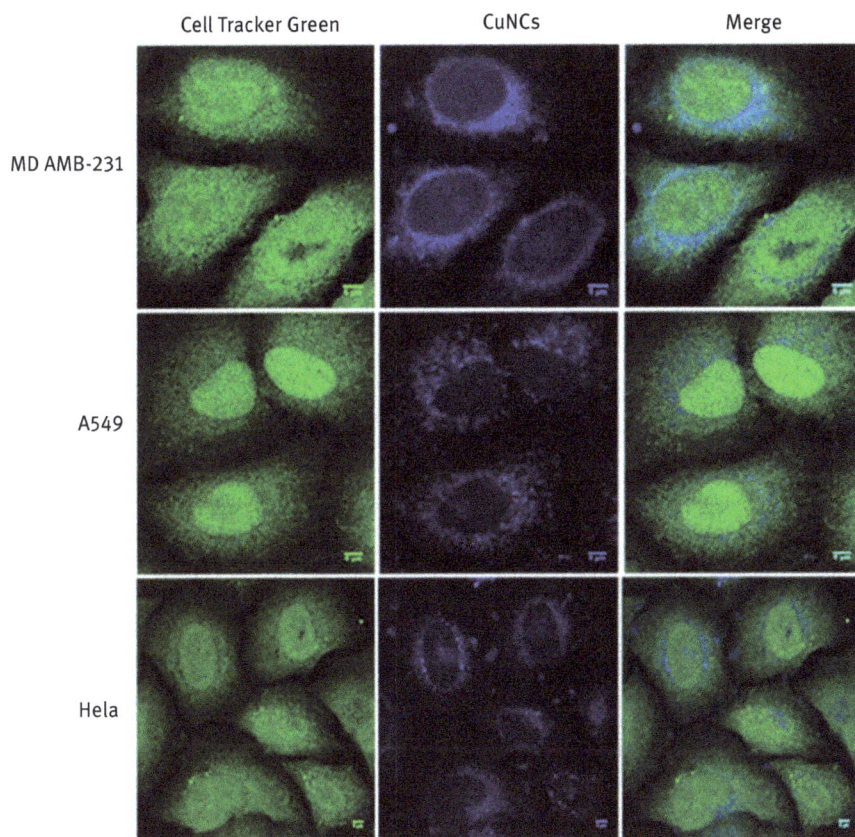

Figure 5.15: Subcellular localization of the GSH-protected CuNCs: HeLa, MDAMB-231 and A549 cells were incubated with CuNCs for 12 h at 37°C. The cells were labelled using Cell Tracker Green CMFDA (5-chloromethylfluorescein diacetate) before fixing in 4 % paraformaldehyde. Images from the fixed cells were acquired confocal microscope. The scale bar of the images represents 5 μM in spatial size (Reference 35, copyright ACS).

Cao et al. synthesized luminescent CuNCs using tannic acid as a capping agent and applied them as selective Fe^{3+} ion sensor in blood serum and live cells using fluorescence imagining techniques [122]. Bhamore et al. reported multicolour imaging (blue when excited at 405 nm and green when excited at 488 nm) in *Bacillus subtilis* cells, based on the excitation dependent emission properties of egg white-templated CuNCs [132].

5.5.3 Nano-thermometry

Mukherjee and co-workers have used HSA-templated CuNCs to serve as a förster resonance energy transfer (FRET) pair with a compatible fluorescence dye, Coumarin-153 [34]. Using temperature-dependent reversible FRET assay, they have established the concept of "nano-thermometry" for detection of temperature of the systems having biological significances. They have investigated the thermal stability of the CuNCs and observed the

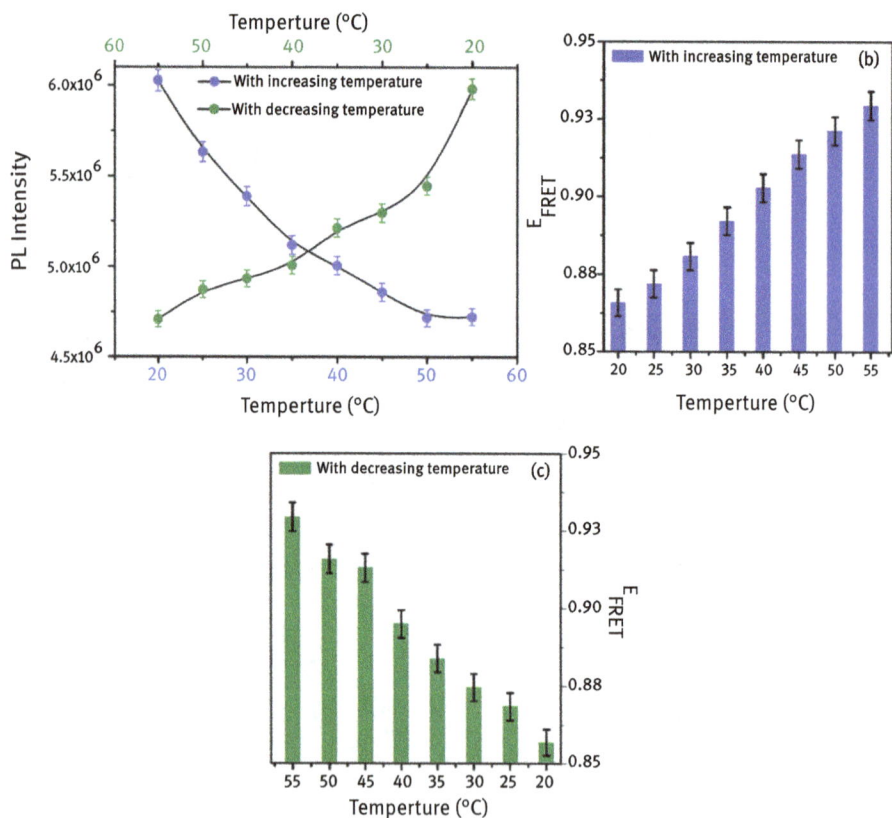

Figure 5.16: (a) The reversible variation of fluorescence intensities of the HSA-templated CuNCs as a function of temperature. The histograms display the variation of FRET efficiencies with (b) increasing temperature and (c) decreasing temperature (Reference 34, copyright ACS).

gradual luminescence quenching upon increasing the temperature within the temperature domain from 25 to 55°C (Figure 5.16).

As HSA gets unfolded upon increasing temperature, the donor–acceptor distance increases, resulting in a decrease in FRET efficiency [34]. Interestingly, they observed that the FRET efficiency as function of temperature is reversible in nature. This concept can be utilized for the detection the temperature in the field of nanoelectronics, nanophotonics and biomedicine [34]. The method of nano-thermometry was further exemplified by Shi et al. using BSA-templated luminescent CuNCs [133]. They have monitored the thermal response of the CuNCs as a function of temperature which has also been proved to be reversible in nature [133].

5.6 Conclusions and future perspective

As seen from this chapter, CuNCs can indeed be considered as the next-generation fluorophores. Although these CuNCs are very rarely explored (as compared to AgNCs and AuNCs) primarily because of their lesser stability, these CuNCs if prepared by proper synthetic protocols can have varied applications spanning almost all branches of science and technology. A rational use of templates (which can serve as a dual role of a stabilizing and reducing agent) can lead to the formation of stable CuNCs defined by a specific atomic composition.

The CuNCs are mostly non-toxic in nature and hence can serve as a substitute to the conventional bio-markers used to study cells in *in vivo* conditions. It has been demonstrated that these NCs can be used in detecting cancerous cells. Hence, as a future prospect along the same lines, CuNCs loaded with anti-cancer drugs can be used as specific drug-delivery vehicles. Thus, there is a plethora of new avenues which can be explored and investigated using these versatile luminescent probes.

Acknowledgements: The authors sincerely thank the research facilities of IISER Bhopal and the Central Instrumentation Facility (CIF), IISER Bhopal. NKD thanks UGC, Govt. of India for research fellowship. SM thanks Indian National Science Academy (Grant No. INSA/CHM/2015045) for providing financial support. SM also thanks the past and present members of the group with a special mention of Dr Subhadip Ghosh and Dr Uttam Anand for their valuable contributions in the works related to the NCs.

References

[1] Kreibig U, Vollmer M. Optical properties of metal clusters. Berlin, Germany: Springer, 1995.
[2] Bhattacharyya K, Mukherjee S. Fluorescent metal nano-clusters as next generation fluorescent probes for cell imaging and drug delivery. Bull Chem Soc Jpn. 2018 3 15;91:447–54. DOI: 10.1246/bcsj.20170377.
[3] Ghosh SK, Pal T. Interparticle coupling effect on the surface plasmon resonance of gold nanoparticles: from theory to applications. Chem Rev. 2007;107:4797–862.

[4]   Schmid G. Clusters and colloids-From theory to applications. Weinheim, Germany: VCH, 1994.

[5]   Zheng J, Nicovich PR, Dickson RM. Highly fluorescent noble-metal quantum dots. Annu Rev Phys Chem. 2007;58:409–13.

[6]   Haberland H. Clusters of atoms and molecules. Berlin, Germany: Springer, 1994.

[7]   Schaaff TG, Knight G, Shafigullin MN, Borkman RF, Whetten RL. Isolation and selected properties of a 10.4 kDa gold:glutathione cluster compound. J Phys Chem B. 1998;102:10643–6.

[8]   Chakraborty I, Pradeep T. Atomically precise clusters of noble metals: emerging link between atoms and nanoparticles. Chem Rev. 2017;117:8208–71.

[9]   Wallace WT, Whetten RL. Coadsorption of CO and O_2 on selected gold clusters: evidence for efficient room-temperature CO_2 generation. J Am Chem Soc. 2002;124:7499–505.

[10]   Campbell CT, Parker SC, Starr DE. The effect of size-dependent nanoparticle energetics on catalyst sintering. Science. 2002;298:811–4.

[11]   Link S, Beeby A, FitzGerald S, Sayed MAE, Schaaff TG, Whetten RL. Visible to infrared luminescence from a 28-atom gold cluster. J Phys Chem B. 2002;106:3410–5.

[12]   Peyser LA, Lee TH, Dickson RM. Mechanism of Ag_n nanocluster photoproduction from silver oxide films. J Phys Chem B. 2002;106:7725–8.

[13]   Félix C, Sieber C, Harbich W, Buttet J, Rabin I, Schulze W, et al. Ag_8 fluorescence in argon. Phys Rev Lett. 2001;86:2992.

[14]   Fedrigo S, Harbich W, Buttet J. Optical response of Ag_2, Ag_3, Au_2, and Au_3 in argon matrices. J Chem Phys. 1993;99:5712–7.

[15]   Ishida Y, Corpuz RD, Yonezawa T. Matrix sputtering method: A novel physical approach for photoluminescent noble metal nanocluster. Acc Chem Res. 2017;50:2986–95.

[16]   Zheng J, Chen Z, Yu M, Liu J. Different sized luminescent gold nanoparticles. Nanoscale. 2012;4:4073–83.

[17]   Johnson PB, Christy RW. Optical constants of the noble metals. Phys Rev. 1972;B6:4370–9.

[18]   March A. Electron theory of metals. Ann Phys. 1916;354:710–24.

[19]   Khandelwal P, Poddar P. Fluorescent metal quantum clusters: an updated overview of the synthesis, properties, and biological applications. J Mater Chem B. 2017;5:9055–84.

[20]   Zheng J, Zhang CW, Dickson RM. Highly fluorescent, water-soluble, size tunable gold quantum dots. Phys Review Lett. 2004;93:077402.

[21]   Johnston RL. Atomic and molecular clusters. London: Taylor & Francis, 2002.

[22]   Knight WD, Clemenger K, De Heer WA, Saunders WA, Chou MY, Cohen ML. Electronic shell structure and abundances of sodium clusters. Phys Review Lett. 1984;52:2141.

[23]   De Heer WA, Selby K, Kresin V, Masui J, Vollmer M, Chatelain A, et al. Collective dipole oscillations in small sodium clusters. Phys Review Lett. 1987;59:1805.

[24]   De Heer WA. The physics of simple metal clusters: experimental aspects and simple models. Rev Mod Phys. 1993;65:611–76.

[25]   Kiejna A, Wojciechowski KF. Metal surface electron physics. Pergamon, United Kingdom 1996.

[26]   Lang N, Kohn W. Theory of metal surfaces: charge density and surface energy. Phys Rev B. 1970;1:4555–68.

[27]   Ashcroft NW, Wood DM. Quantum size effects in the optical properties of small metallic particles. Phys Rev B. 1982;25:6255–74.

[28]   Chattoraj S, Bhattacharyya K. Fluorescent gold nanocluster inside a live breast cell: etching and higher uptake in cancer cell. J Phys Chem C. 2014;118:22339–46.

[29]   Chattoraj S, Amin MA, Bhattacharyya K. Cytochrome C-capped fluorescent gold nanoclusters: imaging of live cells and delivery of cytochrome C. ChemPhysChem. 2016;17:2088–95.

[30]   Chattoraj S, Amin MA, Jana B, Mohapatra S, Ghosh S, Bhattacharyya K. Selective killing of breast cancer cells by Doxorubicin-loaded fluorescent gold nanoclusters: confocal microscopy and FRET. ChemPhysChem. 2016;17:253–9.

[31] Fedrigo S, Harbich W, Buttet J. Recent development in deciphering the structure of luminescent silver nanodots. J Chem Phys. 1993;99:5712–17.

[32] Lin C-AJ, Lee C-H, Hsieh J-T, Wang -H-H, Li JK, Shen J-L, et al. Synthesis of fluorescent metallic nanoclusters toward biomedical application: recent progress and present challenges. J Med Biol Eng. 2009;29:276–83.

[33] Anand U, Ghosh S, Mukherjee S. Toggling between blue- and red-emitting fluorescent silver nanoclusters. J Phys Chem Lett. 2012;3:3605–9.

[34] Ghosh S, Das NK, Anand U, Mukherjee S. Photostable copper nanoclusters: compatible Förster resonance energy-transfer assays and a nanothermometer. J Phys Chem Lett. 2015;6:1293–8.

[35] Das NK, Ghosh S, Priya A, Datta S, Mukherjee S. Luminescent copper nanoclusters as a specific cell-imaging probe and a selective metal ion sensor. J Phys Chem C. 2015;119:24657–64.

[36] Ghosh S, Anand U, Mukherjee S. Luminescent silver nanoclusters acting as a label-free photoswitch in metal ion sensing. Anal Chem. 2014;86:3188–94.

[37] Ghoshal A, Goswami U, Sahoo AK, Chattopadhyay A, Ghosh SS. Targeting Wnt canonical signaling by recombinant sFRP1 bound luminescent Au-Nanocluster embedded nanoparticles in cancer theranostics. ACS Biomater Sci Eng. 2015;1:1256–66.

[38] Leung FC-M, Tam A-Y-Y, Au VK-M, Li M-J, Yam V-W-W. Förster resonance energy transfer studies of luminescent gold nanoparticles functionalized with ruthenium(II) and rhenium(I) complexes: modulation via esterase hydrolysis. ACS Appl Mater Interfaces. 2014;6:6644–53.

[39] Su Y, Qi L, Mubc X, Wang M. A fluorescent probe for sensing ferric ions in bean sprouts based on L-histidine-stabilized gold nanoclusters. Anal Methods. 2015;7:684–9.

[40] Yao Q, Yuan X, Fung V, Yu Y, Leong DT, Jiang D, et al. Understanding seed-mediated growth of gold nanoclusters at molecular level. Nat Commun. 2017;8:927.

[41] Fabris L, Antonello S, Armelao L, Donkers RL, Polo F, Toniolo C, et al. Gold nanoclusters protected by conformationally constrained peptides. J Am Chem Soc. 2005, 2006;128:326–36.

[42] Kundu S, Maheshwari V, Saraf RF. Photolytic metallization of Au nanoclusters and electrically conducting micrometer long nanostructures on a DNA scaffold. Langmuir. 2008;24:551–5.

[43] Lin C-AJ, Yang T-Y, Lee C-H, Sherry HH, Sperling RA, Zanella M, et al. Synthesis, characterization, and bioconjugation of fluorescent gold nanoclusters toward biological labeling applications. ACS Nano. 2009;3:395–401.

[44] Shang L, Brandholt S, Stockmar F, Trouillet V, Bruns M, Nienhaus GU. Effect of protein adsorption on the fluorescence of ultrasmall gold nanoclusters. Small. 2012;8:661–5.

[45] Liu Y-Q, Zhang M, Yin B-C, Ye B-C. Attomolar ultrasensitive microRNA detection by DNA-scaffolded silver-nanocluster probe based on isothermal amplification. Anal Chem. 2012;84:5165–9.

[46] Liu G, Feng D-Q, Chen T, Li D, Zheng W. DNA-templated formation of silver nanoclusters as a novel light-scattering sensor for label-free copper ions detection. J Material Chem. 2012;22:20885–8.

[47] Ghoshal A, Goswami U, Raza A, Chattopadhyay A, Ghosh SS. Recombinant sFRP4 bound chitosan-alginate composite nanoparticles embedded with silver nanoclusters for Wnt/b-catenin targeting in cancer theranostics. RSC Adv. 2016;6:85763–72.

[48] Sarkar S, Chakraborty I, Panwar MK, Pradeep T. Isolation and tandem mass spectrometric identification of a stable monolayer protected silver-palladium alloy cluster. J Phys Chem Lett. 2014;5:3757–62.

[49] Shang L, Dörlich RM, Trouillet V, Bruns M, Nienhaus GU. Ultrasmall fluorescent silver nanoclusters: protein adsorption and its effects on cellular responses. Nano Res. 2012;5:531–42.

[50] Zheng K, Setyawati M, Lim T-P, Leong DT, Xie J. Antimicrobial cluster bombs: silver nanoclusters packed with daptomycin. ACS Nano. 2016;10:7934–42.

[51]   Kawasaki H, Kosaka Y, Myoujin Y, Narushima T, Yonezawa T, Arakawa R. Microwave-assisted polyol synthesis of copper nanocrystals without using additional protective agents. Chem Commun. 2011;47:7740–2.

[52]   Das NK, Chakraborty S, Mukherjee M, Mukherjee S. Enhanced luminescent properties of photo-stable copper nanoclusters through formation of "Protein-Corona"-like assemblies. 2018: DOI: 10.1002/cphc.201800332.

[53]   Guo Y, Cao F, Lei X, Mang L, Cheng S, Song J. Fluorescent copper nanoparticles: recent advances in synthesis and applications for sensing metal ions. Nanoscale. 2016;8:4852–63.

[54]   Salzemann C, Lisiecki I, Brioude A, Urban J, Pileni M-P. Collections of copper nanocrystals characterized by different sizes and shapes: optical response of these nanoobjects. J Phys Chem B. 2004;108:13242–8.

[55]   Salzemann C, Brioude A, Pileni M-P. Tuning of copper nanocrystals optical properties with their shapes. J Phys Chem B. 2006;110:7208–12.

[56]   Salzemann C, Lisiecki I, Urban J, Pileni M-P. Anisotropic copper nanocrystals synthesized in a supersaturated medium: nanocrystal growth. Langmuir. 2004;20:11772–7.

[57]   Liu X, Astruc D. Atomically precise copper nanoclusters and their applications. Co-Ordination Chem Rev. 2018;359:112–26.

[58]   Wang Z, Chen B, Rogach AL. Synthesis, optical properties and applications of light-emitting copper nanoclusters. Nanoscale Horizons. 2017;2:135–46.

[59]   Hu X, Liu T, Zhuang Y, Wang W, Li Y, Fan W, et al. Recent advances in the analytical applications of copper nanoclusters. Trends Anal Chem. 2016;77:66–75.

[60]   Qu X, Li Y, Li L, Wang Y, Liang J, Liang J. Fluorescent gold nanoclusters: synthesis and recent biological application. J Nanomater. 2015;2015:784097.

[61]   Schaaff TG, Whetten RL. Controlled etching of Au: sRcluster compounds. J Phys Chem B. 1999;103:9394–6.

[62]   Luo Z, Yuan X, Yu Y, Zhang Q, Leong DT, Lee JY, et al. From aggregation-induced emission of Au (I)-thiolate complexes to ultrabright Au(0)@Au(I)-thiolate core-shell nanoclusters. J Am Chem Soc. 2012;134:16662–70.

[63]   Rao TUB, Pradeep T. Luminescent Ag₇ and Ag₈ clusters by interfacial synthesis. Angewantde Chemie Int Edition. 2010;49:3925–9.

[64]   Jupally VR, Dass A. Synthesis of Au₁₃₀(SR)₅₀ and Au₁₃₀₋ₓAgₓ(SR)₅₀ nanomolecules through core size conversion of larger metal clusters. Phys Chem Chem Phys. 2014;16:10473–9.

[65]   Wang B, Gui R, Jin H, He W, Wang Z. Red-emitting BSA-stabilized copper nanoclusters acted as a sensitive probe for fluorescence sensing and visual imaging detection of rutin. Talanta. 2018;178:1006–10.

[66]   Wang C, Wang C, Xu L, Cheng H, Lin Q, Zhang C. Protein-directed synthesis of pH-responsive red fluorescent copper nanoclusters and their applications in cellular imaging and catalysis. Nanoscale. 2014;6:1775–81.

[67]   Goswami N, Giri A, Bootharaju MS, Xavier PL, Pradeep T, Pal SK. Copper quantum clusters in protein matrix: potential sensor of Pb²⁺ ion. Anal Chem. 2011;83:9676–80.

[68]   Gao Z, Su R, Qi W, Wang L, He Z. Copper nanocluster-based fluorescent sensors for sensitive and selective detection of kojic acid in food stuff. Sensors Actuators B. 2014;195:359–64.

[69]   Mu͂noz-Bustos C, Tirado-Guízar A, Paraguay-Delgado F, Pina-Luis G. Copper nanoclusters-coated BSA as a novel fluorescence sensor for sensitive and selective detection of mangiferin. Sensors Actuators B. 2017;244:922–7.

[70]   Zhao M, Chen A-Y, Huang D, Zhuo Y, Chai Y-Q, Yuan R. Cu nanoclusters: novel electrochemi-luminescence emitters for bioanalysis. Anal Chem. 2016;88:11527–32.

[71] Xiaoqing L, Ruiyi L, Zaijun L, Xiulan S, Zhouping W, Junkangc L. Fast synthesis of copper nanoclusters through the use of hydrogen peroxide additive and their application for the fluorescence detection of Hg^{2+} in water samples. New J Chem. 2015;39:5240–8.

[72] Ruiyi L, Huiying W, Xiaoyan Z, Xiaoqing L, Xiulanb S, Zaijun L. D-Penicillamine and bovine serum albumin co-stabilized copper nanoclusters with remarkably enhanced fluorescence intensity and photostability for ultrasensitive detection of Ag^+. New J Chem. 2016;40:732–9.

[73] Gao F, Cai P, Yang W, Xue J, Gao L, Liu R, et al. Ultrasmall [^{64}Cu]Cu nanoclusters for targeting orthotopic lung tumors using accurate positron emission tomography imaging. ACS Nano. 2015;9:4976–86.

[74] Ghosh R, Sahoo AK, Ghosh SS, Paul A, Chattopadhyay A. Blue-emitting copper nanoclusters synthesized in the presence of lysozyme as candidates for cell labelling. ACS Appl Mater Interfaces. 2014;6:3822–8.

[75] Wang C, Shu S, Yao Y, Song Q. A fluorescent biosensor of lysozyme-stabilized copper nanoclusters for the selective detection of glucose. RSC Adv. 2015;5:101599–606.

[76] Miao H, Zhong D, Zhoub Z, Yang X. Papain-templated Cu nanoclusters: assaying and exhibiting dramatic antibacterial activity cooperating with H_2O_2. Nanoscale. 2015;7:19066–72.

[77] Zhao T, He X-Y, Li W-Y, Zhangab Y-K. Transferrin-directed preparation of red-emitting copper nanoclusters for targeted imaging of transferrin receptor over-expressed cancer cells. J Mater Chem B. 2015;3:2388–94.

[78] Wang W, Leng F, Zhan L, Chang Y, Yang XX, Lana J, et al. One-step prepared fluorescent copper nanoclusters for reversible pH-sensing. Analyst. 2014;139:2990–3.

[79] Huang H, Li H, Wang A-J, Zhong S-X, Fang K-M, Feng JJ. Green synthesis of peptide-templated fluorescent copper nanoclusters for temperature sensing and cellular imaging. Analyst. 2014;139:6536–41.

[80] Petty JT, Zheng J, Hud NV, Dickson RM. DNA-templated Ag nanocluster formation. J Am Chem Soc. 2004;126:5207–12.

[81] Jia X, Li J, Han L, Ren J, Yang X, Wang E. DNA-hosted copper nanoclusters for fluorescent identification of single nucleotide polymorphisms. ACS Nano. 2012;6:3311–7.

[82] Lan G-Y, Chen W-Y, Chang H-T. Characterization and application to the detection of single-stranded DNA binding protein of fluorescent DNA-templated copper/silver nanoclusters. Analyst. 2011;136:3623–8.

[83] Chen J, Liu J, Fang Z, Zeng L. Random dsDNA-templated formation of copper nanoparticles as novel fluorescence probes for label-free lead ions detection. Chem Commun. 2012;48:1057–9.

[84] Jia X, Yang X, Li J, Li D, Wang D. Stable Cu nanoclusters: from an aggregation induced emission mechanism to biosensing and catalytic applications. Chem Commun. 2014;50:237–9.

[85] Wei W, Lu Y, Chen W, Chen S. One-pot synthesis, photoluminescence, and electrocatalytic properties of subnanometer-sized copper clusters. J Am Chem Soc. 2011;133:2060–3.

[86] Lin Y-J, Chen P-C, Yuan Z, Ma JY, Chang H-T. The isomeric effect of mercaptobenzoic acids on the preparation and fluorescence properties of copper nanoclusters. Chem Commun. 2015;51:11983–6.

[87] Hu X, Mao X, Zhang X, Huang Y. One-step synthesis of orange fluorescent copper nanoclusters for sensitive and selective sensing of Al^{3+} ions in food samples. Sensors Actuators B. 2017;247:312–8.

[88] Huang Y, Liu W, Feng H, Ye Y, Tang C, Ao H, et al. Luminescent nanoswitch based on organic-phase copper nanoclusters for sensitive detection of trace amount of water in organic solvents. Anal Chem. 2016;88:7429–34.

[89] Jia X, Li Z, Wang E. Cu Nanoclusters with aggregation induced emission enhancement. Small. 2013;9:3873–9.

[90] Huang H, Li H, Feng -J-J, Feng H, Wang A-J, Qian Z. One-pot green synthesis of highly fluorescent glutathione-stabilized copper nanoclusters for Fe^{3+} sensing. Sensors Actuators B. 2016;241:292–7.

[91] Ye Y, Dong X, Jiang H, Wang X. An intracellular temperature nano probe based on biosynthesized fluorescent copper nanoclusters. J Mater Chem B. 2017;5:691–6.

[92] Zhou T, Yao Q, Zhao T, Chen X. One-pot synthesis of fluorescent DHLA-stabilized Cu nanoclusters for the determination of H_2O_2. Talanta. 2015;141:80–5.

[93] Guan W, Zhou W, Lu J, Lu C. Luminescent films for chemo and biosensing. Chem Soc Rev. 2015;44:6981–7009.

[94] Cao DH, Stoumpos CC, Farha OK, Hupp JT, Kanatzidis MG. 2D homologous perovskites as light-absorbing materials for solar cell applications. J Am Chem Soc. 2015;137:7843–50.

[95] Hedley GJ, Ruseckas A, Samuel IDW. Light harvesting for organic photovoltaics. Chem Rev. 2017;117:796–837.

[96] Xiao Z, Kerner RA, Zhao L, Tran NL, Lee KM, Koh T-W, et al. Efficient perovskite light-emitting diodes featuring nanometre-sized crystallites. Nat Photonics. 2017;11:108–15.

[97] Zhao L, Yeh Y-W, Tran NL, Wu F, Xiao Z, Kerner RA, et al. In situ preparation of metal halide perovskite nanocrystal thin films for improved light-emitting devices. ACS Nano. 2017;11:3957–64.

[98] Wang Z, Stephen Y, Kershaw V, Chen B, Yang X, Goswami N, et al. In situ fabrication of flexible, thermally stable, large-area, strongly luminescent copper nanocluster/polymer composite films. Chem Mater. 2017;29:10206–11.

[99] Barthel MJ, Angeloni I, Petrelli A, Avellini T, Scarpellini A, Bertoni G, et al. Synthesis of highly fluorescent copper clusters using living polymer chains as combined reducing agents and ligands. ACS Nano. 2015;9:11886–97.

[100] Ghosh R, Goswami U, Ghosh SS, Paul A, Chattopadhyay A. Synergistic anticancer activity of fluorescent copper nanoclusters and cisplatin delivered through a hydrogel nanocarrier. ACS Appl Mater Interfaces. 2015;7:209–22.

[101] Ling Y, Wu JJ, Gao ZF, Li NB, Luo HQ. Enhanced emission of polyethyleneimine-coated copper nanoclusters and their solvent effect. J Phys Chem C. 2015;119:27173–7.

[102] Jin R. Atomically precise metal nanoclusters: stable sizes and optical properties. Nanoscale. 2015;7:1549–65.

[103] Chen P-C, Periasamy AP, Harroun SG, Wu W-P, Chang H-T. Photoluminescence sensing systems based on copper, gold and silver nanomaterials. Coord Chem Rev. 2016;320-321:129–38.

[104] Ghosh S, Anand U, Mukherjee S. Kinetic aspects of enzyme-mediated evolution of highly luminescent meta silver nanoclusters. J Phys Chem C. 2015;119:10776–84.

[105] O' Connor DV, Phillips D. Time correlated single photon counting. New York: Academic Press, 1994.

[106] Lakowicz JR. Principles of fluorescence spectroscopy. USA: Springer, 2006.

[107] Jin R, Zeng C, Zhou M, Chen Y. Atomically precise colloidal metal nanoclusters and nanoparticles: fundamentals and opportunities. Chem Rev. 2016;116:10346–413.

[108] Schaaff TG, Shafigullin MN, Khoury JT, Vezmar I, Whetten RL, Cullen WG, et al. Isolation of smaller nanocrystal au molecules: robust quantum effects in optical spectra. J Phys Chem B. 1997;101:7885–91.

[109] Dass A, Stevenson A, Dubay GR, Tracy JB, Murray RW. Nanoparticle MALDI-TOF mass spectrometry without fragmentation: $au_{25}(SCH_2CH_2Ph)_{18}$ and mixed monolayer $Au_{25}(SCH_2CH_2Ph)_{18-x}(L)_x$. J Am Chem Soc. 2008;130:5940–6.

[110] Franz H, Michael K, Ronald CB, Brian TC. Matrix-assisted laser desorption/ionization mass spectrometry of biopolymers. Anal Chem. 1991;63:1193–203.

[111] Lu Y, Chen W. Application of mass spectrometry in the synthesis and characterization of metal nanoclusters. Anal Chem. 2015;87:10659–67.

[112] Goswami U, Dutta A, Raza A, Kandimalla R, Kalita S, Ghosh SS, et al. Transferrin-copper nanocluster-doxorubicin nanoparticles as targeted theranostic cancer nanodrug. ACS Appl Mater Interfaces. 2018;10:3282–94.

[113] Kong L, Chu X, Wang C, Zhou H, Wu Y, Liu L. D-Penicillamine-coated Cu/Ag alloy nanocluster superstructures: aggregation-induced emission and tunable photoluminescence from red to orange. Nanoscale. 2018;10:1631–40.

[114] Cook AW, Jones ZR, Wu G, Scott SL, Hayton TW. An organometallic Cu_{20} nanocluster: synthesis, characterization, immobilization on silica, and "click" chemistry. J Am Chem Soc. 2018;140:394–400.

[115] Diamond D. Principles of chemical and biological sensors. New York, USA: John Wiley & Sons, 1998.

[116] Sapsford KE, Bradburne C, Detehanty JB, Medintz IL. Sensors for detecting biological agent. Mater Today. 2008;11:38–49.

[117] De M, Ghosh PS, Rotello VM. Applications of nanoparticles in biology. Adv Mater. 2008;20:4225–41.

[118] Thomas SW, Joly GD, Swager TM. Chemical sensors based on amplifying fluorescent conjugated polymers. Chem Rev. 2007;107:1339–86.

[119] Ghosh SK, Rahman DS, Ali AL, Kalita A. Surface plasmon tunability and emission sensitivity of ultrasmall fluorescent copper nanoclusters. Plasmonics. 2013;8:1457–68.

[120] Li D, Li B, Yang SI. A selective fluorescence turn-on sensing system for evaluation of Cu^{2+} polluted water based on ultra-fast formation of fluorescent copper nanoclusters. Anal Methods. 2015;7:2278–82.

[121] Zhong YP, Zhu JJ, Wang QP, He Y, Ge YL, Song CW. Copper nanoclusters coated with bovine serum albumin as a regenerable fluorescent probe for copper (II) ion. Microchimica Acta. 2015;182:909–15.

[122] Cao H, Chen Z, Zheng H, Huang Y. Copper nanoclusters as a highly sensitive and selective fluorescence sensor for ferric ions in serum and living cells by imaging. Biosens Bioelectron. 2014;62:189–95.

[123] Feng J, Ju Y, Liu J, Zhang H, Chen X. Polyethyleneimine-templated copper nanoclusters via ascorbic acid reduction approach as ferric ion sensor. Anal Chim Acta. 2015;854:153–60.

[124] Cui M, Song G, Wang C, Song Q. Synthesis of cysteine-functionalized water-soluble luminescent copper nanoclusters and their application to the determination of chromium (VI). Microchimica Acta. 2015;182:1371–7.

[125] Zhong Y, Wang Q, He Y, Ge Y, Song G. A novel fluorescence and naked eye sensor for iodide in urine based on the iodide induced oxidative etching and aggregation of Cu nanoclusters. Sensors Actuators B. 2015;209:147–53.

[126] Li Z, Guo S, Lu C. A highly selective fluorescent probe for sulfide ions based on aggregation of Cu nanocluster induced emission enhancement. Analyst. 2015;140:2719–25.

[127] Wang CX, Cheng H, Huang YJ, Xu ZZ, Lin HH, Zhang C. Facile sonochemical synthesis of pH-responsive copper nanoclusters for selective and sensitive detection of Pb^{2+} in living cells. Analyst. 2015;140:5634–9.

[128] Liao X, Rui L, Long X, Li Z. Ultrasensitive and wide-range pH sensor based on the BSA-capped Cu nanoclusters fabricated by fast synthesis through the use of hydrogen peroxide additive. RSC Adv. 2015;5:48835–41.

[129] Liu J. DNA-stabilized, fluorescent, metal nanoclusters for biosensor development. Trends Anal Chem. 2014;58:99–111.

[130] Wang XP, Yin BC, Ye BC. A novel fluorescence probe of dsDNA-templated copper nanoclusters for quantitative detection of microRNAs. RSC Adv. 2013;3:8633–6.

[131] Zhang H, Lin Z, Su X. Label-free detection of exonuclease III by using dsDNA-templated copper nanoparticles as fluorescent probe. Talanta. 2015;131:59–63.
[132] Bhamore JR, Jha S, Mungara AK, Singhal RK, Sonkeshariya D, Kailasa SK. One-step green synthetic approach for the preparation of multicolor emitting copper nanoclusters and their applications in chemical species sensing and bioimaging. Biosens Bioelectron. 2016;80:243–8.
[133] Shi Y, Luo S, Ji X, Liu F, Chen X, Huang Y, et al. Synthesis of ultra-stable copper nanoclusters and their potential application as a reversible thermometer. Dalton Trans. 2017;46:14251–5.

Bionotes

Dr. Nirmal Kumar Das did his M. Sc. from IIT Kanpur, Kanpur in 2012. Then, he joined as a PhD scholar in IISER Bhopal under the supervision of Professor Saptarshi Mukherjee. In November 2017, he successfully defended his PhD thesis. He is currently postdoctoral fellow in the same group. His research interests include spectroscopic investigation of organized assemblies, synthesis and spectroscopic applications of luminescent metal nanoclusters.

Professor Saptarshi Mukherjee did his PhD under the supervision of Professor Kankan Bhattacharyya at IACS, Kolkata, and carried out his post-doctoral research with Professor H. Peter Lu at Bowling Green State University, Ohio, USA. He joined the Department of Chemistry, IISER Bhopal in December 2008. His research interests include luminescent metal nanoclusters, protein unfolding and refolding using ultrafast and single-molecule spectroscopy. He is presently as Professor in the Department of Chemistry, IISER Bhopal. He has received the INSA Young Scientist Medal in Chemical Sciences and is also a founding member of the Indian National Young Academy of Science.

Elena Piacenza, Alessandro Presentato, Emanuele Zonaro,
Silvia Lampis, Giovanni Vallini and Raymond J. Turner

6 Selenium and tellurium nanomaterials

Abstract: Over the last 40 years, the rapid and exponential growth of nanotechnology led to the development of various synthesis methodologies to generate nanomaterials different in size, shape and composition to be applied in various fields. In particular, nanostructures composed of Selenium (Se) or Tellurium (Te) have attracted increasing interest, due to their intermediate nature between metallic and non-metallic elements, being defined as metalloids. Indeed, this key shared feature of Se and Te allows us the use of their compounds in a variety of applications fields, such as for manufacturing photocells, photographic exposure meters, piezoelectric devices, and thermoelectric materials, to name a few. Considering also that the chemical-physical properties of elements result to be much more emphasized when they are assembled at the nanoscale range, huge efforts have been made to develop highly effective synthesis methods to generate Se- or Te-nanomaterials. In this context, the present book chapter will explore the most used chemical and/or physical methods exploited to generate different morphologies of metalloid-nanostructures, focusing also the attention on the major advantages, drawbacks as well as the safety related to these synthetic procedures.

This article has previously been published in the journal *Physical Sciences Reviews*. Please cite as:
Piacenza, E., Presentato, A., Zonaro, E., Lampis, S., Vallini, G., Turner, R. J. Selenium and Tellurium Nanomaterials. *Physical Sciences Reviews* [Online] **2018**, 3. DOI: 10.1515/psr-2017-0100

https://doi.org/10.1515/9783110636666-006

Graphical Abstract:

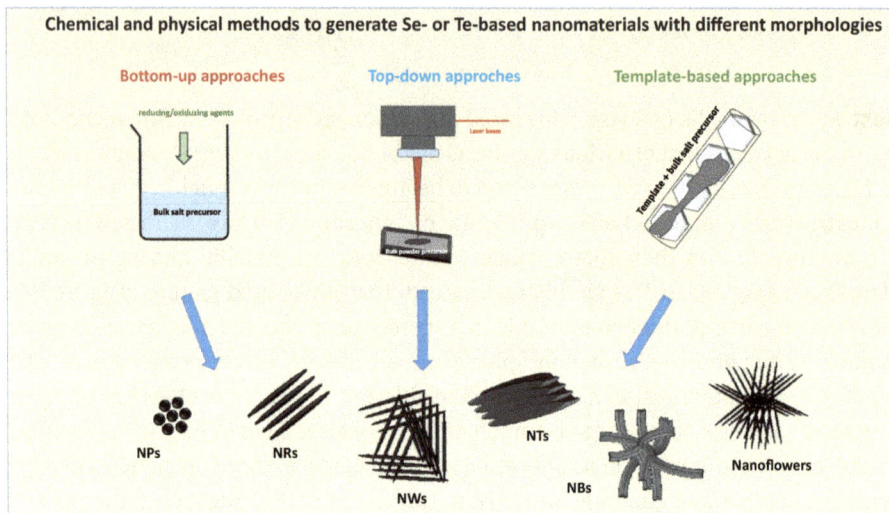

Chemical and physical methods to generate Se- or Te-based nanomaterials with different morphologies

Overview of the chemical and physical methods commonly used to produce various Se- and/or Te-based nanomaterials.

Keywords: selenium, tellurium, nanomaterials, chemical synthesis, physical synthesis, nanoparticles, nanorods, nanowires

6.1 Introduction

Nanoscience and nanotechnology represent the world of the "very small material" (10^{-9} meter), widely impacting at the multidisciplinary level several research fields, such as biomedicine, energy, environmental engineering, chemistry, material science, optoelectronics and life science, to name a few [1, 2]. Indeed, the unique features of nanostructures (e. g., size-dependent, optical, catalytic, electronic and mechanical properties) [3], which derive from their singular physical-chemical characteristics (high surface-volume ratio, large surface energy and spatial confinement), allow the use of nanoscale materials in various application fields. Among the diverse morphologies of existent nanomaterials, nanoparticles (NPs) are considered the building blocks for the generation of various and diverse nanostructures [1], such as: 0-Dimensional (0-D) nanocrystals and quantum dots (QDs); 1-Dimensional (1D) nanorods (NRs), nanowires (NWs), nanotubes (NTs) and nanobelts (NBs); 2-Dimensional (2D) array of nanoparticles (NPs); thin films and 3-Dimensional (3D) structures (superlattices) [4].

Particularly, nanomaterials composed of the metalloid elements selenium (Se) and tellurium (Te) have lately begun to attract a lot of interest in nanoscience, due to their intermediate properties between metals and non-metals [5], which allow their use in a broad spectrum of applications. Indeed, since Se is featured by high photo-conductivity, piezoelectricity, thermoelectricity and spectral sensitivity, it is extensively utilized in photocells, xerography and photographic exposure meters, to name a few [6]. Similarly, Te is a well-known narrow band gap semiconductor, which displays useful physical properties (i. e., nonlinear optical responses, photoconductivity and thermoelectricity), enabling its use to manufacture piezoelectric devices, infrared photoconductive detectors, and thermoelectric materials [7]. Se and Te belong to the chalcogen group (VI) and the p-block of the periodic table, having $[Ar]3d_{10}4s_24p_2$ and $[Kr]4d_{10}\,5s_2\,5p_4$ as chemical configurations, respectively [5, 8]. Further, both these metalloid elements are present as amorphous or crystalline allotropic forms [5, 9, 10]. In this regard, amorphous Se (*a*-Se) exists as either brick-red dust or as black vitreous shape, which derives from the melting of the red state [5]. Generally, reduction reactions or quenching phenomena of Se in vapor phase generate red *a*-Se, which is constituted by Se^8 rings, having the same configuration of elemental sulfur (S^0) [9]. On the other hand, crystalline Se exists as grey form belonging to the hexagonal crystal family (*h*-Se) or as red monoclinic state (*m*-Se), which is usually less stable, and it is generally produced by the evaporation of *h*-Se [5, 9]. Similarly, Te occurs either as brownish *a*-Te or as black crystalline *h*-Te, which is the most stable form at room temperature [10]. Considering that the hexagonal crystal family comprises both the trigonal (*t*) and the hexagonal (*h*) crystal systems, Se and Te can crystallize following both the mentioned space groups [Wendel]. Crystalline Se or Te generates helical chains of atoms covalently bound to each other, which are packed together forming a hexagonal pattern [11]. Although both metalloid elements are featured by this highly organized crystalline structure, in colloidal form Se mainly exists as red *a*-Se [9], while Te results more stable as *t*- or *h*-Te [12]. In this regard, Te crystal structure reveals a high anisotropic nature, which results in the natural tendency of Te-atomic chains to grow along one axis (*c*-axis), forming 1D Te-nanostructures [12].

Considering the wide interest in metalloid-nanomaterial production and application, this book chapter will focus on the chemical-physical procedures lately developed to synthesize these nanostructures, with particular attention toward the strategies aimed to reduce Se- and Te-precursors into their zero-valence states (i. e., Se^0 and Te^0) producing nanosized materials featured by different morphologies and, therefore, diverse properties.

6.2 Preparation methods

Several chemical and physical methods have been recently developed for the synthesis of Se- or Te-based nanostructures (Figure 6.1), having as main focus the generation of high-quality nanomaterials with well-defined and peculiar chemical-physical properties. Considering the fundamental role played by NPs as material at the nanoscale range [1], several synthesis methods have evolved to obtain monodisperse and stable metalloid-NPs. Conversely, 1D metalloid-nanostructures (i. e., NRs, NWs, NTs, NBs) are generally produced exploiting the unique features of Se^0 and Te^0, such as their crystalline nature. In this regard, since Te^0 atoms are featured by an anisotropic crystal structure that results in the formation of helical chains of Te^0 atoms covalently bound with each other, they have the spontaneous tendency to form 1D nanostructures (e. g., NRs, NTs, NBs) [8, 13]. Further, 1D Se- or Te-nanomaterials are obtained by varying crucial parameters of synthesis methods used for NPs generation, being the concentration of reducing agent and/or the reaction time [14].

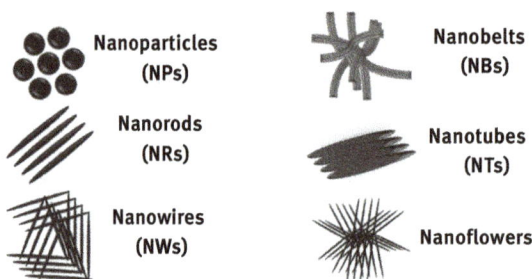

Figure 6.1: Schematic illustration of different Se- or Te-nanostructure morphologies obtained through chemical and physical synthesis methods.

The production of metalloid-nanostructures is generally achieved through: (i) bottom-up approaches, where the conversion of the Se- or Te-precursors – i. e., metalloid salts – into nanostructures is mediated by specific chemical reactions (e. g., reduction, hydrolysis, oxidation); (ii) top-down processes, by which nanomaterials can be obtained upon addition of a bulk precursor to a solvent that under light irradiation mediates the bulk material's melting or transformation and dissociation in smaller entities; (iii) template-based procedures that involve the use of different physical or chemical templates to convert the metalloid-precursor into nanostructures [2, 15]. The choice between the different processes to generate Se- or Te-nanomaterials mainly relies on their ease and versatility, as some of them are typically used to tune the production toward specific nanomorphologies (e. g., NPs, NRs, NWs, NTs or NBs) [14, 16, 17]. Additionally, since both Se and Te are featured by a strong colloidal nature, the production of these metalloid-nanomaterials still has several challenges, such as obtaining of a homogeneous population (monodisperse) of nanostructures featured by thermodynamic stability [14, 18].

6.2.1 Bottom-up approaches for metalloid-nanomaterial generation

High-quality SeNPs are mainly produced exploiting chemical bottom-up approaches (Figure 6.2), due to their ease, competitive cost and short synthesis times, as compared

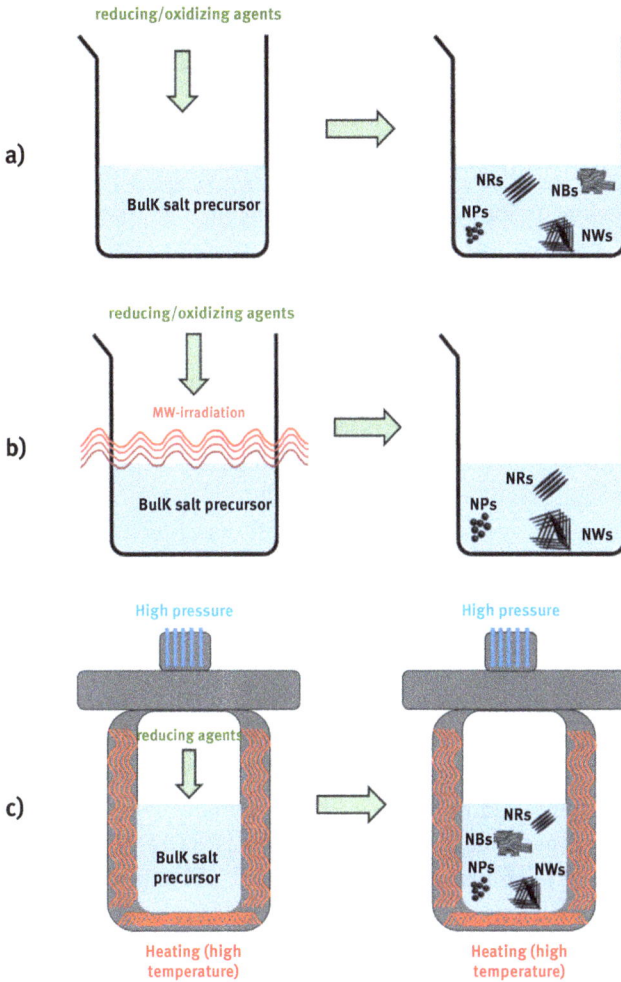

Figure 6.2: Schematic representation of bottom-up systems used to generate different Se- or Te-nanomorphologies, such as: (a) chemical reduction/oxidation reaction, (b) MW-assisted method and (c) hydrothermal process.

to top-down methods [15]. Bottom-up processes for the generation of metalloid-NPs are generally based on reduction reactions occurring between a suitable Se-precursor (i. e., Na_2SeO_4, Na_2SeO_3, SeO_2) and a reducing agent (Figure 6.2a), such as glutathione molecules (GSHs) [19], hydrazine [17, 20], dextrose [21], ascorbic acid [22], acetic acid [23], oxalic acid [23] or sodium ascorbate/thiosulfate [24, 25]. A similar chemical process has been also reported for TeNPs synthesis through a wet chemical reduction reaction, in which sodium tellurite (Na_2TeO_3) salt was exposed to a mixture of phosphoric acid (H_3PO_4), hydrochloric acid (HCl) and a solution of iron(II) acting as reducing agent [26].

Similarly to NPs production, chemical reductions are used to generate also 1D Se- or Te-nanostructures. Indeed, SeNRs or SeNWs are usually produced increasing the concentration of the reducing agent normally utilized to produce SeNPs, improving the reduction kinetics and, therefore, the number of Se^0 atoms available for the nanostructure growth in one direction [27]. Further, SeNRs can be synthesized upon addition of different alcohols (e. g., n-butyl alcohol) to an aqueous system, resulting in the accumulation of SeNPs chains in the water/alcohol interface, which leads to the formation of NRs [28]. On the other hand, considering the natural tendency of Te^0 atoms to deposit along one preferential axis [8, 13], several methods for 1D Te-nanostructures production have been developed based on reduction reactions. Examples include reactions occurring between the precursor orthotelluric acid (H_6O_6Te) or tellurium tetrachloride ($TeCl_4$) and the reducing agent hydrazine (N_2H_4) or potassium borohydride (KBH_4) [29–31]. Further, spontaneous oxidation of sodium hydrogen telluride (NaHTe) with sodium dodecyl benzenesulfonate (SDBS) has also proven its proficiency to generate well-organized TeNRs [32].

Alternatively to chemical approaches, the microwave (MW)-assisted method has shown potential as bottom-up strategy [33] to generate SeNPs [34] (Figure 6.2b). This process occurs through a rapid heating of both Se-precursor and reducing agent, leading to an increase in the reaction kinetics and, therefore, a faster formation of nanomaterials [14, 35, 36]. The superheating effects deriving from the coupling of MWs to a polar organic solvent (e. g., water) [37] are a key component to rapidly generate SeNPs, due to the generation of induced dipole moments in Se-molecules mediated by a high-frequency electromagnetic radiation [38]. In this regard, MW-assisted method was successfully used to produce SeNPs through the reduction of selenous acid (H_2SeO_3) with L-asparagine in polyethylene glycol (PEG) [39], as well as the reaction occurring between selenium chloride ($SeCl_4$) and N_2H_4, in presence of SDS or PEG 600 as surfactants [31]. Since processes based on chemical reductions are still limited in controlling both size and shape of metalloid-nanostructures [40], the MW-assisted mechanism is also exploited to generate 1D metalloid-nanomaterials using polymers [41, 42] or ionic liquids [43], which act as guide for Se^0 or Te^0 atom deposition. As for the chemical synthesis procedures, the same MW-assisted mechanism used to produce SeNPs can lead to the generation of 1D Se-nanostructures (e. g., SeNWs, SeNRs, SeNBs) by changing the concentration of reducing agent along with the MW-irradiation time [37]. Similarly, MW-assisted process was utilized to synthesize TeNRs in water phase using sodium tellurite (Na_2TeO_3) as Te-bulk precursor, N_2H_4 as reducing agent and polyvinylpyrrolidone (PVP) as driving force for Te^0 atoms deposition [44].

To date, one of the most used bottom-up methodologies [33] to generate 1D metalloid-nanomaterials is the hydrothermal synthesis, which is generally described as a super-heated aqueous solution process [45] (Figure 6.2c). This method is based on the use of high pressures and temperatures to favor heterogeneous reactions in aqueous solvents occurring between either Se- or Te-precursors and an appropriate reducing agent [45]. Thus, the hydrothermal route allows to speed up both the rate of bulk precursors dissolution and the reaction kinetics [46], leading to a rapid formation of metalloid NTs and/or NRs with a controllable size and shape [47–51]. Further, the reaction temperature can be easily decreased upon the addition of a suitable solvent [45, 52, 53]. This method was extensively investigated by Cao and coworkers (2011) to synthesize both Se- and TeNRs through the employment of glucose as reducing agent and cetyl trimethylammonium bromide (CTAB) as a structure-directing agent upon heating to 130 °C for 12 hours [50]. Moreover, Te-nanostructures with different morphologies (e. g., NPs, NRs, NWs, NTs) were generated by heating up to 70 °C a mixture of PVP, Na_2TeO_3, sodium peroxide (NaOH), N_2H_4 and ethylene glycol for a timeframe ranging from 15 seconds to 3 hours [50]. Among the several hydrothermal processes developed, solvothermal approaches are increasingly used for 1D Se- or Te-nanostructure production. These synthesis methods involve the use of non-aqueous solvents or solvents near supercritical conditions [45–47], where fluids are featured by high density and non-condensability, as they coexist in thermodynamic equilibrium in two phases (i. e., gas and liquid) [54]. Thus, the use of supercritical fluids allows to obtain SeNWs or TeNRs having defined size and regular shape with high reduction rates and without the necessity to further stabilize them in suspension [45, 50, 52–54]. Variation in the solvents exploited during this synthesis procedure can lead to the production of metalloid-nanomaterials with different morphologies, such as SeNPs or SeNRs upon addition of carbon disulfide (CS_2) or ethanol, respectively [52, 53]. Further, solvothermal methods were also used to synthesize Te flower-like nanomaterials mediated by the continuous supply of 3,3-dimethylbutyl thioacetate (DBTA) as reducing reagent to the system, which resulted in the formation of uniform nanostructures [55].

6.2.2 Top-down approaches for metalloid-nanomaterial generation

Although bottom-up chemical methods are normally preferred for metalloid-NPs generation, several top-down approaches have been recently explored for their synthesis through the exploitation of photocatalysis (PC) [18], pulsed-laser ablation (PLA) [56, 57] and laser induced melting/vaporization (LIM/V) [58–60].

The generation of metalloid-NPs by means of PC processes is based on rapid photocatalytic reactions occurring – upon UV irradiation – between a photocatalyst (e. g., powder of polyoxometalate anions, POM, or titanium dioxide, TiO_2) and a metalloid salt, which is then reduced to its elemental form, forming nanomaterials [18, 61, 62] (Figure 6.3a). This methodology has been used mainly to generate SeNPs compared to Te-nanostructures, as indicated by Triantis and coworkers (2009) [18]. On the other hand, PLA has been recognized as valid alternative to produce monodisperse Se and/or TeNPs. During this physical process, the bulk metalloid-precursor is removed from a solid surface (e. g., silicon wafers, glass or gold films) through irradiation with a

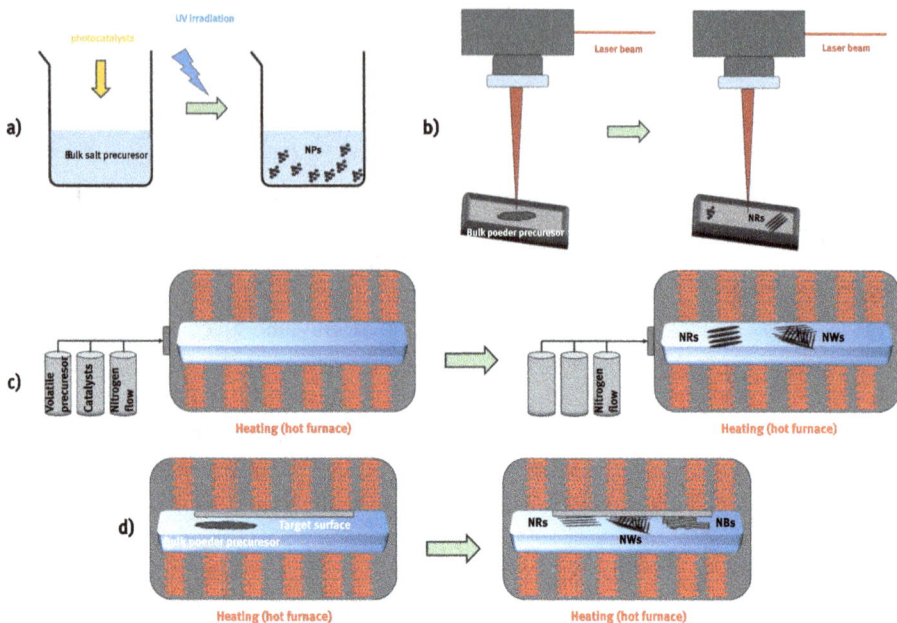

Figure 6.3: Schematic illustration of Se- or Te-nanomaterial synthesis by using top-down approaches of (a) photocatalysis (PC), (b) pulsed laser ablation (PLA), (c) chemical vapor deposition (CVD), (d) physical vapor deposition (PVD).

high intensity pulsed laser source that in turn converts the bulk material into plasma [63, 64] (Figure 6.3b), which contains either Se^0 or Te^0 atoms. Then, metalloid atoms in their zero-valence states start to aggregate forming nucleation seeds upon mutual collision events, allowing the final formation and growth of Se or TeNPs [56, 57, 64, 65]. Similarly to the PLA approach, in the LIM/V process the solid metalloid-precursor is removed and converted into NPs upon laser irradiation in a liquid environment, usually water, resulting in a relatively easy and versatile method for metalloid-NPs synthesis [59, 60, 66, 67]. Indeed, the use of a liquid environment for Se or Te-precursor conversion provides a stronger expansion confinement of the originated plasma as compared to a solid surface, leading to the formation of smaller and more uniform Se or TeNPs [40, 60, 67]. As a result, polycrystalline selenium was used as bulk precursor to produce SeNPs by means of this process [68], while Guisberg et al. (2017) investigated TeNPs synthesis using as Te-precursor a Te-powder and acetone as the solvent [60]. Among the top-down methodologies described above, PC and PLA are commonly used to produce 1D Se-nanostructures [63, 64] varying some parameters of the systems – e. g., temperature, the light source and laser power – to tune the production of SeNPs to 1D nanomaterials [69, 70]. Nevertheless, physical processes other than PC and PLA have recently been investigated to produce these nanomaterials, such as chemical (CVD) or physical vapor deposition (PVD) [14].

In CVD process, a volatile mixture containing the desired precursors along with different gaseous compounds and catalysts (e. g., Cu, Fe, Ni, Si) placed in a quartz tube is heated up by using a hot furnace [71] (Figure 6.3c). The interaction between the three components of this mixture at elevated temperature results in the deposition of pre-cursor elemental atoms as non-volatile solid nanoscale materials [72–74]. Considering the high versatility of this method, CVD approach has been used to synthesize SeNWs [75–77], as well as SeNRs [78]. Moreover, Zhang and coworkers (2005) explored a carbothermal CVD process based on the use of a mixture of Se and carbon powders heated up to 400–500 °C under nitrogen flow to generate SeNWs and SeNRs [79]. On the opposite, PVD procedure is based on the transfer of growth species (i. e., metalloid-nanomaterials) from an appropriate source, which is usually either Se- or Te-powders, to a substrate (e. g., silicon wafers, graphite or aluminum rods) without the occurrence of a chemical reaction [80] (Figure 6.3d). The generation of metalloid elemental forms from the precursors is achieved by two main processes: (i) evaporation, where heating events mediate the detachment of Se^0 or Te^0 atoms, or (ii) sputtering, which is based on the use of gaseous ions (plasma) impacting the metalloid-source, therefore causing its transformation in Se^0 or Te^0 forms, which leads to the formation of nanomaterials with different morphologies [78, 80, 81]. Indeed, PVD method used to synthesize both Se- and TeNBs revealed the presence of several intermediate nanomorphologies (e. g., NRs or NWs) during the synthesis process [82]. Overall, PVD has been more investigated and used to obtain 1D Te-nanomaterials as compared to Se-based ones, mainly due to the spontaneous tendency of Te^0 atoms to anisotropically grow along one axis [80, 83].

6.2.3 Template-based approaches for metalloid-nanostructure generation

Metalloid-NPs can also be synthesize using several template-based strategies, which include the use of block copolymers or surfactants acting as structure-directing as well as reducing and stabilizing agents for the forming nanostructures [84] (Figure 6.4). In this regard, both these agents reversibly adsorb onto the surface of the forming metalloid-NPs in their early stage of generation, therefore providing a chemical tem-plate that allows control of their growth, and also conferring thermodynamic stability [85, 86]. Although surfactants are generally used to produce metalloid-NPs, they can eventually desorb from the nanomaterial surface, resulting in the formation of bigger aggregates [87]. However, since block copolymers are constituted of monomers repeat-ing along their structure, they can irreversibly adsorb onto the nanostructures [88], guarantying the formation of stable metalloid-NPs. Therefore, block copolymers are generally preferred over surfactants to mediate the generation of monodisperse metal-loid-NPs [84, 89]. The most used polymers to obtain stable SeNPs are hydroxyethyl-cellulose (HEC) [90], PVP [91], CTAB [92, 93], polyvinyl alcohols (PVAs) [94], gelatin [94], chitosan [95] and alkane-thiols [96]. Nevertheless, various surfactants have been used since 1990s to produce SeNPs, such as sodium bis(2-ethylhexyl) sulfosuccinate

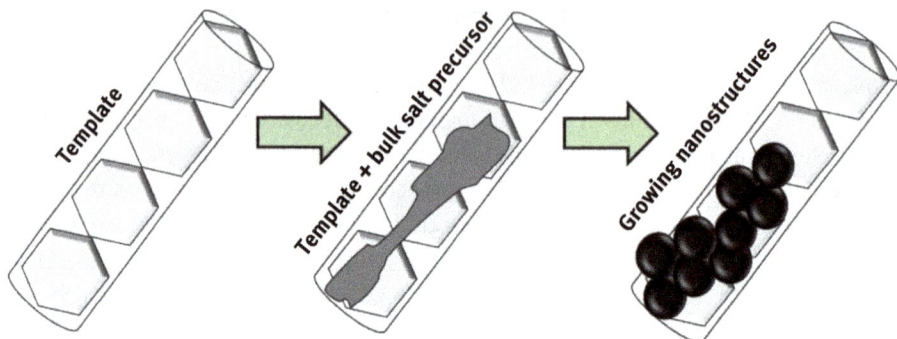

Figure 6.4: Schematic representation of template-based procedures used to produce Se- or Te-nanostructures.

(AOT) reverse micelles [97]. Although several template-based methodologies have been developed for SeNPs synthesis, only few studies explored the use of oleic acid along with ethylene glycol for the production of TeNPs [98].

Surfactants and block copolymers are also used as template to guide the growth of 1D metalloid-nanostructures [14, 50]. The deposition of Se^0 atoms in one direction is achieved drastically increasing either surfactant or block copolymer concentrations in the system, as compared to those used to obtain stabilized SeNPs [13, 14, 99]. The variation in the concentration of template-agents results in their strong binding and adsorption onto the surface of the growing nanostructures [99]. Thus, the electrical charges on Se-nanomaterial surfaces change, leading to the tuning of their morphology into a 1D-nanostructure [14], such as SeNTs [100], SeNWs and SeNRs [101]. Further, SeNRs can be produced by adding to the system an acid solution (e. g., HCl), which speeds up the reduction of Se-precursor into Se^0 that in turn allows their instantaneous deposition along one axis [97]. Although several studies are reported to produce 1D Se-nanomaterials by means of template-based strategies, their use is still limited [14]. Given the natural tendency of Te^0 to deposit and grow in one direction [13, 99], surfactants and block copolymers are mostly exploited as templates to produce TeNRs, TeNWs and TeNTs [13, 50, 99, 100]. Indeed, uniform crystal TeNRs were produced through the reduction occurring between ammonium sulftellurate $((NH_4)_2TeS_4)$ with sodium sulfite (Na_2SO_3) in the presence of sodium dodecyl benzenesulfonate (NaDDBS) as surfactant [102], as well as using sodium tellurate (Na_2TeO_4), ascorbic acid and CTAB as growth-directing polymer [103]. Further, Te-nanomaterials with several morphologies (e. g., TeNRs, TeNWs and Te-feather like nanostructures) were obtained under alkaline conditions (NaOH) using potassium tellurite (K_2TeO_3), sodium hypophosphite monohydrate $(NaH_2PO_2 \cdot H_2O)$ with PVP in different ratios as block copolymer [104].

A general summary of the above described synthesis methods commonly used to generate different morphologies of Se- or Te-nanomaterials is reported in Table 6.1.

Table 6.1: Summary of the synthesis methods used to generate Se- or Te-nanomaterials of various morphologies.

	Se-nanomaterials		Te-nanomaterials	
Bottom-up methods				
Chemical reactions	SeNPs	[15, 17, 19–25]	TeNPs	[26]
	SeNRs	[27, 28]	TeNRs	[29–32]
	SeNWs	[27]	TeNWs	[29–31]
			TeNTs	[29–31]
			TeNBs	[29–31]
MW-assisted process	SeNPs	[31, 34, 37–39]	TeNRs	[41, 43]
	SeNRs	[42]	TeNWs	[41, 44]
	SeNWs	[42]		
Hydrothermal process	SeNRs	[50, 53]	TeNPs	[50, 52]
	SeNTs	[47, 49, 53]	TeNRs	[50]
	SeNBs	[48, 53]	TeNWs	[50]
			TeNTs	[50]
Solvothermal process	SeNPs	[52]	TeNRs	[50, 52]
	SeNRs	[53]	Te-flower like	[55]
	SeNWs	[45, 50]		
Top-down methods				
PC process	SeNPs	[18, 61]		
PLA process	SeNPs	[56]	TeNPs	[57]
	SeNRs	[69, 70]		
LIM/V process	SeNPs	[58, 59, 67, 68]	TeNPs	60
CVD process	SeNRs	[78]		
	SeNWs	[75–77]		
	SeNTs	[71]		
PVD process	SeNBs	[82]	TeNRs	[80, 81]
			TeNWs	[81]
			TeNTs	[80, 81]
			TeNBs	[82]
Template-based methods				
Block copolymers	SeNPs	[90–96]	TeNPs	[98]
	SeNWs	[101]	TeNRs	[103, 104]
			TeNWs	[103, 104]
			TeNTs	[103, 104]
			Te-feather like	[104]
Surfactants	SeNPs	[97]	TeNPs	[100]
	SeNRs	[97, 101]	TeNRs	[13, 50, 103, 104]
	SeNWs	[101]		
	SeNTs	[100]		

6.2.4 Advantages and drawbacks of metalloid-nanostructure synthesis methods

The main advantage of chemical bottom-up approaches used to produce chalcogen-nanomaterials is the possibility to control their final size and morphology by modifying the ratio of the metalloid bulk precursor and the reducing agent used [105], as well as by increasing the temperature of the reaction [29, 105]. In the MW-assisted bottom-up method, changes in the exposure time of chalcogen-precursor to the MWs [14], the reaction time and the pH of the solution [31] are used to generate metalloid nanostructures of different size (e. g., SeNPs) [63] or shapes (e. g., TeNRs, TeNWs, TeNTs) [41]. Additionally, since MW bottom-up method is featured by high reaction rates and shorter production times consequent to the homogenous heating of the system, it is now considered an energy-saving synthesis process [35]. On the other hand, hydrothermal processes are now considered the most efficient and advantageous procedures to obtain highly homogenous nanostructures with defined morphology, crystalline symmetry, narrow size distributions and great purity in composition [45]. Thus, the hydrothermal method is an energy-saving and fairly economic process, which guarantees the control on Se- or Te-nucleation events and, therefore, the nanostructure shape without the use of larger amounts of reagents [45]. Similarly, since solvothermal procedures are based on the use of supercritical fluids with very high reaction rates, dielectric constants and densities [46], they are recognized as highly effective methods for generating homogenous and monodisperse metalloid-nanostructures.

Among the top-down processes exploited for Se- or Te-nanomaterial production, PC is advantageous due to its ease in changing reaction parameters (e. g., concentration of the photocatalyst and ionic strength of the solutions), producing nanostructures featured by different size and morphologies [14, 18]. Moreover, the use of liquid metalloids instead of solid salts as precursors confers a high control on size distribution of the generating nanomaterials [63]. Alternatively, top-down processes as PLA and LIM/V have as their main advantage the possibility to tune the production of metalloid-nanostructures toward different size and shape by simply varying laser source parameters (power density, pulse duration, wavelength) or ambient gas conditions (pressure and flow rate) instead of finding the proper ratio between precursor and reducing agent to obtain the desired product [56, 57, 65]. In the case of CVD process, chemical-physical features of the forming nanostructures depend on the temperature of the substrate used, the nature of the catalysts chosen and their reduction kinetics [73, 75, 76, 79, 106, 107]. Super saturation of gaseous vapors, deposition temperatures, system pressure and carrier gas flow rate are the parameters of PVD procedures responsible for the generation of metalloid-nanostructures of desired chemical-physical characteristics (e. g., size, shape, chemical composition) [81, 82].

Finally, template-based synthesis methods are considered low-cost processes for the generation of Se- or Te-nanomaterials with desired size, shape and chemical composition, upon variation of the concentration of growth-directing agent used (i. e., block copolymers or surfactants), as well as the temperature of the entire system [14, 102, 108]. As a further advantage, these polymers provide hydrophobicity to the metalloid-nanostructures, avoiding their natural tendency to aggregate and, therefore, sterically stabilizing them in solution [109–111].

Although the above described procedures to produce Se- or Te-nanomaterials are effective and advantageous in different ways, the common principal drawback of these synthesis approaches is the generation of metalloid-nanostructures with various size, which is generally caused by: (i) non-homogenous reduction rates or kinetics (bottom-up approaches), (ii) variation in the ionic strength of the solutions (PC top-down process) [14], (iii) splashing phenomenon due to the laser irradiation of Se^0 or Te^0 atoms deposited onto the substrates (PLA and LIM/V top-down procedures) [56, 65], (iv) the presence of impurities in the forming nanomaterials (CVD) [89], or (v) desorption of surfactant molecules from the nanostructure surfaces (template-based methods) [63]. As a consequence, broad Se- or Te-nanostructure distributions (polydispersity) are usually produced by these methods, resulting in the formation of heterogeneous populations of metalloid-nanomaterials in the same system [14, 89]. Additionally, synthesis procedures based on PLA, CVD or PVD processes are not cost-effective to perform, due to the complexity of the entire operating system [14, 112]. Similarly, an important limitation of hydrothermal and solvothermal synthesis processes is the necessity of several empirical trials to find the best conditions for the development of the optimal metalloid-nanomaterial production strategy [45]. Indeed, since these bottom-up approaches are based on the use of high temperature and pressure, designing a suitable apparatus able to resist such hard conditions is still challenging [45].

6.3 Characterization methodologies and instrumentation techniques

Advantages and drawbacks of Se- or Te-nanostructure application in different fields are tightly linked to their peculiar chemical-physical properties, highlighting the necessity to carefully study them [113], with particular attention to size, shape, composition, crystallinity and stability [114].

The size distribution of metalloid-nanomaterials is generally determined by using dynamic light scattering (DLS) technique, which measures the average nanostructure size as the diameter of a hypothetical hard sphere diffusing in the same way as the nanomaterial in the analyzed sample [114, 115]. Thus, DLS is relatively not accurate to monitor size distributions of heterogeneous populations (polydisperse) of nanomaterials, as small structures are generally not detected, due to the dependency of the intensity signal to the 6th power of the radius measured by DLS [115, 116]. Further, this technique results to be inefficient in measuring the dimension of 1D Se- or Te-nanostructures (e. g., NRs, NWs, NTs) [116]. Alternative to DLS analysis, Electron Microscopy (EM) coupled with a suitable software for size measurement (e. g., ImageJ) is now preferred to calculate the actual size of nanomaterials with different shapes. However, the procedures applied to prepare samples for EM imaging, such as dehydration and/or the use of harsh fixative solutions, can lead to agglomeration of the nanomaterials, as well as the generation of artifacts difficult to recognize [117]. Based on the above reported considerations, it is fundamental to firstly perform EM imaging to investigate Se- or Te-nanomaterial morphology, and, subsequently, to measure their size distributions (i) using external software (in the case of polydisperse and not spherical-shaped nanostructures), or (ii) carrying out DLS analysis when the sample is constituted of homogenous NPs.

The chemical composition of metalloid-nanomaterials is usually evaluated exploiting spectroscopy techniques, such as the X-ray Photoelectron Spectroscopy (XPS) or the Energy-Dispersive X-ray Spectroscopy (abbreviated as either EDS or EDX) [114, 118]. Both these techniques are based on the use of relatively low-energy X-rays hitting the samples, therefore causing either an electron ejection from an atom due to photoelectric effect (XPS) or the emission of the energy excess as X-ray (EDS) [118]. Since each element of the periodic table is featured by a specific set of energy levels, the final output of both XPS and EDS is the detection of the element-specific binding energies, which gives the atomic composition of the analyzed sample [118]. Further, X-ray diffraction (XRD) technique can be used to examine both the chemical composition of metalloid-nanomaterials and their crystalline structures [64]. Similarly to XPS and EDS, XRD relies on the use of X-rays to excite the sample [118]. Nevertheless, in XRD analysis the angle at which the X-rays are diffracted by the nanomaterial sample is measured, generating a diffraction pattern that can be then used to identify the crystalline phases and structure of the atoms within the sample [118]. Recently, Selected-Area Electron Diffraction (SAED) has been used as a secondary technique to evaluate crystalline properties of Se- or Te-nanomaterials. Indeed, SAED is based on the measurement of the diffraction of the high-energy electron beams used to excite a small and defined area of samples analyzed by Transmission Electron Microscopy (TEM) [119]. As a result, the generated electron diffraction pattern allows the study of nanostructure crystalline properties [119].

Considering that the application of metalloid-nanomaterials strictly depends on their ability to not aggregate with each other, it is also imperative to evaluate their stability in suspension. This is generally achieved by measuring their surface charges (zeta potential) generated in a liquid environment by means of electrophoresis [115]. Indeed, the motion of nanomaterials in a liquid medium leads to the movement of ions of the medium highly bound to the nanostructure surfaces, generating a charge distribution at the interface [120]. Consequently, the as-produced electrical potential (zeta potential) indicates the degree of electrical repulsion between nanostructures, and, therefore, their stability in suspension [121]. In this regard, by theory nanomaterials having zeta potential values either <−25 or >+25 mV are considered stable, while intermediate zeta potential values are typical of unstable nanostructures in suspension [120].

6.4 Critical safety considerations

Up to now, among the approaches used to generate metalloid-nanomaterials, those based on chemical reactions (e. g., bottom-up methods, PC, CVD and hydrothermal top-down procedures) as well as surfactant or copolymer templates (template-based methodologies) often involve the use of hazardous and harsh chemicals, which, during the synthesis procedures, are converted in toxic waste to be disposed [17, 122]. Moreover, synthesis methodologies such as CVD rely on the addition of metal catalysts to the reaction systems, which can eventually lead to either the generation of an explosion if a spill event occurs, or the incorporation of their residues on the final Se- or Te-nanomaterials, making them potentially toxic for humans [123]. Therefore, scientific research recently has been focused on the development of production methods involving environmental friendly and more biocompatible chemical reagents [45, 122]. Indeed, in the last decade several SeNPs synthesis approaches have been explored using less toxic reducing agents, such as GSHs [19], L-cysteine [124], acetic acid [23], L-asparagine [39], and glucose [125]. Similarly, metalloid-nanomaterials with different morphologies have been synthesized exploiting more biocompatible template-agents, being them chitosan (SeNPs) [49], gelatin (SeNPs) [94], oligosaccharides (SeNWs) [126], proteins (hollow SeNPs) [127] or oleic acid (TeNPs) [98].

Metalloid-nanomaterial production based on physical processes (e. g., PLA, LIM/V, PVD top-down approaches) tend not to use hazardous chemical compounds [14, 122], therefore favoring their exploitation as preferential synthesis routes. Nevertheless, some of these methods require the use of gaseous ions (PVD) to mediate metalloid-nanostructure production [45, 123]. Thus, since the entire PVD process is confined in a close system (e. g., reactor), in the event of a leakage, ions in gaseous phase can be released in the environment, eventually causing explosions [123]. In this regard, small reactors (lab scale) operated under fume hoods result to have a lower possibility of spilling and/or leaking as compared to large-scale systems [128, 129]. Finally, considering that most of the synthesis methods described for Se- or Te-nanomaterial generation are functional under extreme system conditions (e. g., very high temperatures and pressures), there is a serious risk of releasing gaseous compounds in the surrounding environment, which contribute to the development of dangerous explosions as well as the global warming phenomenon [45, 123]. Thus, the new frontier in metalloid-nanostructure production seems to be the use of solvothermal routes based on reduction reactions occurring between Se- or Te-bulk salts and biocompatible reducing agents in the presence of supercritical fluids. Indeed, since these peculiar fluids are featured by a low critical temperature and pressure compared to normal solvents [45], their use in synthesis procedures will result in the possibility to process the precursors under less extreme system conditions.

6.5 Conclusions and future perspectives

Over the last 20 years, various methods aimed to generate Se- or Te-based nanostructures having unique and stronger chemical-physical features as compared to their corresponding bulk materials have been explored. The achievement of this target is dependent on several structural parameters of the as-produced metalloid-nanomaterials, such as size, shape and chemical composition, leading to the exploitation of different methodologies to generate nanostructures with desired characteristics. Indeed, Se- or Te-based nanomaterials have been produced using various procedures, which can be generally summarized as: (i) bottom-up approaches based on chemical reactions occurring between a metalloid precursor and a suitable reducing/oxidizing agent; (ii) top-down methods, where the production of nanostructures is achieved by the dissociation of Se/Te precursor mediated by a hot solvent upon UV/laser irradiation; or (iii) template-based processes, in which growth-directing agents act as templates to generate metalloid-nanomaterials of desired features. Although the developed synthesis methods resulted to be efficacious and advantageous in different fashions, the main two challenges that researchers still must face are the high production costs, and the obtainment of Se- or Te-nanostructure populations having homogenous size (monodisperse) and morphology, in order to assure their uniformity for application purposes. Therefore, a complete comprehension of the mechanisms mediating metalloid-nanomaterial formation and growth over the time should be achieved to rationally design procedures for maximizing the production of uniform Se-or Te-nanostructures. Further, to date most of the methods used to generate such nanomaterials require the use of hazardous and harsh chemicals in the process, which irreversibly results in the formation of toxic waste that can damage both our health and the environment. Thus, it is now fundamental to abandon the use of these harmful substances and to develop new, facile, safe and economic synthetic procedures exploiting environmental friendly and more biocompatible agents to obtain metalloid-nanostructures.

Funding: This work was supported by the Natural Science and Engineering Research Council of Canada (NSERC) [216887-2010]. Natural Science and Engineering Research Council of Canada (NSERC) is gratefully acknowledged for the support of this study (Grant/Award Number: 216887–2010).

References

[1] Horikoshi S, Serpone N. Chapter 1, general introduction to nanoparticles. In: Horikoshi S, Serpone N, editors. Microwaves in nanoparticle synthesis: fundamentals and applications. Weinheim: Wiley-VCH Verlag GmbH & Co. KGaA, 2013:1–24.

[2] Cao C. Chapter 1: introduction. In: Cao G., editor. Nanostructures & nanomaterials, synthesis, properties and applications. London: Imperial College Press, 2004:1–14.

[3] Yuwen L, Wang L. Chapter 11.5, nanoparticles and quantum dots. In: Devillanova F, Du Mont WW, editors. Handbook of chalcogen chemistry: new perspectives in sulfur, selenium and tellurium, 2nd ed. Cambridge: The Royal Society of Chemistry, 2013:232–60.

[4] Rao CN, Muller A, Cheetham AK. Chapter 1, nanomaterials. In: Rao CN, Muller A, Cheetham AK, editors. The chemistry of nanomaterials: synthesis, properties and applications. Weinheim: WILEY-VCH Verlag GmbH & Co. KGaA, 2004:1–11.

[5] Haynes WN. Section 4, properties of the elements and inorganic compounds. In: Haynes WN, Lide DR, Bruno TJ, editors. Handbook of chemistry and physics. 95th ed. Boca Raton: CRC Press/Taylor and Francis, 2014:115–20.

[6] Song XC, Zhao Y, Zheng YF, Yang E, Chen WQ, Fang YQ. Fabrication of Se/C coaxial nanocables through a novel solution process. J Phys Chem C. 2008;112:5352–55.

[7] Song XC, Zhao Y, Zheng YF, Yang E, Chen WQ, Fu FM. Fabrication and characterization of Te/C nanocables and carbonaceous nanotubes. Cryst Growth Des. 2009;9:344–47.

[8] Cooper ED. Tellurium. New York: Van Nostrand Reinhold Co, 1974.

[9] Habashi F. Selenium, physical and chemical properties. In: Kretsinger RH, Uversky VN, Permyakov EA, editors. Encyclopedia of metalloproteins. New York: Springer Science+Business Media, 2013:1924–25.

[10] Habashi F. Tellurium, physical and chemical properties. In: Kretsinger RH, Uversky VN, Permyakov EA, editors. Encyclopedia of metalloproteins. New York: Springer Science+Business Media, 2013:2174–75.

[11] Wendel H. Lattice dynamics of trigonal selenium and tellurium – state of the art. In: Gerlach E, Grosse P, editors. The physics of selenium and tellurium, vol. 13. Berlin: Springer Series in Solid-State Sciences, 1979:47–59.

[12] Li Z, Zheng S, Zhang Y, Teng R, Huang T, Chen C, et al. Controlled synthesis of tellurium nanowires and nanotubes via a facile, efficient, and relatively green solution phase method. J Mater Chem A. 2013;1(47):15046–52.

[13] Gautam UK, Rao NR. Controlled synthesis of crystalline tellurium nanorods, nanowires, nanobelts and related structures by a self-seeding solution process. J Mater Chem. 2004;14:2530–35.

[14] Chaudhari S, Umar A, Mehta SK. Selenium nanomaterials: an overview of recent developments in synthesis, properties and potential applications. Progr Mater Sci. 2016;83:270–329.

[15] Wang Y, Xia Y. Bottom-up and top-down approaches to the synthesis of monodispersed spherical colloids of low melting-point metals. Nano Lett. 2004;4(10):2047–50.

[16] Brus LE. Chemical approaches to semiconductor nanocrystals. J Phys Chem Solids. 1998;59:459–65.

[17] Zhang B, Ye X, Dai W, Hou W, Zuo F, Xie Y. Biomolecule-assisted synthesis of single-crystalline selenium nanowires and nanoribbons via a novel flakecracking mechanism. Nanotechnology. 2006;17:385–90.

[18] Triantis T, Troupis A, Gkika E, Alexakos G, Boukos N, Papaconstantinou E, et al. Photocatalytic synthesis of Se nanoparticles using polyoxometalates. Catal Today. 2009;144:2–6.

[19] Zhang J, Wang H, Bao Y, Zhang L. Nano red elemental selenium has no size effect in the induction of seleno-enzymes in both cultured cells and mice. Life Sciences. 2004;75: 237–44.

[20] Huang B, Zhang J, How J, Chen C. Free radical scavenging efficiency of nano-Se in vitro. Free Radic Biol Med. 2003;35(7):805–13.

[21] De Brouchere L, Watillon A, Van Grunderbuck F. Existence of a nonelectrostatic stabilizing factor in hydrophobic selenium hydrosols. Nature. 1956;178:589.

[22] Gupta KD, Das SR. X-ray study of selenium in the liquid and colloidal state. Indian J Phys. 1941;15:401.

[23] Dwivedi C, Shah CP, Singh K, Kumar M, Bajaj PN. An organic acid-induced synthesis and characterization of selenium nanoparticles. J Nanotechnol. 2011;2011:651971.

[24] Shaker AM. Kinetics of the reduction of Se(IV) to Se-Sol. J Coll Inter Sci. 1996;180:225–31.

[25] Mees DR, Pysto W, Tarcha PJ. Formation of selenium colloids using sodium ascorbate as reducing agent. Colloid Interface Sci. 1995;170:254.

[26] Kurimella VR, Kumar KR, Sanasi PD. A novel synthesis of tellurium nanoparticles using iron (II) as a reductant. Int J Nanosci Nanotechnol. 2013;4(3):209–21.

[27] Chen H, Shin DW, Nam GJ, Kwon KW, Yoo JB. Selenium nanowires and nanotubes synthesized via a facile template-free solution method. Mater Res Bull. 2010;45:699–704.

[28] Song JM, Zhu JH, Yu SH. Crystallization and shape evolution of single crystalline selenium nanorods at liquid-liquid interface: from monodisperse amorphous Se nanospheres toward Se nanorods. J Phys Chem B. 2006;110(47):23790–95.

[29] Mayers B, Xia Y. One-dimensional nanostructures of trigonal tellurium with various morphologies can be synthesized using a solution-phase approach. J Mater Chem. 2002;12:1875–81.

[30] Xia Y, Yang P, Sun Y, Wu Y, Mayers B, Gates B, et al. One-dimensional nanostructures: synthesis, characterization, and applications. Adv Mater. 2003;15(5):353–89.

[31] Panahi-Kalamuei M, Mohandes F, Mousavi-Kamazani M, Salavati-Niasari M, Fereshteh Z, Fathi M. Tellurium nanostructures: simple chemical reduction synthesis, characterization and photovoltaic measurements. Mater Sci Semiconduct Proc. 2014;27:1028–35.

[32] Zheng R, Cheng W, Wang E, Guo S. Synthesis of tellurium nanorods via spontaneous oxidation of NaHTe at room temperature. Chem Phys Lett. 2004;395(4):302–05.

[33] Umer A, Naveed S, Naveed R. Selection of a suitable method for the synthesis of copper nanoparticles. NANO Brief Rep Rev. 2012;7(5):1230005.

[34] Olivas RM, Donard OF. Microwave assisted reduction of SeVI to SeIV and determination by HG/FI-ICP/MS for inorganic selenium speciation. Talanta. 1998;45:1023–29.

[35] Varma RS. Greener approach to nanomaterials and their sustainable applications. Curr Opin Chem Eng. 2012;1(2):123–28.

[36] Gawande MB, Shelke SN, Zboril R, Varma RS. Microwave-assisted chemistry: synthetic applications for rapid assembly of nanomaterials and organics. Acc Chem Res. 2014;47(4):1338–48.

[37] Mingo DM, Baghurst DR. Superheating effects associated with microwave dielectric heating. J Chem Soc Chem Commun. 1992;6:674.

[38] Zhu J, Chen PS, Gedanken A. Microwave assisted preparation of CdSe, PbSe, and $Cu_{2-x}Se$ nanoparticles. J Phys Chem B. 2000;104(31):7344–47.

[39] Yu B, You P, Song M, Zhou Y, Yu F, Zheng W. A facile and fast synthetic approach to create selenium nanoparticles with diverse shapes and their antioxidation ability. New J Chem. 2016;40:1118–23.

[40] Liu J, Liang C, Lin Y, Zhang H, Wu S. Understanding the solvent molecules induced spontaneous growth of uncapped tellurium nanoparticles. Sci Rep. 2016;6:32631.

[41] Liu JW, Zhu JH, Zhang CL, Liang HW, Yu SH. Mesostructured assemblies of ultrathin superlong tellurium nanowires and their photoconductivity. J Am Chem Soc. 2010;132(26):8945–52.

[42] Zhu YJ, Hu XL. Preparation of powders of selenium nanorods and nanowires by microwave-polyol method. Mater Lett. 2004;58:1234–36.

[43] Zhu YJ, Wang WW, Qi RL, Hu XL. Microwave-assisted synthesis of single-crystalline tellurium nanorods and nanowires in ionic liquids. Angew Chem. 2004;116:1434–38.

[44] Liu JW, Chen F, Zhang M, Qi H, Zhang CL, Yu SH. Rapid microwave-assisted synthesis of uniform ultralong Te nanowires, optical property, and chemical stability. Langmuir. 2010;26(13):11372–77.

[45] Byrappa K, Adschiri T. Hydrothermal technology for nanotechnology. Prog Cryst Growth Ch. 2007;53:117e166.

[46] Byrappa K, Yoshimura M. Handbook of hydrothermal technology. New Jersey: Noyes Publications, 2001.

[47] Xi G, Xiong K, Zhao Q, Zhang R, Zhang H, Qian Y. Nucleation-dissolution recrystallization: a new growth mechanism for t-selenium nanotubes. Cryst Growth Des. 2006;6:577–82.

[48] Xie Q, Dai Z, Huang W, Zhang W, Ma D, Hu X, et al. Large-scale synthesis and growth mechanism of single-crystal Se nanobelts. Cryst Growth Des. 2006;6:1514.

[49] Zhang H, Yang D, Ji Y, Ma X, Xu J, Que D. Selenium nanotubes synthesized by a novel solution phase approach. J Phys Chem B. 2004;108:1179–82.

[50] Cao GS, Zhang XJ, Su L, Ruan YY. Hydrothermal synthesis of selenium and tellurium nanorods. J Exp Nanosci. 2011;6(2):121–26.

[51] Qian HS, Yu SH, Luo LB, Gong JY, Fei LF, Liu XM. Synthesis of uniform Te@Carbon-rich composite nanocables with photoluminescence properties and carbonaceous nanofibers by the hydrothermal carbonization of glucose. Chem Mater. 2006;18(8):2102–08.

[52] Lu J, Xie Y, Xu F, Zhu L. Study of the dissolution behaviour of selenium and tellurium in different solvents – a novel route to Se, Te tubular bulk single crystals. J Mater Chem. 2002;12:2755–61.

[53] Ding Y, Li Q, Jia Y, Chen I, Xing J, Qian Y. Growth of single crystal selenium with different morphologies via a solvothermal method. J Cryst Growth. 2002;241:489–97.

[54] Izumi Y, Takeda H, Bamba T. Supercritical fluid. In: Wenk M, editor(s). Encyclopedia of lipidomics. Dordrecht: Springer Science+Business Media B.V, 2016.

[55] Wang S, Guan W, Ma D, Chen X, Wan L, Huang S, et al. Synthesis, characterization and optical properties of flower-like tellurium. Cryst Eng Comm. 2010;12:166–71.

[56] Quintana M, Haro-Poniatowski E, Morales J, Batina N. Synthesis of selenium nanoparticles by pulsed laser ablation. Appl Surf Sci. 2002;195:175–86.

[57] Phae-Ngam W, Kosalathip V, Kumpeerapun T, Limsuwan P, Dauscher A. Preparation and characterization of tellurium nano-particles by long pulsed laser ablation. Adv Mater Res. 2011;214:202–06.

[58] Singh SC, Mishra SK, Srivastava RK, Gopal R. Optical properties of selenium quantum dots produced with laser irradiation of water suspended Se nanoparticles. J Phys Chem C. 2010;114:17374.

[59] Gusbiers G, Wang Q, Khachatryan E, Arellano-Jimenez MJ, Webster TJ, Larese-Casanova P, et al. Anti-bacterial selenium nanoparticles produced by UV/VIS/NIR pulsed nanosecond laser ablation in liquids. Laser Phys Lett. 2015;12:016003.

[60] Guisbiers G, Mimun LC, Mendoza-Cruz R, Nash KL. Synthesis of tunable tellurium nanoparticles. Semicond Sc Technol. 2017;32:04LT01.

[61] Nguyen VN, Amal R, Beydoun D. Photocatalytic reduction of selenium ions using different TiO2 photocatalysts. Chem Eng Sci. 2005;60:5759–69.

[62] Serpone N, Borgarello E, Pelizzetti E. Photocatalysis and environment, trends and applications. London: Kluwer Academic Publishers, 1988.

[63] Chaudhari S, Mehta SK. Selenium nanomaterials: applications in electronics, catalysis and sensors. J Nanosci Nanotechnol. 2014;14:1658–74.

[64] Sarkar B, Bhattacharjee S, Daware A, Tribedi P, Krishnani KK, Minhas PS. Selenium nanoparticles for stress-resilient fish and livestock. Nanoscale Res Lett. 2015;10:371.

[65] Mafune F, Kohno J, Takeda Y, Kondow T, Sawabe H. Formation and size control of silver nanoparticles by laser ablation in aqueous solution. J Phys Chem. 2000;104:9111–17.

[66] Singh SC, Swarnkar RK, Gopal R. Synthesis of titanium dioxide nanomaterial by pulsed laser ablation in water. J Nanosci Nanotech. 2009;9:5367–71.

[67] Singh SC, Mishra SK, Srivastava RK, Gopal R. Optical properties of selenium quantum dots produced by laser irradiation of water suspended Se nanoparticles. J Phys Chem C. 2010;114:17374–84.

[68] Kuzmin PG, Shafeev GA, Voronov VV, Raspopov RV, Arianova EA, Trushina EN, et al. Bioavailable nanoparticles obtained in laser ablation of a selenium target in water. Quantum Electron. 2012;42:1042.

[69] Shi W, Hughes RW, Denholme SJ, Gregory DH. Synthesis design strategies to anisotropic chalcogenide nanostructures. Cryst Eng Commun. 2010;12:641–59.

[70] Jiang ZY, Xie ZX, Xie SY, Zhang XH, Huang RB, Zheng LS. High purity trigonal selenium nanorods growth via laser ablation under controlled temperature. Chem Phys Lett. 2003;368:425–29.

[71] Ren L, Zhang H, Tan P, Chen Y, Zhang Z, Chang Y, et al. Hexagonal selenium nanowires synthesized via vapor-phase growth. J Phys Chem B. 2004;108:4627–30.

[72] Cao G. (editor). Chapter 4: one-dimensional nanostructures: nanowires and nanorods. In: Nanostructures & nanomaterials, synthesis, properties and applications. London: Imperial College Press, 2004:161–62.

[73] Frey H. Chemical vapor deposition (CVD). In: Frey H, Khan HR, editors. Handbook of thin-film technology. Berlin: © Springer-Verlag Berlin Heidelberg, 2015:225–52.

[74] Powel CF, Oxley JH, Blocher JM, Jr. Vapor depositions. New York: Wiley & Sons, 1966.

[75] Engelko V, Mueller GG. Formation of plasma and ion flux on a target, irradiated by an intense electron beam. J Appl Phys Lett. 2005;98:013301–08.

[76] Franklin TC, Adeniyi WK, Nnodimele R. The electro-oxidation of some insoluble inorganic sulfides, selenides, and tellurides in cationic surfactant aqueous sodium hydroxide systems. J Electrochem Soc. 1990;137:480–84.

[77] Mayers B, Gates B, Yin YD, Xia YN. Large-scale synthesis of monodisperse nanorods of Se/Te alloys through a homogeneous nucleation and solution growth process. Adv Mater. 2001;13:1380–84.

[78] Cao X, Xie Y, Zhang S, Li F. Ultra-thin trigonal selenium nanoribbons developed from series-wound beads. Adv Mater. 2004;16:649–53.

[79] Zhang H, Zuo M, Tan S, Li G, Zhang S, Huo J. Carbothermal chemical vapor deposition route to Se one-dimensional nanostructures and their optical properties. J Phys Chem B. 2005;109:10653–57.

[80] Metraux C, Grobety B. Tellurium nanotubes and nanorods synthesized by physical vapor deposition. J Mater Res. 2004;19(7):2159–64.

[81] Sen S, Bhatta UM, Kumar V, Muthe KP, Bhattacharya S, Gupta SK, et al. Synthesis of tellurium nanostructures by physical vapor deposition and their growth mechanism. Cryst Growth Des. 2008;8(1):238–42.

[82] Wang Q, Li GD, Liu YL, Xu S, Wang KJ, Chen JS. Fabrication and growth mechanism of selenium and tellurium nanobelts through a vacuum vapor deposition route. J Phys Chem C. 2007;111:12926–32.

[83] Sen S, Bhandarkar V, Muthe KP, Roy M, Deshpande SK, Aiyer RC, et al. Highly sensitive hydrogen sulphide sensors operable at room temperature. Sensors Actuators B. 2006;115:270–75.

[84] Bockstaller MR, Mickiewicz RA, Thomas EL. Block copolymer nanocomposites: perspectives for tailored functional materials. Adv Mater. 2005;17:1331–49.

[85] Boutonnet M, Kitzling J, Stenius P. The preparation of monodisperse colloidal metal particles from microemulsions. Colloids Surf. 1982;5:209–25.

[86] Eastoe J, Hollamby MJ, Hudson L. Recent advances in nanoparticle synthesis with reversed micelles. Adv Colloid Interface Sci. 2006;128:5–15.

[87] Hotze EM, Phenrat T, Lowry GV. Nanoparticle aggregation: challenges to understanding transport and reactivity in the environment. J Environ Qual. 2010;39:1909–24.

[88] Fleer GJ, Stuart MC, Scheutjens JM. (editors). Chapter 1, polymers in solution. In: Polymers at interfaces. London: Chapman & Hall, 1993:1–25.

[89] Kumar SK, Krishnamoorti R. Nanocomposites: structure, phase behavior, and properties. Ann Rev Chem Biomol Eng. 2010;1:37–58.

[90] Kopeikin VV, Valueva SV, Kipper AI, Borovikova LN, Nazarkina YI, Khlebosolova EN, et al. Adsorption of hydroxyethyl cellulose selenium nanoparticles during their formation in water. Russ J Appl Chem. 2003;76(4):600–02.

[91] Kopeikin VV, Valueva SV, Kipper AI, Borovikova LN, Filippov AI. Synthesis of selenium nanoparticles in aqueous solutions of poly(vinylpyrrolidone) and morphological characteristics of the related nanocomposites. Polym Sci Ser A. 2003;45:374–79.

[92] Gates B, Yin Y, Xia Y. A solution-phase approach to the synthesis of uniform nanowires of crystalline selenium with lateral dimensions in the range of 10–30 nm. J Am Chem Soc. 2000;122:12582.

[93] Mehta SK, Chaudhary S, Kumar S, Bhasin KK, Torigoe K, Sakai H, et al. Surfactant assisted synthesis and spectroscopic characterization of selenium nanoparticles in ambient conditions. Nanotechnology. 2008;19:295601–12.

[94] Shah CP, Kumar M, Bajaj PN. Acid-induced synthesis of polyvinyl alcohol- stabilized selenium nanoparticles. Nanotechnology. 2007;18:385607.

[95] Zhang J, Zhang SY, Xu JJ, Chen HY. A new method for the synthesis of selenium nanoparticles and the application to construction of H2O2 biosensor. Chin Chem Lett. 2004;15(11): 1345–48.

[96] Brust M, Walker M, Bethell D, Schiffrin DJ, Whyman R. Synthesis of thiol-derivatized gold nanoparticles in a two-phase liquid-liquid system. J Chem Soc Chem Commun. 1994;7:801–2.

[97] Liu MZ, Zhang SY, Shen YH, Zhang ML. Selenium nanoparticles prepared from reverse microemulsion process. Chin Chem Lett. 2004;15:1249–52.

[98] He W, Krjci A, Lin J, Osmulski ME, Dickerson JH. A facile synthesis of Te nanoparticles with binary size distribution by green chemistry. Nanoscale. 2011;3(4):1523–25.

[99] Liu Z, Hu Z, Liang J, Li S, Yang Y, Peng S, et al. Size-controlled synthesis and growth mechanism of monodisperse tellurium nanorods by a surfactant-assisted method. Langmiur. 2004;20:214–18.

[100] Zhang SY, Zhang J, Liu Y, Ma X, Chen HY. Electrochemical synthesis of selenium nanotubes by using CTAB soft-template. Electrochim Acta. 2005;50:4365–70.

[101] Bartůněk V, Junková J, Šuman J, Kolářová K, Rimpelová S, Ulbrich P, et al. Preparation of amorphous antimicrobial selenium nanoparticles stabilized by odor suppressing surfactant polysorbate 20. Mat Lett. 2015;152:207–09.

[102] Liu Z, Hu Z, Xie Q, Yang B, Wu J, Qian Y. Surfactant-assisted growth of uniform nanorods of crystalline tellurium. J Mater Chem. 2002;13:159–62.

[103] Li J, Zhang J, Qian Y. Surfactant-assisted synthesis of bundle-like nanostructures with well-aligned Te nanorods. Solid State Sci. 2008;10:1549–55.

[104] Zhu Y, Qian Y. Chapter 7: hydrothermal synthesis of inorganic nanomaterials. In: Prescott WV, Schwartz AI, editors. Nanorods, nanotubes and nanomaterials research progress. New York: Nova Science Publishers, 2008:279–304.

[105] Lin ZH, Wang CR. Evidence on the size-dependent absorption spectral evolution of selenium nanoparticles. Mat Chem Phys. 2005;92:591–94.

[106] Cooper WC. The physics of selenium and tellurium. Montreal: Pergamon Press, 1969:3–20.

[107] Ibragimov NI, Abutalibova ZM, Agaev VG. Electrophotographic layers of trigonal Se in the binder obtained by reduction of SeO2 by hydrazine. Thin Solid Films. 2000;359:125–26.

[108] Ma YR, Qi LM, Ma JM, Cheng HM. Micelle-mediated synthesis of single-crystalline selenium nanotubes. Adv Mater. 2004;16:1023–26.

[109] Ladj R, Bitar A, Eissa M, Mugnier Y, Le Dantec R, Fessi H, et al. Individual inorganic nanoparticles: preparation, functionalization and in vitro biomedical diagnostic applications. J Mater Chem B. 2013;1:1381–96.

[110] Lu Z, Yin Y. Colloidal nanoparticle clusters: functional materials by design. Chem Soc Rev. 2012;41:6874–87.

[111] Xu H, Cao W, Zhang X. Selenium-containing polymers: promising biomaterials for controlled release and enzyme mimics. Acc Chem Res. 2013;46:1647–58.

[112] Habiba K, Makarov VI, Weiner BR, Morell G. Chapter 10, fabrication of nanomaterials by pulsed laser synthesis. In: Waqar A, Nasar A, editors. Manufacturing nanostructures. Altrincham: One Central Press, 2014:263–92.

[113] Lin PC, Lin S, Wang PC, Sridharb R. Techniques for physicochemical characterization of nanomaterials. Biotechnol Adv. 2014;32(4):711–26.

[114] Thompson M. The characterization of nanoparticles. London: The Royal Chemistry Society. AMCTB 48 2010:1–3.

[115] Cho EJ, Holback H, Liu KC, Abouelmagd SA, Park J, Yeo Y. Nanoparticle characterization: state of the art, challenges, and emerging technologies. Mol Pharmac. 2013;10:2093–2110.

[116] Hoo CM, Starostin N, West P, Mecartney ML. comparison of atomic force microscopy (AFM) and dynamic light scattering (DLS) methods to characterize nanoparticle size distributions. J Nanopart Res. 2008;10:89–96.

[117] Mahl D, Diendorf J, Meyer-Zaika W, Epple M. Possibilities and limitations of different analytical methods for the size determination of a bimodal dispersion of metallic nanoparticles. Colloids Surf A. 2011;377:386–392.

[118] Cao G. (editor). Chapter 8: characterization and properties of nanomaterials. In: Nanostructures & nanomaterials, synthesis, properties and applications. London: Imperial College Press, 2004:329–90.

[119] Egerton RF. (editor). Chapter 3, the transmission electron microscope. In: Physical principles of electron microscopy: an introduction to TEM, SEM, and AEM. New York: Springer Science+Business Media, Inc., 2005:57–92.

[120] Hiemenz P, Rajagopalan R. Chapter 12, electrophoresis and other electrokinetic phenomena. In: Principles of colloidal and surface chemistry. 3rd ed. New York: Marcel Dekker; 1997. p. 534–71.

[121] Hunter RJ. Zeta potential in colloid science: principles and applications. Oxford: Academic Press, 1981.

[122] American Chemistry Society (ACS) NANO. ACS Green Chemistry Institution. Green nanotechnology challenges and opportunities. Available at: https://greennano.org/sites/greennano2.uoregon.edu/files/GCI_WP_GN10.pdf. Accessed: Jun 2011.

[123] Centers for Disease Control and Prevention National Institute for Occupational Safety and Health (CDC). Department of Health and Human Services, Centers for Disease Control and Prevention, National Institute for Occupational Safety and Health, DHHS (NIOSH). Current strategies for engineering controls in nanomaterial production and downstream handling processes. 2014 Available at: https://www.cdc.gov/niosh/docs/2014-102/pdfs/2014-102.pdf. Accessed: 9 Nov 2017.

[124] Li Q, Chen T, Yang F, Liu J, Zheng W. Facile and controllable one-step fabrication of selenium nanoparticles assisted by l-cysteine. Mater Lett. 2010;64(5):614–17.

[125] Jiang F, Cai W, Tan G. Facile synthesis and optical properties of small selenium nanocrystals and nanorods. Nanoscale Res Lett. 2017;12(1):401.

[126] Lin ZH, Wang CR. Evidence on the size-dependent absorption spectral evolution of selenium nanoparticles. Mat Chem Phys. 2005;92:591–94.

[127] Lee S, Kwon C, Park B, Jung S. Synthesis of selenium nanowires morphologically directed by Shinorhizobial oligosaccharides carbohydrate. Research. 2009;344:1230–34.

[128] Tsai SJ, Hoffman M, Hallock MF, Ada E, Kong J, Ellenbecker MJ. Characterization and evaluation of nanoparticle release during the synthesis of single-walled and multiwalled carbon nanotubes by chemical vapor deposition. Environ Sci Technol. 2009;43:6017–23.

[129] Yeganeh B, Kull CM, Hull MS, Marr LC. Characterization of airborne particles during production of carbonaceous nanomaterials. Environ Sci Technol. 2008;42(12):4600–06.

Bionotes

MSc Elena Piacenza received her BS degree in Biotechnology from the University of Verona (Italy) in 2013, and her MS degree in Sciences and Technologies of Bio Nanomaterials from Ca'Foscari University (Venice, Italy) in 2015. Subsequently she joined the Biological Sciences Department at the University of Calgary (Canada) as PhD student in the Biophysical Chemistry program, under the supervision of Prof. Raymond J. Turner. Currently she is a graduate visiting student in the group of Prof. Giovanni Vallini (Environmental Microbiology and Microbial Biotechnology) at the University of Verona (Italy). Her research and PhD project is focused on the production of biocompatible metalloid-nanomaterials by using microorganisms as cell factories, their chemical-physical characterization and future applications.

Dr. Alessandro Presentato received his BS degree in Biological Sciences from the University of Palermo (Italy) in 2008, and his MS degree in Cellular and Molecular Biology from the University of Palermo in 2011. He then joined the Department of Pharmacy and Biotechnology of the University of Bologna as graduate student and he obtained his PhD in Cellular and Molecular Biology under the supervision of Prof. Davide Zannoni in 2015. Subsequently, he worked as Post-Doctoral fellow in the Microbial Biochemistry group of Prof. Raymond J. Turner at the University of Calgary (Canada) between 2015 and 2017. He is currently working as Post-Doctoral fellow in Prof. Giovanni Vallini's group of Environmental Microbiology and Microbial Biotechnology at the University of Verona (Italy). His research is focused on the production of biogenic nanomaterials using microorganisms as cell factories and their potential applications.

Dr. Emanuele Zonaro obtained his BS and MS degrees in Molecular and Industrial Biotechnology from the University of Verona (Italy). Then, he obtained his PhD in Molecular, Industrial and Environmental Biotechnology at University of Verona in 2016. Currently, he is working as Post-Doc in Prof. Giovanni Vallini's group of Environmental Microbiology and Microbial Biotechnology at University of Verona. His research is mainly focused on the study of the interactions between microorganisms and the metalloids Selenium and Tellurium, the production of nanomaterials through biological synthesis and the applications of these nanomaterials as antimicrobial and antibiofilm agents.

Dr. Silvia Lampis got her PhD in 'Molecular, Environmental and Industrial Biotechnology' with a thesis entitled 'New insights in bacterial selenium transformation' in 2006 at University of Verona (Italy). From November 2007 she got a permanent position as Assistant Professor in General Microbiology, BIO19, at the Department of Biotechnology at Verona University. She has a scientific expertise on environmental microbiology with a special interest on microbial interactions with metals in terrestrial ecosystems and relative applications. She is author and co-author of many scientific works published in internationally scientific journals focused mainly on the bacterial transformation of metals and metalloids and toxic organic compounds. The scientific interest ranges from bacteria involved in selenium and tellurium cycle to all metal resistant microorganisms; the study of bacteria isolated from industrial site with ability to degrade aromatic and aliphatic hydrocarbons; the study of bacterial community involved in water treatment processes.

Prof. Giovanni Vallini graduate magna cum laude in Agricultural Sciences at University of Pisa (Italy) and he spent the first part of his scientific career at the Soil Microbiology Center - Italian National Research Council (CNR) in Pisa. He is currently chair professor of General Microbiology and head of the Environmental Microbiology and Microbial Biotechnology group at the Department of Biotechnology of the University of Verona (Italy), where he was formerly even appointed as Head of the Department. In the last twenty years, Prof. Vallini has carried out studies on the microbial biodegradation of xenobiotic organic compounds in soils and waters, in the frame of activities dealing with the development of bioremediation strategies for the clean-up of contaminated sites. His interest also has focused on how interactions between plants and rhizosphere microorganisms can affect phytoextraction mechanisms of heavy metals and metalloids, with special reference to Lead, Selenium and Arsenic. Starting from this specific investigation on microbial strains capable of transforming chemical forms of metals and metalloids, Prof. Vallini attention has then moved to the exploitation of bacterial species for the synthesis of elemental nanomaterials with special antimicrobial capabilities and contrast activity towards biofilm formation in medical, industrial and environmental fields. Prof. Vallini can list more than 130 papers published in scientific journals and contributions in monographies together with science divulgation articles.

Prof. Raymond J. Turner Turner joined the University of Calgary in 1998 in the Department of Biological Sciences and is presently a Professor of Biochemistry an d Microbiology (full professor since 2006). He has held the post of Associate Department Head and Graduate program director from 2013 – 2016. He has also been chair of various research cluster units over the past 10 years. He is presently funded by grants from Canadian granting councils of NSERC and CIHR, as well as industry partners. He has received excellence in research and excellence in graduate student supervision awards. Research interests are multi-disciplinary from bioinorganic and environmental chemistry of chalcogen metals, metal toxicity mechanisms towards bacteria, microbiology of metal nanoparticles, bacteria growing as biofilm, membrane protein structural biology, multidrug resistance, molecular microbiology of protein transporters and translocators and photochemistry of novel fluorophores.

Index

https://doi.org/10.1515/9783110636666-007

www.ingramcontent.com/pod-product-compliance
Lightning Source LLC
Chambersburg PA
CBHW080907220326
41598CB00034B/5501